ATOMS AND MOLECULES
IN ASTROPHYSICS

ATOMS AND MOLECULES IN ASTROPHYSICS

Proceedings of the Twelfth Session of the Scottish
Universities Summer School in Physics, 1971.

A NATO Advanced Study Institute

Scottish Universities Summer School, 12th,
University of Stirling, 1971.

Edited by

T. R. CARSON

*Department of Astronomy, University of St. Andrews,
St. Andrews, Scotland*

and

M. J. ROBERTS

*Department of Physics, University of Stirling,
Stirling, Scotland*

1972

ACADEMIC PRESS · LONDON AND NEW YORK

ACADEMIC PRESS INC. (LONDON) LTD.
24/28 Oval Road,
London NW1

United States Edition published by
ACADEMIC PRESS INC.
111 Fifth Avenue
New York, New York 10003

Copyright © 1972 By Scottish Universities Summer School

All Rights Reserved
No part of this book may be reproduced in any form by photostat, microfilm, or any other means, without written permission from the publishers

Library of Congress Catalog Card Number: 72-84455
ISBN: 0-12-161050-0

PRINTED IN GREAT BRITAIN BY
ROYSTAN PRINTERS LIMITED
Spencer Court, 7 Chalcot Road
London NW1

List of Participants

ORGANIZING COMMITTEE

PERCIVAL, PROFESSOR I. C. *University of Stirling, Stirling, Scotland (Director).*
DICKINSON, DR. A. S. *University of Stirling, Stirling, Scotland (Secretary).*
KOLLATH, MR. K. *University of Stirling, Stirling, Scotland (Secretary).*
LAWLEY, DR. K. P. *University of Edinburgh, Edinburgh, Scotland (Treasurer).*
CARSON, DR. T. R. *University of St. Andrews, St. Andrews, Scotland (Editor).*
ROBERTS, DR. M. J. *University of Stirling, Stirling, Scotland (Editor).*
DUNCAN, DR. A. J. *University of Stirling, Stirling, Scotland (Steward).*
WALKER, DR. ISOBEL C. *University of Stirling, Stirling, Scotland (Steward)*

LECTURERS

BURKE, PROFESSOR P. G. *The Queen's University, Belfast, Northern Ireland.*
FOLEY, DR. H. *Columbia University, New York, U.S.A.*
LITVAK, DR. M. *Harvard Observatory, Cambridge, U.S.A.*
PERCIVAL, PROFESSOR I. C. *University of Stirling, Stirling, Scotland.*
VAN REGEMORTER, DR. H. *Observatoire de Meudon, Paris, France.*
SEATON, F.R.S., PROFESSOR M. J. *University College, London.*
SPEER, DR. R. *Imperial College, London.*

SEMINAR SPEAKERS

CHURCHWELL, DR. E. *Max-Planck-Institut für Radioastronomie, Bonn, West Germany.*
GABRIEL, DR. A. *Culham Laboratory, Berkshire, England.*
DE JONG, DR. T. *The Observatory, Leiden, Holland.*

MALLOW, DR. J. V. *University of Jerusalem, Israel.*
OLTHOF, DR. H. *University of Groningen, Groningen, Holland.*
PAGEL, DR. B. E. J. *Greenwich Observatory, England.*
SCHUTTE, DR. A. *University of Leiden, Leiden, Holland.*
SMITH, DR. S. *University of Colorado, Boulder, Colorado, U.S.A.*

PARTICIPANTS

ALLARD, NICOLE, *Observatoire de Paris, 92 — Meudon, France.*

ANDERSEN, S. *Copenhagen University Observatory, Østervold 3, Copenhagen.*

AUBÈS, M. *Laboratoire de Physique des Milieux Ionisés, Université Paul Sabatier, 31 Toulouse, France.*

BARLOW, M. *University of Sussex, Brighton, England.*

BARWIG, P. *University of Bonn, West Germany.*

BERGER, P. *Department of Earth and Space Sciences, S.U.N.Y., Stonybrook, N.Y. 11790.*

BLACK, J. H. *Harvard College Observatory, 60 Garden Street, Cambridge, Massachusetts 62138, U.S.A.*

BROCKLEHURST, M. *Physics Department, University College, Gower Street, London, W.C.1.*

CAMPOS, J. *Catedra de Fisica Atomica y Nuclear, Facultad de Ciencias, Universidad de Madrid, Spain.*

CATO, T. *Onsala Space Observatory, Onsala, Sweden.*

CHANDRA, N. *Department of Applied Maths, The Queen's University, Belfast, Northern Ireland.*

CHEN, P. C. *Department of Earth and Space Sciences, S.U.N.Y., Stonybrook, N.Y. 11790.*

CHURCHWELL, E. *Max-Planck-Institut für Radioastronomie, 53 Bonn, Argelanderstrasse 3, West Germany.*

CLAEYS, WILLY J. *Institut de Physique Nucleaire, Avenue Cardinal Mercier, (Parc d'Arenberg), Heverlee (Louvain), Belgium.*

CLEGG, R. *University of Maryland, U.S.A.*

COHEN, J. S. *Physics Department, Rice University, Houston, Texas 77001, U.S.A.*

COHEN, R. J. *Nuffield Radio Astronomy Laboratories, Jodrell Bank, Macclesfield, Cheshire, England.*

LIST OF PARTICIPANTS vii

DANKS, A. *Institut d'Astrophysique, Université de Liège, Ave. de Cointe 5, B-4200 Cointe (Ougree), Belgium.*

DROWART, A. *Aeronomy Institute, Bruxelles, Belgium.*

FERTEL, JEANNE. *Max-Planck Institut für Radioastronomie, 53 Bonn, Argelanderstrasse 3, West Germany.*

FON, WAI-CHU. *Department of Mathematics, University of Malaya, Kuala Lumpur, Malaysia.*

GAIL, H. P. *Institut für Theoretische Astrophysik, 69 Heidelberg, Berlinstrasse 19, Germany.*

HALL, R. *Université de Paris VI T 12E5, 9 quai St. Bernard, Paris V.*

HILBORN, R. *Physics Department, S.U.N.Y. Stonybrook, N.Y., U.S.A. 11790.*

HOFSÄSS, D. *D-23 Kiel, Institut für Theoret. Physik der Universitat, Olshausehstrasse, Keil, West Germany.*

HUTTON, N. *University of Stirling, Stirling, Scotland.*

DE JONG, T. *The Observatory, Leiden, Holland.*

KANAI, M. *University of Pisa, Italy.*

KENNEDY, E. T. *Physics Department, U.C.D., Belfield, Dublin, Eire.*

KNAPP, H. F. P. *Kamerlingh Onnes Laboratorium, University of Leiden, Leiden, Holland.*

LARSEN, A. *Heriot-Watt University, Edinburgh, Scotland,*

MACADAM, K. *University of Stirling, Stirling, Scotland.*

MCILVEEN, A. *Department of Applied Mathematics, The Queen's University, Belfast, Northern Ireland.*

MALLOW, J. *The Hebrew University of Jerusalem, Racah Institute of Physics, Israel.*

MARSHALL, P. *Heriot-Watt University, Edinburgh, Scotland.*

MASON, HELEN. *Department of Physics, University College, Gower Street, London, W.C.1.*

NACCACHE, P. *University of Stirling, Stirling, Scotland.*

OLTHOF, H. *Department of Space Research, University of Groningen, P.O. Box 800, Groningen, Holland.*

PFENNIG, H. *Max-Planck Institut für Radioastronomie, 53 Bonn, Argelanderstrasse 3, West Germany.*

PREIST, T. *University of Exeter, Exeter, England.*

VAN RAAN, A. *Physics Laboratory of The State University of Utrecht, Holland.*

ROBINS, E. S. *University of Stirling, Stirling, Scotland.*

ROTTMAN, G. *Physics Department, The John Hopkins University, Baltimore, Maryland, U.S.A. 2128.*

SALEM, M. *Depto de Fisica, Fac. Cs. Exactas, Universidad de Buenos Aires, Peru, 272, Capital Federal, Argentina.*

SCHUTTE, A. *Kamerlingh Onnes Laboratorium, University of Leiden, Leiden, Holland.*

SEGRE, E. *Observatoire de Paris, 1 Place Jules Jansen, 92—Meudon, France.*

SIMIEVIC, A. *Racah Institute of Physics, Hebrew University of Jerusalem, Israel.*

STERN, B. *Observatoire de Paris, 1 Place Jules Jansen, 92—Meudon, France.*

SUME, A. *Onsala Space Observatory, Onsala, Sweden.*

SUSSMAN, B. *Physics Department, S.U.N.Y. Stonybrook, N.Y., U.S.A. 11790.*

TAYLOR, I. R. *The Queen's University, Belfast, Northern Ireland.*

THOMAS, B. *University of Pittsburgh, U.S.A. 15213.*

TRAN-MINH, N. *Observatoire de Paris, 1 Place Jules Jansen, 92—Meudon, France.*

TURNBULL, R. S. *Natural Philosophy Department, University of Strathclyde, Glasgow, Scotland.*

VOIGHT, B. *University of Copenhagen, Copenhagen.*

WALMSLEY, M. *Max-Planck Institut für Radioastronomie, 53 Bonn, Argelanderstrasse, 3, West Germany.*

WEBER, K. W. *Institut für Astrophysik und Extraterrestrische Forschung, D-53, Bonn 1, Poppelsdorfer Allee 49, West Germany.*

WENDIN, G. *Institute of Theoretical Physics, Frack S-40220, Göteborg, 5, Sweden.*

WHITWORTH, A. P. *Department of Mathematics, Manchester University, England.*

Director's Preface

In August 1971 the Twelfth Scottish Universities Summer School in Physics took place at the University of Stirling. The lecture courses were divided almost equally between atomic and molecular physics and astrophysics. In view of recent developments there was an emphasis on phenomena which are observable only at radio, ultraviolet and X-ray wavelengths and on states of atoms, molecules and ions which produce—or which are believed to produce —radiation at these wavelengths.

In this volume it is shown how detailed observation of the radiation from atoms and molecules is combined with basic theory to obtain the abundances, temperatures and densities in stellar atmospheres and interstellar regions.

I am glad to be able to acknowledge here the work of the editors, Dick Carson and Mike Roberts, of the School secretaries, Alan Dickinson and Klaus Kollath, for their meticulous attention to every detail of the running of the School, and to the rest of the Organizing Committee and all our assistants, official and unofficial, who helped to make the School such a success.

The organisers and participants of the Summer School would like to express their deepest sympathy to the relatives and colleagues of Dr. A. Schutte who tragically died in a motor accident in Genoa on March 30th, 1972. He will always be remembered at Stirling for his active participation in the School and his untimely death will be a great loss to fundamental physics research.

The contributions of the lecturers speak for themselves in this volume.

I. C. Percival.

Editors' Preface

The contributions in the present volume, representing essentially the lectures as delivered at the Summer School, have been prepared by the authors, with the exception of Professor Seaton's lectures which have been written up by Dr. M. Brockelhurst assisted by Mr. P. Naccache. In every case the editing has been kept to a minimum, consistent with accuracy (by the correction of obvious errors) and with some measure of uniformity within the volume (by minor changes in presentation). For these alterations and all other changes from the contributions as received, the Editors accept full responsibility.

The material in the first six chapters was presented in the form of series of lectures, while that in the remaining chapters was presented in the form of seminars, with the exception of Dr. Speer's contribution which embodies both his lectures and his two seminars. A further seminar by Dr. S. Smith on "Cross–Beam Collision Techniques" has not been included in these proceedings.

T. R. Carson
M. J. Roberts

Contents

Atomic Processes
P. G. Burke

I	Potential Scattering	1
II	Multichannel Scattering	8
III	Electron Scattering by Hydrogen Atoms and Hydrogen-like Ions	15
IV	Electron Scattering by Complex Atoms and Ions	26
V	Photoionization and Recombination	39
VI	Molecular Processes	49

Highly Excited Atoms
Ian C. Percival

I	Introduction	65
II	Elementary properties	65
III	Correspondence principle for intensity	68
IV	Classical collisions	69
V	Discrete states	70
VI	Scaling laws	71
VII	Initial conditions	72
VIII	Binary encounter or classical impulse approximation	72
IX	Orbit integration and Monte Carlo method	74
X	Correspondence principles	75
XI	Heisenberg's form	76
XII	Strong coupling	77
XIII	Ranges of validity	77
XIV	Cross-sections (quantal)	79
XV	Cross-sections (correspondence)	80
XVI	Literature	81

Spectral Line Broadening
H. Van Regemorter

	INTRODUCTION	85
I	Basic Principles	86
II	The Semi-classical Approach	93
III	Quantum Theory of Line Broadening	102
IV	Applications to a Few Cases	110

The Spectra of Gaseous Nebulae
M. J. Seaton

I	Introduction	121
II	Physical Processes	122
III	Ionization Equilibrium	122
IV	Observed Line Intensities	125
V	Electron Temperatures and Forbidden Lines	125
VI	Model Calculations	128
VII	Optical Observations of Recombination Lines	130
VIII	Interpretation of Line Observations	131
IX	The Spectra of Neutral Helium	131
X	The Optical Continuum	132
XI	The Radio Continuum	133
XII	Radio Recombination Lines	135
XIII	Helium and Carbon Recombination Lines	147
	Appendix	149

Introduction to Molecular Spectra
H. M. Foley

I	The Molecular Wave Equation	156
II	Properties of the Electronic States	158
III	Rotation and Vibration	162
IV	Symmetry and Statistics	163
V	Electronic Dipole Matrix Element-Selection Rules	165
VI	Vibration–Rotation and Rotation Spectra	168
VII	Spin–Orbit Interaction and Higher Order Rotational Effects	170
VIII	Rotation of Polyatomic Molecules	175

IX	Molecular Hyperfine Structure							183
X	Emission and Absorption							186
XI	Astrophysical Observations							187
XII	The Astrophysical Molecules							189
XIII	Literature							198

Non-Equilibrium Processes in Interstellar Molecules
M. M. Litvak

I	Introduction							201
II	Observation of Maser Molecules							206
III	Non-Equilibrium Conditions							221
IV	Nonlinear Properties of Masers							244
V	Formation of Molecules							259

Radio Recombination Lines: An Observer's Point of View
E. Churchwell

I	General Considerations							277
II	Importance of Radio Recombination Lines in Astronomy				278

Some Recent Aspects of Spectroscopy at UV and X-Ray Wavelengths
R. J. Speer

I	Introduction							285
II	Ultraviolet Spectroscopy at Total Solar Eclipse					293
III	Light Sources							298
IV	Diffraction Gratings							302

Spectral Intensities from Helium-Like Ions
A. H. Gabriel

I	Introduction							311
II	Temperature and Density Measurement						311
III	The Helium-like Ion							313
IV	Low Density							316
V	Intermediate Density							317
VI	High Density							317
VII	Conclusions							318

Abundances in the Solar Corona
H. Olthof
321

The Formation of H_2 Molecules in Dark Interstellar Clouds
T. De Jong

I	Introduction	327
II	The Reactions	328
III	Results	331
IV	Discussion	332
V	Conclusion	334

Formation of Molecular Hydrogen on Cold Surfaces
A. Schutte
337

Emission-line Spectra as Probes of Dust Clouds
B. E. J. Pagel

I	M 82	341
II	Seyfert Galaxies	342
III	η Carinae	343

The Investigation of UV Oscillator Strengths in C, N and O Ions
Jeffry V. Mallow

I	Introduction	347
II	Relevant theoretical formulae and experimental procedure	347
III	Results	350
IV	Conclusion	351

Atomic Processes[†]

P. G. BURKE

The Queen's University, Belfast, Northern Ireland

I. POTENTIAL SCATTERING

1.1. The Scattering Cross-section

We shall start by considering a beam of electrons incident upon a scattering centre. The appropriate equation for the electron wave function $\psi(\mathbf{r})$ is the time independent Schrödinger equation

$$\left[-\frac{\hbar^2}{2m}\nabla^2 + V(\mathbf{r})\right]\psi(\mathbf{r}) = E\psi(\mathbf{r}) \tag{1.1}$$

where $V(\mathbf{r})$ is the potential field exerted by the scattering centre and E is the energy eigenvalue. It is convenient to use atomic units ($1 = m = \hbar = e$) so that (1.1) becomes

$$[-\tfrac{1}{2}\nabla^2 + V(\mathbf{r})]\psi(\mathbf{r}) = E\psi(\mathbf{r}) \tag{1.2}$$

where r is now in units of $a_0 = \hbar^2/me^2$ while V and E are units of e^2/a_0.

We shall suppose that the wave function of the incident beam is an eigenfunction of energy and momentum and thus, by the uncertainty principle, is not localized in space. At large distances the wave function for the incident and scattered beams becomes

$$\psi(\mathbf{r}) \underset{r\to\infty}{\sim} e^{ikz} + f(\theta)\frac{e^{ikr}}{r} \tag{1.3}$$

where $f(\theta)$ is the scattering amplitude and $k^2 = 2mE/\hbar^2 = 2E$ (in atomic units).

The flux of particles passing through an area $d\mathbf{S}$ follows from the time derivative of the probability density integrated over a fixed volume

$$\mathbf{j} \cdot d\mathbf{S} = \frac{\hbar}{2mi}(\psi^* \operatorname{grad} \psi - \psi \operatorname{grad} \psi^*) \cdot d\mathbf{S} \tag{1.4}$$

[†] References 1–9 provide a general background to these lectures.

Substituting the second term in (1.3) into (1.4) gives

$$\mathbf{j} \cdot d\mathbf{S} = \frac{k}{r^2} |f(\theta)|^2 r^2 \, d\Omega \tag{1.5}$$

which is the flux passing through an area $r^2 \, d\Omega$ at a radius r.

The differential cross-section $\sigma(\theta)$ defined by

$$\sigma(\theta) = \frac{\text{scattered flux/unit solid angle}}{\text{incident flux/unit area}}$$

$$\sigma(\theta) = 1/k \, . \, k |f(\theta)|^2 \, d\Omega = |f(\theta)|^2 \, d\Omega \tag{1.6}$$

and the total cross-section, obtained by integrating over all scattering angles is

$$\sigma = 2\pi \int_0^\pi |f(\theta)|^2 \sin \theta \, d\theta \tag{1.7}$$

1.2. Partial Wave Analysis

In order to calculate the scattering amplitude and consequently the cross-section we must solve the Schrödinger equation (1.1). At this point in our development it is assumed that $V = V(\mathbf{r})$ is spherically symmetric. The equation is then most conveniently solved by carrying out a partial wave expansion. We write

$$\psi(\mathbf{r}) = \frac{1}{r} \sum_{l=0}^{\infty} A_l(k^2) P_l(\cos \theta) u_l(r) \tag{1.8}$$

where the energy dependent constants $A_l(k^2)$ are to be obtained. Substituting (1.8) into (1.2), multiplying by $P_{l'}(\cos \theta)$ and integrating over all angles gives

$$\left(\frac{d^2}{dr^2} - \frac{l(l+1)}{r^2} - U(r) + k^2 \right) u_l(r) = 0 \tag{1.9}$$

where $U(r) = 2V(r)$.

For potentials which are less singular than r^{-2} at the origin, we find two solutions at the origin

$$u_l(r) \propto r^{l+1} \tag{1.10}$$

and

$$u_l(r) \propto r^{-l} \tag{1.11}$$

Only the first satisfies the physical condition of regularity.

For potentials which vanish faster than r^{-1} asymptotically (we will consider Coulomb potentials separately later) we find that apart from an arbitrary normalization factor

$$u_l(r) \underset{r \to \infty}{\sim} \sin(kr - \tfrac{1}{2}l\pi + \delta_l) \tag{1.12}$$

which defines the phase shift δ_l.

In order to determine the scattering amplitude we must choose $A_l(k^2)$ in (1.8) so that $\psi(r)$ has the asymptotic form given by (1.3). We first expand the plane wave

$$e^{ikz} = \sum_{l=0}^{\infty} (2l+1) i^l j_l(kr) P_l(\cos\theta) \tag{1.13}$$

where

$$j_l(x) = \left(\frac{\pi}{2x}\right)^{\tfrac{1}{2}} J_{l+\tfrac{1}{2}}(x)$$

and has the asymptotic form

$$j_l(kr) \underset{r \to \infty}{\sim} (kr)^{-1} \sin(kr - \tfrac{1}{2}l\pi) \tag{1.14}$$

Equating the coefficient of the ingoing spherical wave $\exp[-i(kr - \tfrac{1}{2}l\pi)]$ in (1.3) and (1.8) gives

$$A_l(k^2) = \frac{1}{k}(2l+1) i^l e^{i\delta_l} \tag{1.15}$$

and it follows immediately that

$$f(\theta) = \frac{1}{2ik} \sum_{l=0}^{\infty} (2l+1)(e^{2i\delta_l} - 1) P_l(\cos\theta) \tag{1.16}$$

and

$$\sigma = \frac{4\pi}{k^2} \sum_{l=0}^{\infty} (2l+1) \sin^2 \delta_l \tag{1.17}$$

We see that the phase shifts completely determine the cross section. For most potentials of interest in atomic physics the phase shifts must be determined by numerically integrating (1.9) subject to the boundary conditions (1.10) and (1.12). In practice, at low energies, only a small number of phase shifts are significant and the summations in (1.16) and (1.17) truncate.

It is of interest to rewrite (1.12) in two alternative forms

$$u_l(r) \underset{r \to \infty}{\sim} \sin(kr - \tfrac{1}{2}l\pi) + \cos(kr - \tfrac{1}{2}l\pi) K_l \tag{1.18}$$

and
$$u_l(r) \underset{r \to \infty}{\sim} \exp[-i(kr - \tfrac{1}{2}l\pi)] - \exp[i(kr - \tfrac{1}{2}l\pi)]S_l \qquad (1.19)$$

which differ from each other and from (1.12) just in normalization factors. These forms define the K-matrix and the S-matrix given in terms of the phase shift by
$$K_l = \tan \delta_l \qquad (1.20)$$
and
$$S_l = \exp(2i\delta_l). \qquad (1.21)$$

They are obviously related by
$$S_l = \frac{1 + iK_l}{1 - iK_l}. \qquad (1.22)$$

It is also convenient to introduce the T-matrix through
$$T_l = S_l - 1. \qquad (1.23)$$

1.3. Integral Equation Formalism

The partial wave expansion is only convenient at low energies when the number of phase shifts contributing to the summations (1.16) and (1.17) is small. At high energies it is necessary to obtain a direct expression for the scattering amplitude. This can be achieved by writing the Schrödinger equation as an integral equation.

The time independent Schrödinger equation.
$$(H - E)\psi = 0. \qquad (1.24)$$

In most situations of interest the Hamiltonian can be written as
$$H = H_0 + V \qquad (1.25)$$

where H_0 is the unperturbed Hamiltonian and V is the potential interaction, so (1.24) then becomes
$$(E - H_0)\psi = V\psi \qquad (1.26)$$

which can be formally solved to give
$$\psi = \phi + \frac{1}{E - H_0} V\psi \qquad (1.27)$$

where ϕ is a solution of the unperturbed equation
$$(E - H_0)\phi = 0. \qquad (1.28)$$

If H_0 is chosen to be the kinetic energy operator in (1.7) we see immediately that (1.27) is equivalent to (1.3).

Equation (1.27) is called the Lippmann–Schwinger equation. We can develop an explicit expression for the scattered wave term by writing the free particle Green's function as

$$G_0 = \frac{1}{E - H_0} = \left(\frac{1}{2\pi}\right)^3 \int \frac{|\phi_{\mathbf{k}'}(\mathbf{r})\rangle\langle\phi_{\mathbf{k}'}(\mathbf{r}')|}{E - E'} d\mathbf{k}' \quad (1.29)$$

where we have introduced the complete set of solutions of (1.28)

$$\phi_{\mathbf{k}} = \exp(i\mathbf{k}\cdot\mathbf{r}) \quad (1.30)$$

The G_0 is only defined uniquely when we specify the path of integration past the singularity at $E' = E$. It is conventional to introduce a small positive or negative part into E and to define

$$G_0^{\pm} = \lim_{\varepsilon \to 0+} \frac{1}{E \pm i\varepsilon - H_0}. \quad (1.31)$$

The physical significance of these two paths of integration can be determined by evaluating $G_0^{\pm}$ in the coordinate representation using (1.30). We find directly that

$$G_0^{\pm}(\mathbf{r}, \mathbf{r}') = -\frac{1}{2\pi} \frac{\exp(\pm ik|\mathbf{r} - \mathbf{r}'|)}{|\mathbf{r} - \mathbf{r}'|} \quad (1.32)$$

and thus the + and − signs are associated with outgoing and ingoing wave solutions respectively. We write the corresponding solution of (1.24)

$$\psi^{\pm} = \phi + \lim_{\varepsilon \to 0+} \frac{1}{E \pm i\varepsilon - H_0} V\psi^{\pm}. \quad (1.33)$$

1.4. The Born Approximation

Equation (1.33) can be written explicitly in the form

$$\psi^+(\mathbf{r}) = e^{ikz} - \frac{1}{2\pi} \int \frac{\exp(ik|\mathbf{r} - \mathbf{r}'|)}{|\mathbf{r} - \mathbf{r}'|} V(r')\psi^+(\mathbf{r}')d\mathbf{r}'. \quad (1.34)$$

The scattering amplitude is obtained by considering the asymptotic form of the outgoing wave term using

$$|\mathbf{r} - \mathbf{r}'| \underset{r \to \infty}{\sim} r - \hat{\mathbf{n}}\cdot\mathbf{r}' \quad (1.35)$$

where $\hat{n}$ is a unit vector in the direction of $\mathbf{r}$. Comparing with (1.3) and setting $\mathbf{k}' = k\hat{n}$ we obtain

$$f(\theta) = -\frac{1}{4\pi}\int \exp(-i\mathbf{k}'\cdot\mathbf{r}')U(r')\psi_k^+(\mathbf{r}')\,d\mathbf{r}' \tag{1.36}$$

which can be conveniently written as

$$f(\theta) = -\frac{1}{2\pi}\langle\phi_{k'}, V\psi_k^+\rangle \tag{1.37}$$

In both (1.36) and (1.37) ψ_k^+ is defined by (1.33) with ϕ_k replacing ϕ on the right hand side. In a similar way, starting from ψ^-, we can show that

$$f(\theta) = -\frac{1}{2\pi}\langle\psi_{k'}^-, V\phi\rangle. \tag{1.38}$$

The first Born approximation is obtained by approximating ψ_k^+ in (1.37) or $\psi_{k'}^-$ in (1.38) by ϕ_k or $\phi_{k'}$ respectively. This gives

$$f_B(\theta) = -\frac{1}{4\pi}\int \exp[i(\mathbf{k}' - \mathbf{k})\cdot\mathbf{r}]U(r)\,d\mathbf{r}. \tag{1.39}$$

Higher order Born approximations can be obtained by iterating (1.33) to give

$$\psi^\pm = (1 + G_0^\pm V + G_0^\pm V G_0^\pm V + \ldots)\phi. \tag{1.40}$$

The convergence of the Born expansion has been investigated by a number of authors, including Klein and Zemach,[10] but the situation is still uncertain in some cases of practical interest.

1.5. Effective Range Theory

We now discuss briefly the analytic behaviour of the S-matrix in the neighbourhood of zero energy. Blatt and Jackson[11] and Bethe[12] showed that when the potential falls off exponentially the phase shift could be parametrized according to

$$k^{2l+1}\cot\delta_l = -\frac{1}{a_l} + \tfrac{1}{2}r_{0l}k^2 + O(k^4) \tag{1.41}$$

where a_l is the scattering length and r_{0l} is the effective range.

In atomic physics we are interested in potentials $V(r) \underset{r\to\infty}{\sim} r^{-n}$, and Levy and Keller[13] have derived general expressions for the analytic behaviour of the scattering amplitude. For an atom in a non-degenerate S-state $V(r) \underset{r\to\infty}{\sim} -\tfrac{1}{2}\alpha r^{-4}$, where α is the polarizability, and O'Malley et al.[14] have shown that

$$k \cot \delta_0 = -\frac{1}{a_0} + \frac{\pi \alpha}{3 a_0{}^2} k + \frac{2\alpha}{3 a_0} k^2 \log\left(\frac{\alpha k^2}{16}\right) + O(k^2) \qquad (1.42)$$

and

$$k^2 \cot \delta_l = \frac{8(l + \tfrac{3}{2})(l + \tfrac{1}{2})(l - \tfrac{1}{2})}{\pi \alpha} + \ldots, \qquad l > 0. \qquad (1.43)$$

The effect of the long range potential on the phase shift is, as expected, most marked for the higher partial waves where the phase shift has a linear dependence on energy.

For electron scattering by a positive ion $V(r) \sim -Zr^{-1}$, where Z is the charge on the ion. In this case we obtain the usual bound state spectrum

$$k_{nl}^2 = -\frac{Z^2}{(n - d_{nl})^2} \qquad (1.44)$$

where n has integer values and d_{nl} is the quantum defect. This is a measure of the departure of the potential from pure Coulomb at short distances and is a slowly varying function of energy. Seaton[15–16] showed that the phase shift at positive energies can be related to the extrapolated quantum defect by

$$\tan \delta_l = \frac{\tan \pi d_l(k^2)}{(1 - e^{2\pi \eta})} \qquad (1.45)$$

where $\eta = -Z/k$ and $d_l(k^2)$ is an analytic function of energy which has the value d_{nl} at the bound state energies given by (1.44).

The asymptotic form of the wave function is defined in terms of this phase shift by

$$u_l(r) \underset{r \to \infty}{\sim} \sin(kr - \tfrac{1}{2}l\pi - \eta \log 2kr + \sigma_l + \delta_l) \qquad (1.46)$$

where

$$\sigma_l = \arg \Gamma(l + 1 + i\eta).$$

This important relation allows the cross-section close to the threshold to be determined from a knowledge of the bound state spectrum.

1.6. The Eikonal Approximation

This is obtained in going to the classical limit of the Schrödinger equation and has been studied in detail by Glauber[17]. It has recently found extensive application in electron atom scattering. Writing the wave function in the form

$$\psi(\mathbf{r}) = \exp[iS(\mathbf{r})/\hbar] \qquad (1.47)$$

and substituting into the time independent Schrödinger Eqn (1.1) gives

$$\frac{1}{2m}(-i\hbar\nabla^2 S + (\nabla S)^2) = E - V(r). \tag{1.48}$$

The classical limit is obtained when the first term on the left hand side is negligible compared with the second term. This is equivalent to taking the limit $\hbar \to 0$. We now assume that the scattering is confined to small angles, and it is therefore a good approximation to carry out the integral in (1.48) along a straight line path parallel to the incident beam direction. This gives

$$S(z) \simeq \hbar k z + \int_{-\infty}^{z} \sqrt{2m(E-V)}\, dz' \tag{1.49}$$

where the integration constant has been chosen to give a plane incident wave. The scattering amplitude is then obtained by substituting the expression for ψ^+ given by (1.47) and (1.49) into (1.37). After further reduction we find (once more in atomic units) that

$$f(\theta) = \frac{k}{i} \int_0^\infty J_0(Kb)(e^{i\chi(b)} - 1) b\, db \tag{1.50}$$

where b is the impact parameter, K is the momentum transfer and the phase or eikonal $\chi(b)$ is defined by

$$\chi(b) = -\frac{1}{2k} \int_{-\infty}^\infty U(\sqrt{b^2 + z^2})\, dz. \tag{1.51}$$

Unlike the Born approximation, this method satisfies unitarity and therefore, when it is extended to inelastic scattering, gives better results at low energies.

II. MULTICHANNEL SCATTERING

1.1. The Scattering Cross-section

The scattering of an electron by a complex atom is more complicated than the situation discussed above. Firstly, the atom has structure and can be excited during the collision. Thus the scattered electron can lose energy. Secondly, since the incident and target electrons are identical, the total wave function must satisfy the Pauli exclusion principle. This gives rise to exchange potentials which have a non-local character.

ATOMIC PROCESSES

The wave function describing an electron incident on an atom in state ϕ_i, and scattered leaving the atom in any state ϕ_j, is

$$\Psi \underset{r \to \infty}{\sim} \phi_i \chi_{\frac{1}{2}}^{m_i} e^{ik_i z} + \sum_j \phi_j \chi_{\frac{1}{2}}^{m_j} f_{ji}(\theta, \phi) \frac{e^{ik_j r}}{r} \qquad (2.1)$$

where $\chi_{\frac{1}{2}}^{m_i}$ is the spin function for the scattered electron, and where the wave number k_i is related to the total energy of the system E and the energy of the atomic state E_i through

$$E = E_i + \tfrac{1}{2} k_i^2. \qquad (2.2)$$

Of course, scattering is only possible into those states which are energetically allowed, for which $k_i^2 > 0$. The remaining states of the atom, for which $k_i^2 < 0$, can only exist virtually for a short time during the collision, but can, in some cases, play a significant role in the collision.

The scattering cross-section is defined as in §1.1, and in this case is

$$\sigma_{ji}(\theta, \phi) = \frac{k_j}{k_i} |f_{ji}(\theta, \phi)|^2 \qquad (2.3)$$

and the total cross-section is obtained by integrating over all angles.

2.2. The Radial Equations

As in §1.2, we can obtain the scattering amplitude by solving the time independent Schrödinger equation

$$H_{N+1} \Psi = E \Psi \qquad (2.4)$$

where the Hamiltonian H_{N+1}, describing the scattering of an electron by an atom or ion with nuclear charge Z containing N electrons, is

$$H_{N+1} = \sum_{i=1}^{N+1} \left(-\tfrac{1}{2} \nabla_i^2 - \frac{Z}{r_i} \right) + \sum_{i>j=1}^{N+1} \frac{1}{r_{ij}}. \qquad (2.5)$$

This can be written conveniently as

$$H_{N+1} = H_N + T_{N+1} + V_{N+1} \qquad (2.6)$$

where T_{N+1} is the kinetic energy of the $(N+1)$th electron, and where V_{N+1} is the interaction of this electron with the rest of the system.

In order to carry out the partial wave analysis, we make use of the fact that for light atomic systems, where the neglect of spin dependent terms in the Hamiltonian is a good approximation, L the total orbital angular momentum and S the total spin and their z components are conserved. We can therefore define a channel by the quantum numbers

$$\Gamma_i = \alpha_i k_i L_i S_i l_i s_i L S M_L M_S \pi \quad (2.7)$$

where L_i and S_i specify the atomic state, l_i is the orbital angular momentum of the scattered electron and s_i is its spin, π is the parity of the system and α_i specifies any other quantum numbers required to define the atomic state. L and S therefore satisfy the triangular relations

$$|L_i - l_i| \leq L \leq L_i + l_i \quad (2.8)$$
$$|S_i - \tfrac{1}{2}| \leq S \leq S_i + \tfrac{1}{2}$$

We now form channel functions which are eigenstates of LSM_LM_S and π:

$$\Phi_{\Gamma_i}(x_1, x_2, \ldots x_N, \hat{\mathbf{r}}_{N+1}, \sigma_{N+1})$$
$$= \sum_{\substack{M_{L_i} m_{l_i} \\ M_{S_i} m_i}} C_{L_i l_i}(LM_L; M_{L_i} m_{l_i}) C_{S_i \frac{1}{2}}(SM_S; M_{S_i} m_i)$$
$$\times \phi_i(x_1, \ldots x_N) Y_{l_i}^{m_{l_i}}(\hat{\mathbf{r}}_{N+1}) \chi_{\frac{1}{2}}^{m_i}(\sigma_{N+1}) \quad (2.9)$$

where $x_i = (\mathbf{r}_i, \sigma_i)$ are the space-spin variables of the ith electron, and we expand our total wave function in terms of a complete set of these functions:

$$\Psi = \sum_{k=1}^{N+1} (-)^{N+1-k} \sum_{\Gamma_i} \Phi_{\Gamma_i}(x_1, \ldots x_{k-1}, x_{k+1}, \ldots x_{N+1}, \hat{\mathbf{r}}_k, \sigma_k) r_k^{-1} u_i(r_k) \quad (2.10)$$

where the summation over k ensures that the total wave function satisfies the Pauli exclusion principle.

Substituting this expression for Ψ into (2.4), premultiplying by $\Phi_{\Gamma_i}(x_1 \ldots x_N \hat{\mathbf{r}}_{N+1} \sigma_{N+1})$ and integrating over the coordinates of all the electrons except r_{N+1} gives

$$\left(\frac{d^2}{dr^2} - \frac{l_i(l_i+1)}{r^2} + k_i^2\right) u_i(r) = 2 \sum_j \left[V_{ij}(r) u_j(r) + \int_0^\infty K_{ij}(r, r') u_j(r') \, dr'\right]$$

where (2.11)

$$V_{ij}(r_{N+1}) = \langle \Phi_{\Gamma_i}(x_1, \ldots x_N, \hat{\mathbf{r}}_{N+1}, \sigma_{N+1}), V_{N+1} \Phi_{\Gamma_j}(x_1, \ldots x_N, \hat{\mathbf{r}}_{N+1}, \sigma_{N+1})\rangle$$

and (2.12)

$$\int_0^\infty K_{ij}(r_{N+1}, r_N) u_j(r_N) \, dr_N$$
$$= -N\langle \Phi_{\Gamma_i}(x_1, \ldots x_N, \hat{\mathbf{r}}_{N+1}, \sigma_{N+1})(H_{N+1} - E)$$
$$\times \Phi_{\Gamma_j}(x_1, \ldots x_{N-1}, x_{N+1}, \hat{\mathbf{r}}_N, \sigma_N) u_j(r_N)\rangle. \quad (2.13)$$

The evaluation of the direct potential V_{ij} and exchange potential K_{ij} for any atom, except the simplest, is complex, and can only satisfactorily be carried out by a computer. However it is of interest to note here that

$$V_{ij}(r) \underset{r\to\infty}{\sim} \frac{N-Z}{r}\delta_{ij} + \sum_{\lambda=1}^{\lambda_{max}} a_{ij}^\lambda r^{-\lambda-1}. \tag{2.14}$$

The dipole coefficients $(a_{ij}^1)^2$ are directly proportional to the oscillator strength of the transition from state i to state j. Further, since V_{N+1} is spin independent, V_{ij} is zero unless $S_i = S_j$.

As in the case of potential scattering, the solution of (2.11) must be carried out numerically subject to the boundary conditions

$$u_{ij}(r) \propto r^{l_i+1} \tag{2.15}$$

at the origin, and

$$u_{ij}(r) \underset{r\to\infty}{\sim} k_i^{-\frac{1}{2}}[\sin(k_i r - \tfrac{1}{2}l_i\pi)\delta_{ij} + \cos(k_i r - \tfrac{1}{2}l_i\pi)K_{ij}], \quad \text{open channels}$$

$$u_{ij}(r) \underset{r\to\infty}{\sim} 0, \quad \text{closed channels} \tag{2.16}$$

for neutral atomic systems. The second index j on u_{ij} has been introduced to denote the possible linearly independent solutions of (2.11). If there are n_a channels open (i.e., have $k_i^2 > 0$) then we can find n_a linearly independent solutions satisfying (2.15) and (2.16). The matrix K_{ij} is the multi-channel generalization of (1.20) and can be shown to be real (from unitarity) and symmetric (from time reversal invariance).

The unitary and symmetric S-matrix and also the T-matrix are defined, in analogy with (1.22) and (1.23), by the $n_a \times n_a$ dimensional matrix equations

$$\mathbf{S} = \mathbf{I} + \mathbf{T} = \frac{\mathbf{I} + i\mathbf{K}}{\mathbf{I} - i\mathbf{K}} \tag{2.17}$$

where $\mathbf{I}$ is the unit matrix. It is now necessary to relate the S-matrix obtained by solving (2.11) with the scattering amplitude $f_{ji}(\theta, \phi)$. To do this we expand the plane wave in (2.1) in terms of the channel functions defined by (2.9)

$$\phi_i \chi_{\frac{1}{2}}^{m_i} \exp(ik_i z_{N+1}) \underset{r_{N+1}\to\infty}{\sim} \frac{i\pi^{\frac{1}{2}}}{k_i r_{N+1}} \sum_{LSl_i} C_{L_il_i}(LM_L; M_L 0)C_{S_i\frac{1}{2}}(SM_S; M_{S_i}m_i)$$

$$\times i^{l_i}(2l_i+1)^{\frac{1}{2}}\Phi_{\Gamma_i}(x_1, \ldots x_N, \hat{\mathbf{r}}_{N+1}, \sigma_{N+1})$$

$$\times \{\exp[-i(k_i r_{N+1} - \tfrac{1}{2}l_i\pi)] - \exp[i(k_i r_{N+1} - \tfrac{1}{2}l_i\pi)]\}. \tag{2.18}$$

Equating the ingoing wave term in this equation with that in (2.10), where we assume the asymptotic form (2.16), gives

$$f_{ji}(\theta, \phi) = i \left(\frac{\pi}{k_i k_j}\right)^{\frac{1}{2}} \sum_{LSl_i l_j} i^{(l_i - l_j)} (2l_i + 1)^{\frac{1}{2}} C_{L_i l_i}(LM_L; M_{L_i} 0)$$

$$\times C_{S_i \frac{1}{2}}(SM_S; M_{S_i} m_i) C_{L_j l_j}(LM_L; M_{L_j} m_j) C_{S_j \frac{1}{2}}(SM_S; M_{S_j} m_j)$$

$$\times (\delta_{ij} - S_{ij}) Y_{l_j}^{m_j}(\theta, \phi). \qquad (2.19)$$

The total cross-section for the transition $\alpha_i L_i S_i \to \alpha_j L_j S_j$ is obtained by integrating (2.3) over all angles, averaging over initial spin directions and summing over final spin directions:

$$\sigma(\alpha_i L_i S_i \to \alpha_j L_j S_j) = \frac{\pi}{k_i^2} \sum_{\pi L S l_i l_j} \frac{(2L + 1)(2S + 1)}{2(2L_i + 1)(2S_i + 1)} |T_{ji}|^2. \qquad (2.20)$$

2.3. The Structure of the Scattering Amplitude

Before proceeding further with the theoretical development it is useful to pause and consider the likely form of the solution of (2.11) as the total energy of the system changes. In Fig. 1 we show the spectrum of the Hamiltonian H_{N+1}, and we identify four distinct energy regions.

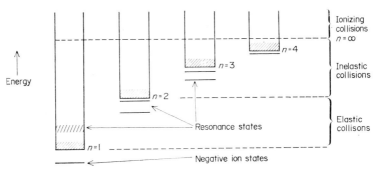

FIG. 1. Spectrum of the atomic Hamiltonian.

Region 1. Here all channels are closed and the only allowed solutions of (2.11) correspond to decaying waves in all channels. This is the well known bound state region which will not concern us further in these lectures.

Region 2. One channel is open and the corresponding K-matrix has dimension one. Only elastic scattering is possible. In certain circumstances bound states would exist in the absence of coupling to the open channel. These produce resonances in the elastic cross-section when the coupling between the open and closed channels is included. A typical example is

$$e^- + \text{He}^+ (1s) \to \text{He}^* (2s^2\, {}^1S) \to e^- + \text{H}(1s) \tag{2.21}$$

The transition via the intermediate state interferes with the direct process and produces a resonance. In this case there are a number of Rydberg series of resonances $2s\,ns\,{}^1S$, $2s\,np\,{}^1P$, ... $n = 2, 3, ...\infty$ converging to the $n = 2$ threshold. Low energy shape or potential resonances may also occur.

Region 3. More than one channel is open and the K-matrix becomes multi-dimensional. Inelastic scattering is possible as well as resonances converging to the higher thresholds. In many cases these resonances dominate the excitation cross-sections close to threshold.

Region 4. An infinite number of channels become open and ionization as well as excitation becomes energetically possible. This region is very difficult to treat adequately although, at higher energies, the Born approximation can be used.

2.4. Approximate Methods

The methods which have been used for the direct solution of equation (2.11) will be discussed in Section III. Here we describe very briefly some approximate methods that have been developed.

For approximate wave functions Ψ_i and Ψ_j defined by (2.10), with the asymptotic boundary condition given by (2.16), we obtain the variational principle (Kohn)[18]

$$\delta(\mathbf{L} - \tfrac{1}{2}\mathbf{K}) = 0 \tag{2.22}$$

where

$$L_{ij} = \langle \Psi_i, (H_{N+1} - E)\Psi_j \rangle. \tag{2.23}$$

Thus, given approximate trial values $\mathbf{L}^t$ and $\mathbf{K}^t$, an improved Kohn estimate is given by

$$\mathbf{K}^K = \mathbf{K}^t - 2\mathbf{L}. \tag{2.24}$$

If we allow the radial wave function in the variational principle (2.22) to have arbitary variations, we can show that it must satisfy (2.11) and thus $\mathbf{L} = 0$ in (2.24). On the other hand, if we start off with some trial form for the wave function, (2.24) gives us an approximate (correct to second order in the error in the wave function) estimate for $\mathbf{K}$. This approximation is often called the Kohn method. The distorted wave approximation is obtained by calculating the u_i in Ψ_i and Ψ_j using central potentials, and neglecting coupling, and then using (2.24).

The Coulomb–Born approximation (CB) is obtained by taking the trial functions $u_i = k_i^{-\frac{1}{2}} F_i(r)$ where $F_i(r)$ are the regular Coulomb wave functions.

The integrals in (2.24) are calculated neglecting exchange. The Coulomb–Born–Oppenheimer (CBO) approximation uses the same functions but includes exchange. Two variants of the above approximations have been developed by Seaton.[19] In the CB I and CBO I one assumes that $\mathbf{K} \ll 1$ and thus from (2.17) one can write

$$\mathbf{S} = \mathbf{I} + 2i\mathbf{K} \tag{2.25}$$

In the CB II and CBO II the matrix relation (2.17) is used.

Recent work by Bely[20] has also made extensive use of a method introduced by Ochkur[21] to estimate the exchange amplitude. Ochkur was able to show that the exchange amplitude was approximately related to the direct amplitude through

$$g(\theta, \phi) \approx \frac{K^2}{k^2} f(\theta, \phi) \tag{2.26}$$

where K^2 is the square of the momentum transfer. This result follows by retaining just the first term in the expansion of the Born–Oppenheimer (BO) amplitude in inverse powers of the energy and does not have the unpleasant divergent features of the BO method at low energies. Bely calls this approach the Coulomb exchange approximation. Further approximate methods will be introduced in connection with the explicit processes discussed in Sections III and IV.

2.5. Multichannel Effective Range Theory

When the potentials in (2.11) are short range we can define a derivative matrix

$$\mathbf{R} = \mathbf{u}(a) \left[a \frac{d\mathbf{u}}{da} \bigg|_{r=a} \right]^{-1} \tag{2.27}$$

where a is chosen so that the potential is effectively zero for $r \geq a$. It can be shown that $\mathbf{R}$ is a meromorphic function of energy with only poles on the real energy axis (Lane and Thomas).[22] It can be expanded as

$$\mathbf{R} = \frac{1}{2a} \sum_i \frac{\gamma_i \times \gamma_i}{E_i - E} \tag{2.28}$$

where the γ_i are reduced-width amplitudes. The $\mathbf{R}$-matrix is then simply related to the $\mathbf{K}$-matrix by means of the regular and irregular Bessel functions. It is found that

$$\mathbf{K} = \mathbf{k}^{l+\frac{1}{2}} \mathbf{M} \, \mathbf{k}^{l+\frac{1}{2}} \tag{2.29}$$

where **k** and *l* are diagonal matrices, and where the real and symmetric matrix **M**, introduced by Ross and Shaw,[23] is analytic through all the thresholds and can be expanded:

$$\mathbf{M}(k^2) = \mathbf{M}_0 + \mathbf{M}_1 k^2 + \mathbf{M}_2 k^4 + \dots . \tag{2.30}$$

If we have information, experimental or theoretical or both, about scattering in one region of energy we can calculate **M** in this region and extrapolate it to neighbouring regions of energy.

Gailitis[24] has extended this **M**-matrix approach to the situation where there are Coulomb potentials; this has been used to calculate the resonances in $e^- - \mathrm{He}^+$ scattering, using information on the scattering above the $n = 2$ threshold.

An alternative but equivalent approach has been to generalize the quantum defect theory (see Section I.5) to coupled channels. Seaton[25] shows that the **K**-matrix, defined by

$$\mathbf{u} \underset{r \to \infty}{\sim} \mathbf{k}^{-\frac{1}{2}}(\sin \theta \cdot \mathbf{1} + \cos \theta \cdot \mathbf{K}) \tag{2.31}$$

where

$$\theta = \mathbf{k}r - \tfrac{1}{2}\pi \mathbf{l} - \boldsymbol{\varepsilon} + \arg \Gamma(\mathbf{l}+1+i\boldsymbol{\eta})$$

is a slowly varying function of the energy (**ε** is the vector with components $\eta_i \log(2k_i r)$ and **1** is the vector with all components unity). The **S**-matrix is then given by

$$\mathbf{S} = \frac{i\mathbf{l} - \mathbf{K}}{\mathbf{T} + \mathbf{K}} \tag{2.32}$$

where **T** is a diagonal matrix with elements

$$T_{ij} = \begin{cases} i, & k_j^2 > 0 \\ \tan \pi v_j, & k_j^2 < 0 \end{cases} \tag{2.33}$$

where $k_j^2 = -Z^2/v_j^2$. When all the channels are open (2.32) reduces to (2.17). When all the channels are closed we have bound states at the energies where

$$|\mathbf{T} + \mathbf{K}| = 0 \tag{2.34}$$

This reduces to the result obtained in Section I in the one channel case.

III. ELECTRON SCATTERING BY HYDROGEN ATOMS AND HYDROGEN-LIKE IONS

This system has been the subject of the most extensive study, both because it provides a model for more complex atomic systems and also because it is important in its own right.

Most attempts to calculate cross sections at low energies have been based on the solution of (2.11). These must be truncated to a numerically manageable set which, at the same time, is physically meaningful. Unfortunately this is not always possible. The usual procedure is to retain in (2.11) those states between which the transition is required, and to include, as well, those states which are strongly coupled either to the initial, or to the final state. This gives the so-called close coupling approximation. In some cases, e.g., transitions between highly excited states, this gives too many coupled equations to be handled numerically. However, for transitions involving the low lying levels, satisfactory results can often be obtained.

When the $1s, 2s$ and $2p$ states are included in the close coupling approximation, we get the following coupled channels for each L and S

Table I

$\pi = (-)^L$	$\pi = (-)^{L+1}$
$1s, \quad l_i = L$	$2p, \quad l_i = L$
$2s, \quad l_i = L$	
$2p, \quad l_i = L+1, L-1$	

This gives 4 coupled channels when $L \geq 1$, and 3 coupled channels when $L = 0$. Note that the single channel, when the parity is $(-)^{L+1}$, is not of interest for excitation from the $1s$ state.

In Fig. 2 we show the ^1S phase shift calculated in a number of approximations below the $n = 2$ excitation threshold. This illustrates a number of very important features of electron–atom scattering and it is therefore worth spending some time discussing them.

(i) *Zero Energy Limit of the Phase Shift*

This is π radians, which is a special case of the generalized Levinson theorem

$$\delta_l(0) - \delta_l(\infty) = (n_b + n_p)\pi \qquad (3.1)$$

where n_b is the number of true bound states, n_p is the number of bound states excluded by the Pauli principle, and we ignore the effect of resonances at the higher thresholds. This was first discussed by Swan.[28] Examples of excluded bound states are

$$1s^2\, {}^3S \quad \text{in} \quad e^- - H \qquad (3.2)$$
$$1s^3\, {}^2S \quad \text{in} \quad e^- - He$$

In both these cases the phase shift starts from π although there is no true bound state. In our example $n_p = 0$ but the $1s^2\,{}^1S$ state is a true bound state of H^- so $n_b = 1$.

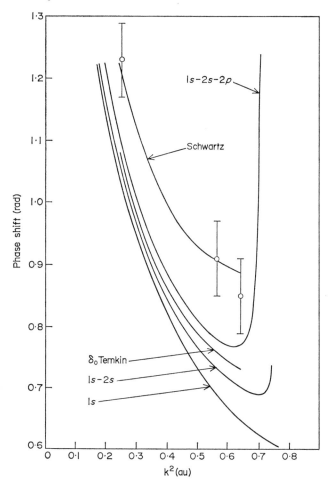

FIG. 2. ^1S phase shift for elastic $e^- - H$ scattering in various approximations. Schwartz[27] results are from a 50 parameter variational calculation. The points, with their associated error bars, refer to the calculations of Temkin.[26]

(ii) *Minimum Principle for the Phase Shift*

We see that

$$\delta_{\text{static}} \leqslant \delta_{1s2s2p} \leqslant \delta_{\text{variational}} \tag{3.3}$$

at all energies considered. Here, the static approximation is obtained by retaining just the $1s$ state in (2.11) and the variational result was obtained by Schwartz[27] using the Kohn variational method and a trial function with up to 50 terms. His phase shift results are probably correct to four significant figures, and can be assumed to be essentially exact in our discussion.

The basic theory has been developed by Hahn et al.,[29–30] Gailitis[24], McKinley and Macek,[31] and Sugar and Blankenbecler.[32] A convenient way of discussing this principle is to introduce, following Feshbach,[33] the projection operators P, onto those states retained in the close coupling expansion, and $Q = 1 - P$. Then the Schrödinger Eqn (2.4) can be written

$$P(H_{N+1} - E)(P + Q)\Psi = 0$$
$$Q(H_{N+1} - E)(P + Q)\Psi = 0. \quad (3.4)$$

The second equation gives

$$Q\Psi = -Q \frac{1}{Q(H_{N+1} - E)Q} QH_{N+1}P\Psi \quad (3.5)$$

which can be substituted into the first equation to give

$$P\left[H_{N+1} - PH_{N+1}Q \frac{1}{Q(H_{N+1} - E)Q} QH_{N+1}P - E\right]P\Psi = 0. \quad (3.6)$$

This equation is identical in content to the original equation. It must be compared with the close coupling equations which can be written in this notation as

$$P[H_{N+1} - E]P\Psi^P = 0. \quad (3.7)$$

The difference between (3.6) and (3.7) is contained in the optical potential term

$$V_{opt} = -PH_{N+1}Q \frac{1}{Q(H_{N+1} - E)Q} QH_{N+1}P. \quad (3.8)$$

Introducing the eigenstates of QHQ

$$QH_{N+1}Q\Phi_n = \varepsilon_n \Phi_n \quad (3.9)$$

then

$$V_{opt} = -\sum_n \frac{PH_{N+1}Q\Phi_n \rangle \langle \Phi_n QH_{N+1}P}{\varepsilon_n - E} \quad (3.10)$$

where the summation includes an integration over the continuous spectrum of $QH_{N+1}Q$.

If $E < \varepsilon_1$, then $V_{opt} < 0$. That is, if E lies below the spectrum of the channels omitted from the close coupling expansion, then the optical potential is attractive. But the monotonicity theorem (Spruch[34]) says that, if one potential is everywhere more attractive than another, then it gives a larger phase shift. It follows that the static phase shift is less than the $1s\,2s\,2p$ close coupling phase shift below the energy of the resonances below the $n = 2$ threshold (see below). In fact the result applies above these resonances if we define the phase shift to be continuous in energy through the resonances. In a similar way the $1s\,2s\,2p$ phase shift lies below the exact phase shift.

(iii) *Resonances*

Just below the $n = 2$ threshold, we see that the $1s\,2s\,2p$ approximation has a resonance. Its position is now known to be 9·56 eV and its width approximately 0·0475 eV. It is the first of an infinite series of resonances converging onto the $n = 2$ threshold. To see how these resonances arise, consider the asymptotic form of the Eqns (2.11) coupling the $2s$ and $2p$ states of hydrogen:

$$\left(\frac{d^2}{dr^2} + k_2^2\right) u_{2s}(r) = \frac{6}{r^2} u_{2p}(r)$$

$$\left(\frac{d^2}{dr^2} - \frac{2}{r^2} + k_2^2\right) u_{2p}(r) = \frac{6}{r^2} u_{2s}(r) \quad (3.11)$$

where we have neglected the quadrupole term coupling, which is diagonal in the $2p$ channel. The matrix of coefficients of the r^{-2} terms in this equation can be diagonalized, and Eqn (3.11) written

$$\left(\frac{d^2}{dr^2} - \frac{\lambda_i(\lambda_i + 1)}{r^2} + k_2^2\right) u_i(r) = 0, \quad i = 1, 2 \quad (3.12)$$

where

$$\lambda_1 = -\tfrac{1}{2} + i\sqrt{(\sqrt{37} - \tfrac{5}{4})}$$

$$\lambda_2 = -\tfrac{1}{2} + \sqrt{(\sqrt{37} + \tfrac{5}{4})} \quad (3.13)$$

and u_1 and u_2 are linear combinations of u_{2s} and u_{2p}. In the first solution, the modified centrifugal term is attractive, and gives an infinite number of bound states. The corresponding solution has the zero energy asymptotic form

$$u_1 \underset{r \to \infty}{\sim} r^{\frac{1}{2}} \sin\left(\operatorname{Im} \lambda_1 \log r + \delta\right) \quad (3.14)$$

which is oscillatory, confirming the statement about bound states. When the coupling with the $1s$ channel is included, these bound states broaden into resonances. It can be shown that similar resonance series occur below all the higher thresholds. Since all atoms become hydrogenic at their higher thresholds, this affect is common to all atoms, although the infinite number in each series is a direct consequence of the orbital degeneracy which occurs only in hydrogen. This degeneracy is lifted by the Lamb shift in hydrogen.

3.2. $e^- - H$ $1s - 2s$ and $1s - 2p$ Excitation

We show in Fig. 3 several calculations of the $1s-2s$ excitation cross-section, close to the threshold. These are the three state $1s - 2s - 2p$ close coupling approximation (Burke et al.[35]), the six state $1s-2s-2p-3s-3p-3d$ close coupling approximation (Burke et al.[36]) and the correlation approximation (Taylor and Burke[37]). The latter work solved (3.6) where P projected onto the $1s, 2s$, and $2p$ states, and the optical potential term was approximated by a set of 20 states of the form

$$\chi_j(\mathbf{r}_1, \mathbf{r}_2) = \sum_i r_1^{p_i} r_2^{q_i} r_{12}^{s_i} \exp\left[-\kappa(r_1 + r_2)\right] a_{ij} \qquad (3.15)$$

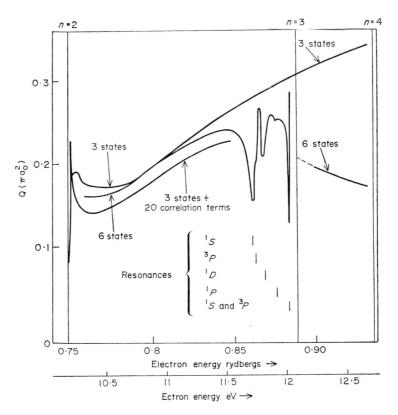

FIG. 3. The $1s - 2s$ $e^- - H$ excitation cross-section Q close to threshold.

chosen to diagonalize

$$\langle \chi_j, QH_{N+1}Q\chi_k \rangle = \varepsilon_j^{Tr} \delta_{jk}. \tag{3.16}$$

The resultant ε_j^{Tr} are upper bounds on the ε_j defined by (3.9).

We note that all three approximations are in good agreement below the $n = 3$ threshold. In this region they all have the correct asymptotic form since they all contain all open channels. We also note the resonances converging onto the $n = 3$ threshold in the six state calculation. A further resonance with the configuration $2s\,2p\,^1P$ occurs just above the $n = 2$ threshold. This resonance also appears in the H^- photodetachment cross-section. Finally, we note that above the $n = 3$ threshold the three state and the six state calculations diverge, mainly because the latter approximation allows for loss of flux omitted from the former.

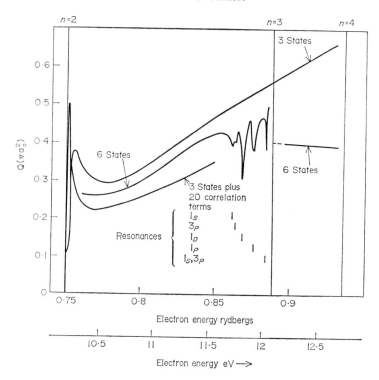

FIG. 4. The $1s-2p\ e^-$—H excitation cross-section Q close to threshold.

In Fig. 4 we show the $1s-2p$ excitation cross-section close to the threshold. This has very similar features to the $1s-2s$ cross-section.

In Fig. 5 we show the $1s-2s$ excitation cross-section over an energy range from threshold to 1000 eV. The experimental measurements were made by Kauppila et al.[38] They include direct excitation and indirect excitation by cascade from the higher states. After normalizing to the Born approximation above about 200 eV, it is found that the measurements are about 20% lower than the six state results below the $n = 3$ threshold. Above the $n = 3$ threshold, both the Born approximation and the three state ($1s\,2s\,2p$ close coupling) calculations are in error.

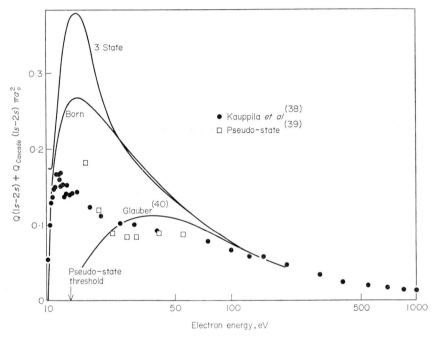

Fig. 5. The $1s$–$2s$ e^-–H excitation cross-section up to 1000 eV.

Part of the difficulty with the close coupling approximation is that it can only include a finite number of channels. The continuum terms in the sum Γ_i in (2.10) are omitted completely. The pseudo-state method includes the atomic states of interest together with some pseudo-states which are not eigenstates of the target Hamiltonian, e.g.,

$$\Phi_{\Gamma_1}, \Phi_{\Gamma_2}, \ldots \Phi_{\Gamma_n}, \quad \Phi_{\Gamma_{n+1}}^{\text{P.S.}}, \ldots \Phi_{\Gamma_{n+m}}^{\text{P.S.}}. \qquad (3.17)$$

The pseudo-states expansion, by a suitable choice of states, will converge much faster than the eigenstate expansion. In the pseudo-state calculation in Fig. 5 carried out by Burke and Webb [39] the following states were included

$$1s\,2s\,2p\,3s^{\text{P.S.}}\,3p^{\text{P.S.}} \qquad (3.18)$$

The radial parts of $3s^{P.S.}$ and $3p^{P.S.}$ had the form

$$P_{3s}(r) = (a_1 r + a_2 r^2 + a_3 r^3) e^{-\mu_1 r} \tag{3.19}$$

$$P_{3p}(r) = (b_2 r^2 + b_3 r^3) e^{-\mu_2 r}$$

where μ_1 and μ_2 were chosen to confine these states to the same range as the $2s$ and $2p$ states, and the coefficients a_1, a_2, a_3, b_2 and b_3 to ensure that these states were orthogonal to the $1s$, $2s$, and $2p$ states, and were normalized.

The other calculation presented in Fig. 5 is based on the Glauber approximation (Tai et al.[40]). This has been discussed briefly in Section I.6. It is seen that the Glauber approximation is applicable at much lower energies than the Born approximation, while tending to the Born approximation at higher energies. However since exchange is not included in this approximation it cannot be extended down to the threshold region with any reliability.

3.3. $e^- - H$ Transitions Involving Excited States

The close coupling method fails for transitions between excited states because of the number of states which would have to be coupled gets too large.

Seaton[41] and Saraph[42] have discussed a semi-classical or impact parameter method which has been used to calculate $n \to n + 1$ transitions in hydrogen. The incident electron is assumed to follow a classical path, which is valid for large impact parameters, when the velocity is not too low, i.e. when

$$R \gg a \quad \text{and} \quad R \gg \lambdabar \tag{3.20}$$

where R is the distance of closest approach, a is the atomic dimensions and $\lambdabar = k^{-1}$. The cross-section is then obtained either from

$$\sigma_{i \to j}^w = 2\pi \int_{R_0}^{\infty} P_{ji}(R) R \, dR \tag{3.21}$$

in the case of weak coupling, where R_0 is a cut-off distance chosen to give the Born approximation at high energies, or by

$$\sigma_{i \to j}^s = \tfrac{1}{2}\pi R_1^2 + 2\pi \int_{R_1}^{\infty} P_{ji}(R) R \, dR \tag{3.22}$$

in the case of strong coupling, where R_1 is chosen by the condition $P_{ji}(R_1) = \tfrac{1}{2}$ and $P_{ji}(R)$ is oscillatory for $R < R_1$.

The transition probability is taken from first order time dependent perturbation theory:

$$P_{ji} = \frac{1}{\omega_i} \sum_{\substack{\text{degenerate} \\ \text{levels}}} \left| \int_{-\infty}^{\infty} \exp[i(E_j - E_i)t] \langle \Psi_j | V | \Psi_i \rangle \, dt \right|^2 \quad (3.23)$$

where V is the time dependent interaction between the incident electron and the atom and ω_i is the statistical weight of the initial state. If, in the evaluation of this integral, we consider just the dipole term we obtain

$$P_{ji} = \frac{R_{ji}}{3\omega_i} \left| \int_{-\infty}^{\infty} \frac{\exp[i(E_j - E_i)t] \, \mathbf{r}'(t)}{r'^3(t)} \, dt \right|^2 \quad (3.24)$$

where

$$R_{ji} = \sum_{\substack{\text{degenerate} \\ \text{levels}}} |\int \Psi_j^* \mathbf{r} \Psi_i \, d\mathbf{r}|^2 = \frac{3\omega_i f_{ij}}{k_i^2 - k_j^2} \quad (3.25)$$

and thus the transition probability is proportional to the oscillator strength of the transition. The best cross-section estimates were obtained from the smaller of σ^w and σ^s, and in Fig. 6 the results for $n = 5 \to 6$ transitions are compared with the Born and the classical results of Gryzinski.[43]

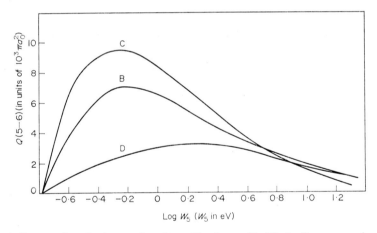

FIG. 6. Cross-sections for the $n = 5 \to 6$ transition in e^-–H. B is the Born approximation, C the classical approximation of Gryzinski,[43] and D the dipole approximation of Saraph. (After Saraph[42]).

Recently a significant advance has been made by Percival and Richards[44] in the study of transitions involving highly excited states of atoms. It is just in this situation that the correspondence principle leads us to expect that there is a region where classical mechanics can be applied. Percival and Richards have been able to generalize Heisenberg's correspondence principle to the case of strongly coupled states where the change in principal quantum number is small compared with itself. This is discussed further by Percival in his lectures.

3.4. Electron Scattering by Hydrogen-like Ions

We conclude this section by considering briefly the scattering of electrons by hydrogen-like ions. By making a transformation to new variables $\rho_i = Zr_i$ and $\varepsilon = EZ^{-2}$, the Schrödinger Eqn (2.4) can be written

$$\left[-\sum_{i=1}^{N+1} \left(\tfrac{1}{2}\nabla_i^2 + \frac{1}{\rho_i} \right) + \frac{1}{Z} \sum_{i>j} \frac{1}{\rho_{ij}} \right] \Psi = \varepsilon \Psi. \quad (3.26)$$

As Z increases, the electron–electron repulsion term decreases in importance, and consequently the close coupling approximation might be expected to

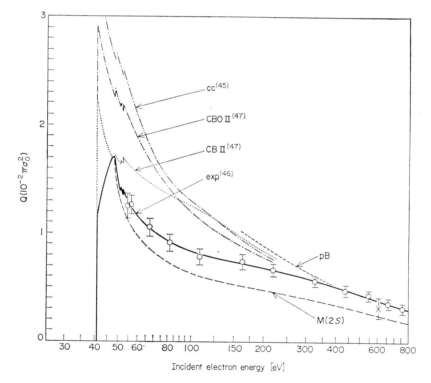

FIG. 7. The $1s-2s$ excitation cross-section of e^-–He$^+(1s)$ in various theoretical calculations compared with experiment. The curve $p.B$ is the plane-wave Born approximation, and the curve $M(2s)$ shows the experimentally determined cross-section for direct excitation to He$^+(2s)$.

increase in accuracy provided all states which become degenerate in the limit of infinite nuclear charge are included in the expansion.

In Fig. 7 we show the $1s-2s$ excitation cross section of $He^+(1s)$, calculated in the $1s\,2s\,2p$ close coupling approximation (Burke et al.[45]) and compared with the experiments of Dance et al.[46] Also shown are the Coulomb–Born calculations of Burgess et al.[47] The agreement between theory and experiment is still not satisfactory but is better than for $e^- - H$, bearing out our general discussion given above.

IV. ELECTRON SCATTERING BY COMPLEX ATOMS AND IONS

4.1. Electron Scattering by He and He-Like Ions

After $e^- - H$ scattering, the scattering of electrons by neutral helium provides the next simplest test of our theories. This system is also of great astrophysical importance.

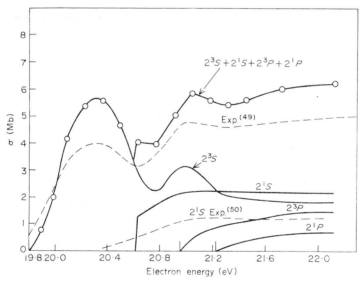

FIG. 8. Total metastable excitation cross-sections for electron impact on He, and the individual 2^3S, 1^1S, 2^3P, and 2^1P contributions.

The most accurate method of calculating the excitation cross-section of the $n = 2$ levels close to threshold, is using the close coupling approximation. This has been carried out including the following atomic states

$$1s^2\,{}^1S,\ 1s\,2s\,{}^3S,\ 1s\,2s\,{}^1S,\ 1s\,2p\,{}^3P,\ \text{and}\ 1s\,2p\,{}^1P \tag{4.1}$$

by Burke et al.[48] This gives five coupled integro-differential equations for $L = 0$ and seven for $L \geq 1$. Figure 8 shows the calculated total metastable excitation cross-section compared with measurements of Schulz and Fox.[49] We see that the cross-section exhibits peaks at ≈ 20.3 eV, 21.0 eV and perhaps a broader peak at higher energies. These calculations, and also angular distribution measurements by Andrick and Ehrhardt[51] show that the $e^- - $He system has the following resonant states,

$$(1s\,2s^2 + 1s\,2p^2)\,^2S\,(E_r = 19.3 \text{ eV}), \quad 1s\,2s\,2p\,^2P\,(E_r \approx 20.3 \text{ eV}),$$

$$(1s\,2s\,3d + 1s\,2p^2)\,^2D\,(E_r \approx 21.0 \text{ eV}), \quad (1s\,2s\,4f + 1s\,2p\,3d)\,^2F\,(E_r \approx 22 \text{ eV}),$$

(4.2)

where the most significant configurations are indicated. The ^2S resonance lies below the 2^3S threshold at 19.8 eV and contributes only to elastic scattering. The others, as shown in Fig. 8, strongly affect the excitation cross-sections. The dipole polarizability α of the 2^3S state is $\approx 313\,\alpha_0^3$ and of the 2^1S state is $\approx 788\,\alpha_0^3$ mainly arising from the first terms in the oscillator strength summation

$$\alpha_i = \sum_j \frac{f_{ij}}{(E_j - E_i)^2}. \quad (4.3)$$

It is in large part due to the resultant attractive αr^{-4} potential, coupled with the centrifugal repulsion $-l(l+1)/r^2$ at large distances, which is the cause of these resonances.

Another way of looking at this problem is to observe that the orbital degeneracy in H, which is broken in He, allows resonances, which would have occurred below threshold, to appear above or between thresholds in $e^- - $He. Experiments indicate that the $n = 3$ and 4 excitation cross-sections are dominated by resonances close to threshold, but no detailed calculations have yet been carried out. As in $e^- - $H scattering, it is doubtful if the $n = 1 \to 2$ close coupling calculations have too much validity above the $n = 3$ threshold, if only the states given in (4.1) are included.

Finally we note that calculations have been carried out by Vainshtein and Vinogradov,[52] using the impact parameter method, for the $2^3P - 3^3S$ and the $3^3S - 3^3P$ transitions.

Plasma measurements for ionization, excitation of resonance lines, and transitions between $n = 2$ singlet and triplet states in CV and OVII have been made by Kunze et al.,[53] and similar measurements at higher temperatures have been made by Elton and Köppendörfer.[54] The intensity ratio $I(2^1P - 1^1S)/I(2^3P - 1^1S)$ has been calculated by Bely[55] allowing for cascade, using the Coulomb–Born approximation and the Coulomb exchange approximation.

4.2. Electron Scattering by Li and Li-like Ions

The scattering of electrons by Li has been calculated by Karule,[56] Karule and Peterkop,[57] Burke and Taylor,[58] using the close coupling approximation, and by Vo Ky Lan[59] and other workers using the polarized orbital approximation. The elastic scattering calculations are summarized in Fig. 9, where they are compared with measurements of Perel *et al.*[60] (recent experimental work reported by Bederson has indicated that these should be re-normalized downwards by about a factor of two).

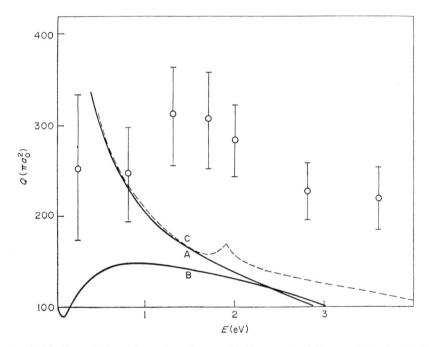

FIG. 9. Elastic scattering criss-sectoon for $e^- -$Li. Curves A and B are polarized orbital calculations of Vo Ky Lan,[59] curve C close coupling calculations of Burke and Taylor,[58] and the experimental points are those of Perel *et al.*[60] (After Vo Ky Lan.[59])

The close coupling calculations, carried out with the inclusion of the $1s^2 2s\,^2S$ and the $1s^2 2p\,^2P$ states, are given by curve C, while the single channel polarized orbital calculations are given by curves A and B. The polarized orbital method replaces the undistorted atomic state in (2.10) by

$$\Phi_\Gamma + \Phi_\Gamma^{pol} \tag{4.4}$$

where Φ_Γ^{pol} is the distorted part of the atomic wave function, calculated in the static field of the incident electron. This gives rise to additional direct and exchange potentials, the leading direct potential being asymptotic to αr^{-4}, where α is the atomic dipole polarizability. The additional exchange potential is important and its neglect leads to curve B. Curve A is in excellent agreement with the close coupling calculations below the $1s^2\,2p\,^2P$ threshold.

Transitions in Li-like ions are important in the measurement of electron temperatures (Heroux[61–62]). For example, in Fig. 10, we give the level scheme of NV. In coronal conditions the population of the $2p$ and $3p$ levels will depend on a balance between electronic excitation from the ground state and radiative decay back to the ground state. Because of the large energy difference between the two excitation energies, the relative population of the levels will be a very sensitive function of the electron temperature in the range 10–100 eV.

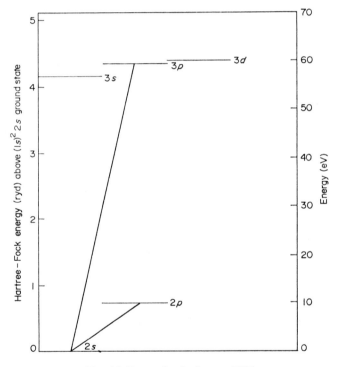

FIG. 10. Energy level scheme of NV.

Burke et al.[63] have obtained cross-sections for all transitions involving the levels $2s$, $2p$, $3s$, $3p$ and $3d$ in NV. Their results for the $2s-2p$ and $2s-3p$ transitions are given in Figs. 11 and 12. The strong coupling results are obtained including just the initial and final states in (2.18). The Bethe–Seaton approximation follows directly from the Bethe approximation. It gives

$$\sigma_{i \to j} = \frac{4\pi^2}{k_i^2 \sqrt{3}} \frac{f_{ji}}{E_j - E_i} g(k_j, k_i) \tag{4.5}$$

where f_{ji} is the optical oscillator strength and $g(k_j, k_i)$ is the Kramers–Gaunt factor which is $\sim \log k_i^2$ at high energies, but which is taken by Seaton[64] to be about 0.2 at low energies. Recent plasma measurements by Boland et al.[65] have confirmed the close coupling results.

Bely[66] has now calculated the cross-sections for $2s-ns, 2s-np$, and $2s-nd(n \leq 7)$ transitions in Li-like ions using the Coulomb–Born approximation.

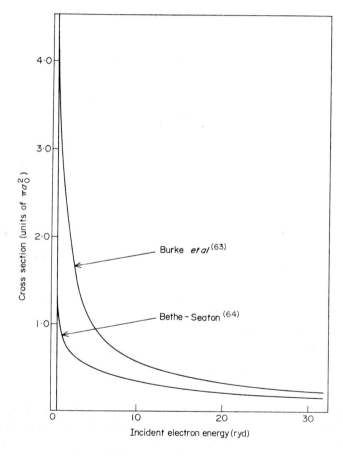

FIG. 11. The $2s-2p$ cross-section for $e^- - NV$.

4.3. Electron Scattering by other Alkalis and Alkali-like Ions

Karule[56] and Karule and Peterkop[57] have extended their 2 state close coupling calculations to Na, K, and Cs. The calculations for $e^- - K$ are particularly interesting, in view of recent measurements of the spin exchange cross-sections for elastic scattering (Collins et al.[67]) and for excitation (Rubin et al.[68]). These provide a very sensitive test of theory and the agreement of theory with experiment illustrated for elastic scattering in Fig. 13 is satisfactory.

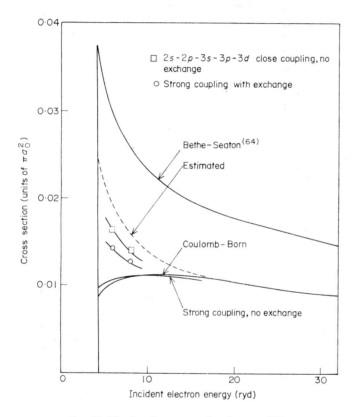

FIG. 12. The $2s-3p$ cross-section for $e^- - NV$.

The close coupling equations for $e^- - $Mg II ($3s$, $3p$ and $3d$) and for $e^- - $Ca II ($4s$, $3d$ and $4p$) have been solved by Burke and Moores.[69] The calculated positions and widths of the resonances are in moderate agreement with experiments of Newsom.[70] Other calculations in the Coulomb–Born II approximation for $e^- - $Mg II have been carried out by Bely et al.[71]

4.4. Electron Scattering by Beryllium-like Ions

Osterbrock[72] has carried out close coupling calculations for members of the isoelectronic sequence from BI to Ne VII. The results for the lowest $^1S-^1P$ and $^1S-^3P$ transitions are in moderate agreement with Born–Oppenheimer calculations of Beigman and Vainshtein,[73] after an error in the latter work had been corrected.

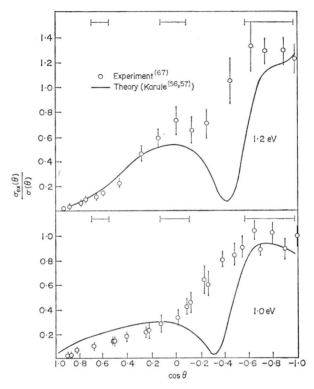

FIG. 13. The ratio of the differential spin-exchange cross-section $\sigma_{ex}(\theta)$ to the differential cross-section $\sigma(\theta)$ for elastic $e^- - K$ scattering. (After Collins et al.[67])

4.5. Transitions Involving np^q Atoms and Ions

A most important application is the calculation of transitions between ground state terms in atoms and ions with configurations

$$1s^2 \, 2s^2 \, 2p^q$$

and

$$1s^2 \, 2s^2 \, 2p^6 \, 3s^2 \, 3p^q$$

with $q = 2, 3$ and 4. The terms of interest are

$$
\begin{array}{llll}
q = 2 & {}^3P & {}^1D & {}^1S \\
q = 3 & {}^4S & {}^2D & {}^2P \\
q = 4 & {}^3P & {}^1D & {}^1S
\end{array}
\quad (4.6)
$$

which gives rise to forbidden lines in the visible spectrum e.g., in atomic oxygen where we have the spectrum shown in Fig. 14.

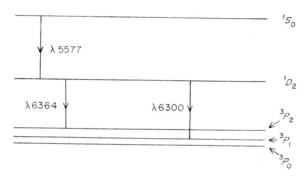

FIG. 14. Forbidden lines in O I.

The transition probabilities are small, since they are forbidden to electric dipole transitions, and we have for example

$$
\begin{array}{l}
\text{O I} \quad {}^1S \text{ life-time} \approx 0.7 \text{ sec, electric quadrupole.} \\
\text{O I} \quad {}^1D \text{ life-time} \approx 100\text{--}200 \text{ sec, magnetic dipole.}
\end{array}
\quad (4.7)
$$

The forbidden lines in O I are observed in aurorae and the night air glow. The same transitions in O III form the strongest lines in the spectra of many gaseous nebulae.

The close coupling equations describing the collision with the inclusion of the ground state, have been formulated by Smith et al.,[74] Henry et al.[75] and Saraph et al.[76] and were discussed in an early work by Seaton,[77] who gave the first formulation in terms of coupled integro-differential equations. The formalism follows closely that described in Section II. However it is convenient to impose on our solution the orthogonality conditions

$$(u_i, P_{nl}) = 0 \quad \text{when} \quad l = l_i. \quad (4.8)$$

For the closed shells this implies that

$$(u_i, P_{1s}) = (u_i, P_{2s}) = 0 \quad \text{when} \quad l_i = 0 \quad (4.9)$$

which is certainly satisfied, without approximation, if the total wave function is antisymmetric, since we cannot get more than two electrons into an s shell.

However, for the open $2p$ shell (we assume that we are here discussing the first row atoms and ions) (4.8) is not true for the general solution. In general we can write

$$u_i = \tilde{u}_i + a_i P_{2p} \delta_{l_i 1} \tag{4.10}$$

where $\tilde{u}_i$ satisfies (4.8). This suggests that our general expansion (2.10) should be replaced by

$$\Psi = \sum_{k=1}^{N+1} (-1)^{N+1-k} \sum_{\Gamma_i} \Phi_{\Gamma_i}(x_1, \ldots x_{k-1}, x_{k+1}, \ldots x_N, \hat{r}_k \sigma_k) r_k^{-1} \tilde{u}_i(r_k)$$
$$+ C\Psi_0(1s^2 2s^2 2p^{q+1} LS; x_1, \ldots x_{N+1}). \tag{4.11}$$

The second term occurs whenever $2p^{q+1}$ contains a term LS, the total orbital and spin angular momenta of the collision. We can now derive coupled integro-differential equations for the $\tilde{u}_i$ coupled to a linear equation for the coefficient C, where the evaluation of the potential matrix elements is considerably simplified, since the orbitals occuring, both bound and continuum, are orthogonal.

It should be remarked at this point that this technique of forcing all the orbitals to be orthogonal, has been the basis of recent general formulations of the electron-atom collision problem (Smith and Morgan,[78] and Burke et al.[79]). It is in no sense an approximation but must be viewed in terms of its usefulness in allowing the complex two electron integrals which occur to be evaluated.

For neutral atomic systems it is probably a good approximation to neglect all target configurations, except the lowest, when the incident energy is below the threshold for these higher configurations. Results for $e^- - C$, $e^- - N$ and $e^- - O$ scattering retaining just the ground state configuration have been given by Henry et al.[75]

For ions, Rydberg series of resonances converging to these higher thresholds can lie close to the thresholds of interest, and care must therefore be taken in the calculations. Consider for example the $e^- -$ O III system discussed by Eissner et al.[80] In this case the lowest lying configuration is $2s^2 2p^2$ and the first excited configuration is $2s 2p^3$. The energy levels for the carbon isoelectronic sequence are shown in Fig. 15. The energy differences vary as linear functions of Z in accordance with the theory of Layzer.[81] If a $3s$ electron is now added to the ion in an excited state the binding energy is

$$\frac{Z^2}{(3-d)^2} \tag{4.12}$$

where d is the quantum defect which behaves like Z^{-1} in the limit of large Z. It follows that terms found from the configuration $2s\,2p^3\,3s$ will be bound for large Z and indeed have been observed. In Fig. 16 we show the energy levels $2s^2\,2p^2\,{}^3P,\,{}^1D,\,{}^1S$ and $2s\,2p^3\,({}^5S)\,3s\,{}^4S,\,2s\,2p^3\,({}^3D)\,3s\,{}^2D$ and $2s\,2p^3\,({}^3P)\,3s\,{}^2P$ as a function of Z. For e^- − O III scattering we see that the ${}^2P$ and ${}^2D$ states lie in the continuum just above the ${}^1S$ and ${}^1D$ thresholds, respectively, giving rise to resonances which can be expected to play significant roles in the excitation cross-sections. In Fig. 17 we show the total collision strengths defined as

$$\Omega(i \to j) = k_i^2 (2S_i + 1)(2L_i + 1)\sigma_{i \to j} \qquad (4.13)$$

for the transitions ${}^3P-{}^1S$ and ${}^1D-{}^1S$ obtained by Eissner et al.[80] using semi-empirical methods and compared with calculations neglecting the resonance obtained by Saraph et al.[76]

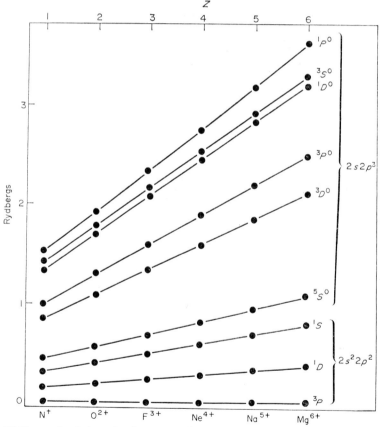

FIG. 15. Energy levels in $2s^2\,2p^2$ and $2s\,2p^3$ configurations in the carbon isoelectronic sequence, relative to $2s^2\,2p^2\,{}^3P$. (After Eissner et al.[80])

4.6. Fine Structure Transitions

In many problems of astrophysical interest it is important to know the cross-sections for transitions involving the fine structure levels of the atom or ion. For example Saraph and Seaton[82] have discussed the dependence of the ratio

$$R = \frac{I(^2D_{3/2} \to {}^4S)}{I(^2D_{5/2} \to {}^4S)} \qquad (4.14)$$

upon electron temperature for O II. We are thus interested in calculating the excitation cross-sections

$$e^- + O^+({}^4S) \to e^- + O^+({}^2D_{3/2})$$
$$e^- + O^+({}^4S) \to e^- + O^+({}^2D_{5/2}). \qquad (4.15)$$

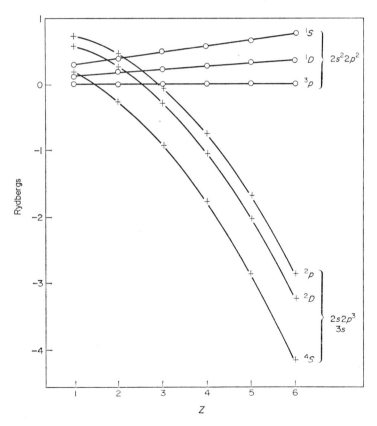

FIG. 16. Energy levels, relative to $2s^2\,2p^2\,{}^3P$, for $2s^2\,2p^2\,{}^1D$ and ${}^1S$, and $2s\,2p^3\,({}^5S)\,3s\,{}^4S$, $2s\,2p^3\,({}^3D)\,3s\,{}^2D$ and $2s\,2p^3\,({}^3P)\,3s\,{}^2P$ (After Eissner et al.[80]).

If the spin–orbit interaction was zero, and thus the $^2D_{3/2}$ and $^2D_{5/2}$ levels were exactly degenerate, we could obtain the S-matrix in this new representation from the representation defined by (2.7), by a unitary transformation. We would need to transform between the two coupling schemes

$$(S_i\tfrac{1}{2})S(L_i l_i)LJM_J \qquad (4.16)$$

and

$$(S_i L_i)J_i(\tfrac{1}{2}l_i)j_i JM_J. \qquad (4.17)$$

This is simply achieved by the Wigner $9-j$ coefficient (Edmonds[83]).

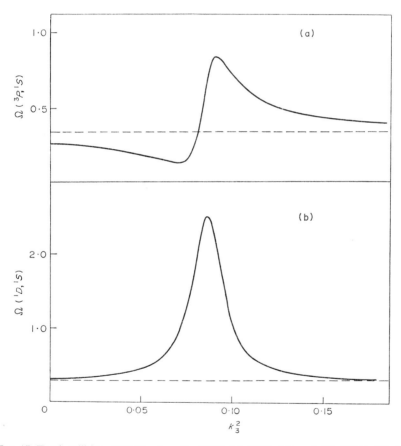

FIG. 17. Total collision strengths for $e^- - $O III: (a) $\Omega(^3P, {}^1S)$ and (b) $\Omega(^1D, {}^1S)$. The broken lines are the results of Sataph et al.[76] (After Eissner et al.[80]).

In the case interest of the spin–orbit interaction is small and this method of proceeding can still be used. We find that

$$S(\alpha_i L_i S_i J_i l_i j_i J, \alpha_j L_j S_j J_j l_j j_j J)$$

$$= \sum_{LS} \begin{pmatrix} S_i & L_i & J_i \\ \tfrac{1}{2} & l_i & j_i \\ S & L & J \end{pmatrix} S(\alpha_i L_i S_i l_i SL, \alpha_j L_j S_j l_j SL)$$

$$\times \begin{pmatrix} S_j & L_j & J_j \\ \tfrac{1}{2} & l_j & J_j \\ S & L & J \end{pmatrix}. \qquad (4.18)$$

This expression can be simplified if one of the two terms contains only one fine structure level as is the case in (4.15). We find then that

$$\Omega(\alpha_i L_i S_i, \alpha_j L_j S_j J_j) = \frac{2J_j + 1}{(2S_j + 1)(2L_j + 1)} \Omega(\alpha_i L_i S_i, \alpha_j L_j S_j) \qquad (4.19)$$

if $L_i = 0$ or if $S_i = 0$.

4.7. Ionization

In this section we discuss very briefly recent theoretical work on electron–atom ionization. This is treated fully in a review article by Rudge.[8] First a few general remarks are in order. From a purely theoretical point of view ionization is very much more difficult than excitation. This arises from the difficulty of writing down a wave function with the correct asymptotic form for the two outgoing electrons in the final state. Considerable effort is being given to this problem but so far no calculational procedures have been introduced which will lead necessarily to the correct result even for the electron–hydrogen atom system. On the other hand since in most cases of astrophysical interest we are interested only in cross-sections integrated over the relative energies of the outgoing electrons this tends to correct errors in the approximations and reliable results can be obtained.

It is convenient when discussing ionization to define a direct scattering amplitude $f(\mathbf{k}, \mathbf{k}')$ an an exchange scattering amplitude $g(\mathbf{k}, \mathbf{k}')$, here $f(\mathbf{k}, \mathbf{k}')$ describes the process where the incident electron is scattered with final momentum $\mathbf{k}'$ and $g(\mathbf{k}, \mathbf{k}')$ describes the process where the incident electron has final velocity $\mathbf{k}$. It is therefore not unexpected that

$$f(\mathbf{k}, \mathbf{k}') = g(\mathbf{k}', \mathbf{k}). \qquad (4.20)$$

Peterkop[84–87] and Rudge and Seaton[88] have derived the following integral expression for the scattering amplitude

$$f(\mathbf{k}, \mathbf{k}') = -\left(\frac{1}{2\pi}\right)^{5/2} e^{i\Delta} \int \phi(-\mathbf{k}, z, \mathbf{r}) \phi(-\mathbf{k}', z', \mathbf{r}')(H - E)\Psi^+(\mathbf{r}, \mathbf{r}') \, d\mathbf{r} \, d\mathbf{r}' \quad (4.21)$$

where we have restricted consideration for convenience to the two-electron system. Here Ψ^+ is the total wave function satisfying outgoing wave boundary conditions and the ϕ are Coulomb wave functions normalized to unit ingoing wave amplitude corresponding to nuclear charge Z. Finally the phase

$$\Delta = \frac{2Z}{k} \log\left(\frac{k}{\sqrt{k^2 + k'^2}}\right) + \frac{2Z'}{k'} \log\left(\frac{k'}{\sqrt{k^2 + k'^2}}\right). \quad (4.22)$$

The cross-section for ionization, neglecting symmetrization, is

$$\sigma = \frac{1}{k_0} \int_0^E dE \, k k' \int\int d\hat{\mathbf{k}} \, d\hat{\mathbf{k}}' \, |f(\mathbf{k}, \mathbf{k}')|^2 \quad (4.23)$$

where k_0 is the incident electron wave number. Using $Z = 1$ and $Z' = 0$ in (4.21) and taking a plane incident wave to define Ψ gives the Born approximation. This overestimates the cross-section by as much as 50%. A better result is obtained by replacing f in (4.23) by

$$\tfrac{1}{8}|f(\mathbf{k}, \mathbf{k}') + f(\mathbf{k}', \mathbf{k})|^2 + \tfrac{3}{8}|f(\mathbf{k}, \mathbf{k}') - f(\mathbf{k}', \mathbf{k})|^2$$

where we have used (4.20) to define the exchange amplitude. Since exact wave functions are not used in calculating $f(\mathbf{k}, \mathbf{k}')$ the phase between $f(\mathbf{k}, \mathbf{k}')$ and $f(\mathbf{k}', \mathbf{k})$ is not correct and this has been taken as disposable in versions of the so called Born exchange approximations. Different choices of phase give results differing by up to 20% but agreement with theory for H(1s) can be satisfactory (Geltman et al.[89]). A considerable number of calculations have now been carried out for complex atoms using the Born and Born exchange approximations.

One important feature which is not given correctly by these simple approximations is the threshold behaviour of the cross-section. Wannier[90] using classical mechanical arguments showed that $\sigma \propto E^{1.127}$ for neutral atomic systems, where E is the total energy measured from threshold. Recent work by Peterkop[91] and Rau[92] has indicated that this result is correct.

V. PHOTOIONIZATION AND RECOMBINATION

5.1. Theory of Photoionization

The primary physical processes in the gas surrounding a star is often the absorption of ultra-violet radiation in atomic photoionization

$$A + h\nu \rightarrow A^+ + e^- \quad (5.1)$$

and the inverse process known as radiative recombination. The cross-section for abosorption of light frequency v is

$$\sigma_v = \frac{8\pi^3 v}{3c} \frac{1}{\omega_i} \sum_{if} \left| \langle \Psi_f^-, \sum_\mu e_\mu \mathbf{r}_\mu \Psi_i \rangle \right|^2 \tag{5.2}$$

in the dipole length approximation (Bethe and Salpeter[93]). Here Ψ_i is the initial state normalized to unity and Ψ_f^- is the final continuum state with ingoing wave boundary conditions, normalized to unit energy internal in the continuum

$$\langle \Psi_f^-(E), \Psi_f^-(E') \rangle = \delta(E - E') \tag{5.3}$$

ω_i is the statistical weight of the initial atomic state $(2L_i + 1)(2S_i + 1)$ and the factor three arises from the three possible orientations of the position vector $\mathbf{r}_\mu$. It is often convenient to write (5.2) in the equivalent forms

$$\sigma_v = \frac{8\pi^3 e^2 v}{c\omega_i} \sum_{if} \left| \langle \Psi_f^- \sum_\mu z_\mu \Psi_i \rangle \right|^2 \tag{5.4}$$

or

$$\sigma_v = \frac{e^2 h^2}{2\pi m^2 c v \omega_i} \sum_{if} \left| \langle \Psi_f^- \sum_\mu \frac{\partial}{\partial z_\mu} \Psi_i \rangle \right|^2 \tag{5.5}$$

known as the length and the velocity formulae respectively. A third form known as the acceleration form is occasionally used. If exact initial and final wave functions are used then (5.4) and (5.5) give the same result. If approximate wave functions are used then (5.4) and (5.5) give results which in general are different. The length formulation emphasises the wave functions at larger distances than the velocity formulation and is preferred if the wave functions are known best in that region.

It is usual when calculating the photoionization cross-section of a complex atomic system to adopt the expression defined by (2.10) for the final state wave function. The radial wave function $\mathbf{u}$ must then satisfy the boundary condition

$$\mathbf{u}^- \underset{r\to\infty}{\sim} \left(\frac{2m}{\pi\hbar^2 k}\right)^{\frac{1}{2}} (\sin\theta + \cos\theta \mathbf{K})(1 + i\mathbf{K})^{-1} \tag{5.6}$$

which is consistent with (5.3). See also (2.31). This form is obtained by taking appropriate linear combinations of the solutions defined by (2.16). The cross-section for photoionization of the jth channel is then given by using the jth column of $\mathbf{u}^-$ in defining the Ψ_f^- in (5.4) or (5.5). Further discussion of multichannel photoionization cross-sections is given by Henry and Lipsky.[94]

5.2. Photoionization of Hydrogen-like Ions

The cross-section for photoionization of a hydrogen-like ion of charge Z. from level nl can be obtained analytically. Averaging over the l quantum number we obtain

$$\sigma_n(Z, \varepsilon) = \left(\frac{2^6 \alpha \pi a_0^2}{3\sqrt{3}}\right) \frac{n}{Z^2} \frac{1}{(1+n^2\varepsilon)^3} g_{II}(n, \varepsilon) \tag{5.7}$$

where

$$\frac{2^6 \alpha \pi a_0^2}{3\sqrt{3}} = 7 \cdot 907 \times 10^{-18} \text{ cm}^2 \tag{5.8}$$

$\alpha = e^2/\hbar c$ being the fine structure constant $\approx 137^{-1}$ and a_0 the Bohr radius (Seaton[95]). The photon energy is

$$h\nu = Z^2 I_H (1/n^2 + \varepsilon) \tag{5.9}$$

$I_H (= 13 \cdot 605 \text{ eV})$ being the binding energy of the ground state of hydrogen. The ejected energy is $Z^2 \varepsilon$ in units of 13·6 eV. In (5.7) g_{II} is the Kramers–Gaunt factor which has the asymptotic expansion (Menzel and Pekeris[96])

$$g_{II}(n, \varepsilon) = 1 + 0 \cdot 1728 n^{-2/3}(u+1)^{-2/3}(u-1)$$
$$- (0 \cdot 0496) n^{-4/3}(u+1)^{-4/3}(u^2 + 4u/3 + 1) + \ldots \tag{5.10}$$

where $u = n^2 \varepsilon$.

It is also important to know the absorption cross-section for a $n \to n'$ transition. Assuming thermal Dopler broadening this is

$$K_\nu(n, n') = K_0(n, n') \exp\left(-b(\nu - \nu_{nn'})^2\right) \tag{5.11}$$

where

$$K_0(n, n') = \frac{\pi e^2}{m \nu_{nn'}} \left(\frac{M}{2\pi kT}\right)^{\frac{1}{2}} f(n, n') \tag{5.12}$$

and

$$b = \left(\frac{c}{\nu_{nn'}}\right)^2 \frac{M}{2kT} \tag{5.13}$$

where m is the electron mass, M is the ion mass and $f(n, n')$ is the oscillator strength. Hydrogenic f-values are given by

$$f(n, n') = \frac{2^5}{3\pi\sqrt{3}} \frac{nn'}{n'^2 - n^2} g_I(n, n') \tag{5.14}$$

where g_I is given by (5.10) with $u = -(n/n')^2$.

5.3. Photoionization of H^-

The negative hydrogen ion is the dominant source of opacity in the solar atmosphere and is important in the outer atmospheres of numerous other stars. Early work on the photoionization of H^- from the ground state was carried out by many workers using the dipole length approximation. Chandrasekhar[97] introduced the dipole velocity and acceleration formalisms which are not so sensitive to the long range part of the H^- wave function which was difficult to calculate accurately. In more recent work, Doughty et al.[98] used a 70-parameter wave function for H^- calculated by Schwartz and a $1s\,2s\,2p\,3s\,3p\,3d$ close coupling wave function for the final state.

The length, velocity and acceleration results of Doughty et al.'s calculation are compared with the experiments of Smith and Burch[99-100] in Fig. 18. The velocity results are seen to be in excellent agreement with experiment. One important feature which is missing from these curves is a sharp peak at 1129·5 Å with a width of 1·55 Å which is due to the intermediate $2s\,2p\,^1P$ resonance state of H^-. This state lies just above the $n = 2$ state of H and decays most of the time leaving the hydrogen atom in the $2s$ or $2p$ state. It has already been noted in connection with $1s-2s$ and $1s-2p$ excitation in $e^- - H$ scattering where it gives rise to a peak close to threshold (see Figs 3 and 4). Macek[101] has shown that the oscillator strength of the transition from the H^- ground state to this resonance state is approximately 0·044.

The absorption by H^- ions due to electron transitions from one state of the continuum to another state of the continuum can also be important in the stellar opacity. These free–free, transition matrix elements have been calculated by, amongst others, John[102] using the $1s$ exchange approximation for the $e^- - H$ continuum system. In this work $S \leftrightarrow P$ transitions were considered.

5.4. Photoionization of He

The spectrum of He is considerably more complex than that of H^- due to the presence of the Coulomb interaction seen by the ejected electron giving rise to Rydberg series of resonances. The absorption spectrum of He in its ground state was obtained by Madden and Codling[103] using the continuous light from the 180 MeV electron synchrotron at the National Bureau of Standards. Madden and Codling observed the resonant process

$$\text{He} + h\nu \to \text{He}^*(2s\,np\,^1P, 2p\,ns\,^1P) \to \text{He}^+ + e^- \quad (5.15)$$

interfering with the direct process

$$\text{He} + h\nu \to \text{He}^+ + e^-. \quad (5.16)$$

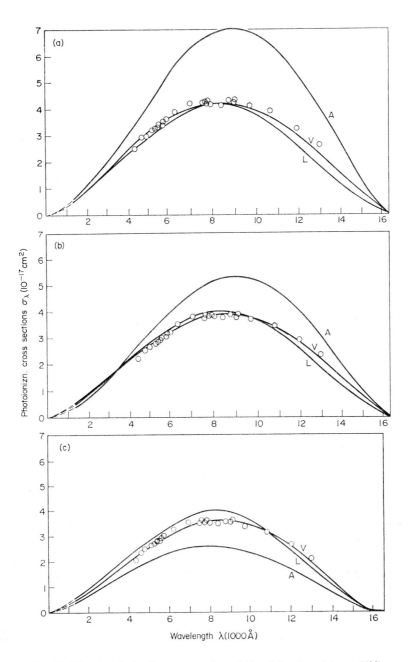

FIG. 18. The photoionization cross-section of H^-. (After Doughty et al.[98])

This gives rise to Rydberg series of resonances converging onto the $2s$, $2p$ level of the ion as n varies from 2 to ∞. Further faint series can also be seen converging onto the $n = 3$ and $n = 4$ levels of the ion. Only one broad series is actually seen converging onto the $n = 2$ threshold rather than two as implied by (5.15). This was explained by Cooper, Fano and Prats[104] who noted that the correct zeroth-order states are

$$(2s\,np \pm 2p\,ns)\,{}^1P. \tag{5.17}$$

The states with the + signs will be more strongly excited and decay faster than those with the − signs. The − series can only be seen under good resolution. A further series with the configuration $2p\,2pnd\,{}^1P$, expected in LS coupling, has not yet been observed but has been calculated (Burke and McVicar[105]).

An important observable in the discussion of photoionization into resonant continua is the shape of the resonance called by Fano[106] the line profile index. Figure 19 illustrates the varying absorption shapes found in different rare gas spectra. Each plot shows a partially resolved Rydberg series converging to a threshold for a new photoionization process (a) He near the $n = 2$ level of He^+ (see (5.15) and (5.16)). (b) Kr near the N_I edge (Samson[107]) and (c) Ar near the K edge (Schnopper et al.[108]). Fano and Cooper[109] show that the most general expression for the photoionization cross-section in the neighbourhood of a resonance is

$$\sigma = \sigma_a \frac{(q + \varepsilon)^2}{1 + \varepsilon^2} + \sigma_b \tag{5.18}$$

where σ_a and σ_b are smooth functions of energy and ε is given by

$$\varepsilon = \frac{E - E_r}{\Gamma/2} \tag{5.19}$$

where E_r is the resonance position and Γ its width. The line profile index can have any value between $-\infty$ and $+\infty$ and determines the shape of the resonance. In Fig. 19(a) q has a value close to -2.8 while in Fig. 19(b) the window type resonances correspond to a value of q close to zero.

If we define $\psi_j(E)$ as that linear combination of the continuum states at energy E which interacts with the discrete resonance state ϕ

$$\langle \psi_j(E) | H | \phi \rangle \neq 0 \tag{5.20}$$

$$\langle \psi_i(E) | H | \phi \rangle = 0 \qquad i \neq j$$

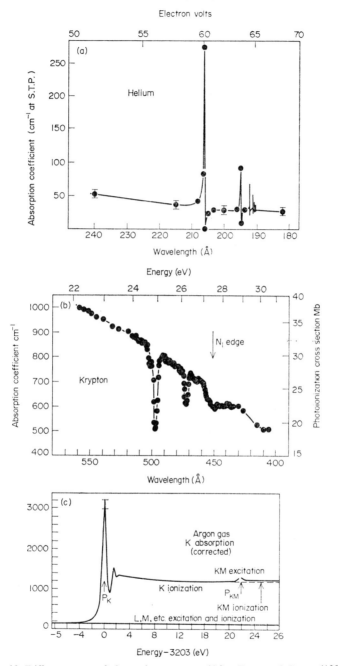

FIG. 19. Different types of absorptions spectra. (After Fano and Cooper[109])

then Fano and Cooper show that

$$q = \frac{\langle \Phi | z | \psi_0 \rangle}{\pi \langle \phi | H | \psi_j \rangle \langle \psi_j | z | \psi_0 \rangle} \quad (5.21)$$

where ψ_j is normalized to unit energy interval in the continuum, Φ is obtained from ϕ by admixing the appropriate amount of continuum required to diagonalize H and ψ_0 is the initial state of the atom or ion. When only one continuum channel is open, then ϕ interacts with the full continuum and σ_b is zero. This is the situation in He below the $n = 2$ threshold of He$^+$. We see that the value of the line profile index depends on subtle considerations based upon the moduli and phases of three matrix elements, and no general rules are yet available.

Our considerations up until now have been concerned with photoionization from the ground state of He. Photoionization from the metastable 2^3S and 2^1S states may be a significant mechanism for depopulating the 2^3S metastable state in some planetary nebulae. This process has recently been considered by Jacobs[110] and by Norcross.[111] It is found that there is a high probability that the He$^+$ ion will be left in an excited state if the incident photon energy is high enough. The processes are

$$\text{He}(2\,^1S) + h\nu \to \text{He}^+(2s, 2p) + e^-$$

$$\text{He}(2\,^3S) + h\nu \to \text{He}^+(2s, 2p) + e^- \quad (5.22)$$

5.5. Photoionization of Complex Atoms and Ions

A considerable amount of work has been carried out on the photoionization of atoms and ions with incomplete outer $2p$ and $3p$ subshells. Important amongst these calculations are the photodetachment cross-sections of C$^-$ and O$^-$ (e.g., Henry[112]) and of Si$^-$, S$^-$ and Cl$^-$ (Conneely et al.[113]). Work on the photoionization of Al, Si, P, S and Cl has also been reported by the latter workers. The absorption of radiation by some of these systems makes a major contribution to the bound-free opacity in many common astrophysical plasmas.

As an example the photoionization cross-sections of $1s^2\,2s^2\,2p^6\,3s^2\,3p^2\,^3P$ ^1D and ^1S states of Si calculated by Conneely et al.[113] are shown in Fig. 20 where they are compared with the measurements of Rich.[114] The calculations are carried out using close coupling continuum wave functions and the transition amplitude is evaluated in the dipole-length approximation. The dipole velocity approximation results were about half the size of the length values and are not given. The final states, allowed by angular momentum and parity considerations, are given in the following table

Table II

States allowed in the photoionization of Si:

Initial State	Final State	No. of coupled eqns.
$2p^2$ ^3P	$2p\,ks\,^3P^0$, $2p\,kd\,^3P^0$	2
	$2p\,kd\,^3D^0$	1
$2p^2$ ^1D	$2p\,ks\,^1P^0$, $2p\,kd\,^1P^0$	2
	$2p\,kd\,^1D^0$	1
	$2p\,kd\,^1F^0$, $2p\,kg\,^1F^0$	2
$2p^2$ ^1S	$2p\,ks\,^1P^0$, $2pkd^1P^0$	2

We see that in order to calculate the photoionization cross-sections shown in the figure it was necessary to solve two single equations and four sets of two coupled equations for the energies of interest. The spectrum of Si does not involve resonances in the approximation considered by Conneely *et al.* where only terms of the ground state configuration are retained in the final state. However the spectra of P, S and Cl are complicated by resonances converging to higher thresholds of the residual ion.

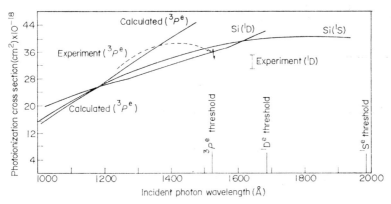

FIG. 20. Comparison of the close-coupling photoionization cross-section of Conneely *et al.*[113] for Si with the experiment of Rich[114]. (After Conneely *et al.*[113])

5.6. Recombination of Electrons and Ions

The recombination coefficient α is defined by the equation

$$\frac{dn(A^+)}{dt} = -\alpha n(e)n(A^+) \qquad (5.23)$$

where $n(A^+)$ is the number density of positive ions and $n(e)$ is the number density of free electrons. In plasmas of low density the recombination is dominated by radiative recombination. If the photoionization cross-section σ_{ij} for the process

$$A_i + h\nu \rightarrow A_j^+ + e^- \tag{5.24}$$

is known then the radiative recombination coefficient α_{ji} for the inverse process may be obtained from the standard relation of Milne[115] in the form

$$\alpha_{ji}(T) = \frac{\omega_i}{\omega_j} \left[\frac{2^{1/2} \exp(I_i/kT)}{c^2 \pi^{1/2} (mkT)^{3/2}} \right] \int_{I_i}^{\infty} (h\nu)^2 \sigma_{ij}(\nu) \exp\left(-\frac{h\nu}{kT}\right) d(h\nu) \tag{5.25}$$

where I_i is the ionization potential and ω_i and ω_j are the statistical weights of the states A_i and A_j^+ respectively.

If the photoionization cross-section times ν^2 is approximately constant over a range of energy appreciable compared with kT then this equation can be written as (Bates and Dalgarno[116])

$$\alpha_{ji}(T) = \frac{1\cdot 3 \times 10^6 I_i^2}{T^{1/2}} \frac{\omega_i \sigma_{ij}^T}{\omega_j} \tag{5.26}$$

where σ_{ij}^T is the threshold value of the cross-section in cm^2, I_i is in Rydberg's and T is in °K. The result is in cm^3/sec.

We have seen that the cross-section for photoionization of hydrogen-like ions from level n is known analytically (see 5.7). The corresponding recombination coefficient at temperature T into level n is then

$$\alpha_n(Z, T) = \mathscr{D} Z x_n^{3/2} S_n(\Lambda) \tag{5.27}$$

where, if α_s is the (Sommerfeld) fine structure constant $e^2/\hbar c$,

$$\mathscr{D} = \frac{2^6}{3} \left(\frac{\pi}{3}\right)^{1/2} \alpha_s^4 c a_0^4 = 5\cdot 197 \times 10^{-14} \text{ cm}^3/\text{sec}$$

$$\Lambda = hcRZ^2/kT = 157,890 Z^2/T \tag{5.28}$$

$$x_n = \Lambda/n^2$$

$$S_n(\Lambda) = \int_0^\infty \frac{g_{II}(n, \varepsilon) \exp(-x_n u)}{1 + u} du.$$

Retaining the leading term in the Kramers–Gaunt factor (5.10) we find that

$$S_n(\Lambda) = \exp(x_n) \mathscr{E}_i(x_n) \tag{5.29}$$

where $\mathscr{E}_i$ is the exponential integral. A table of radiative recombination coefficients for H^+ for $n = 1$ to 12 and $T = 250°K (\times 2) 64000K$ has been given by Bates and Dalgarno.[116]

At higher densities three body recombination will dominate radiative recombination. This process is the inverse of ionization

$$A + e^- \to A^+ + e^- + e^-. \tag{5.30}$$

The second electron takes away the excess energy released in the recombination of the first electron and the ion. This process has been considered in considerable detail by Bates et al.[117–118] and Bates and Kingston.[119]

The final process which we consider in this Section is dielectronic recombination. Stabilization here is effected through an intermediate resonance state

$$A_i^+ + e^- \rightleftharpoons A_j^* \to A + h\nu \tag{5.31}$$

and plays an important role in many high temperature plasmas such as solar corona (Burgess[120]). A statistical equilibrium is built up in the resonance state state A_j^* through collisions between the electrons and the ions. Most of the time the resonance state decays back into the continuum through autoionization with a typical life-time of 10^{-13} sec. Occasionally, however, decay will occur via radiative decay leaving the atom in an excited but bound state A. The typical life-time of this mode of decay is 10^{-8} sec. The recombination coefficient is

$$\alpha = \frac{A_s A_a}{A_s + A_a} \frac{\omega_j}{2\omega_i} \frac{h^3}{(2\pi mkT)^{3/2}} \exp(-\varepsilon/kT) \tag{5.32}$$

where A_s is the probability of radiative decay of A_j^*, A_a is the probability of autoionization and ε is the energy of the resonance state relative to the state A_i^+. The total recombination coefficient obtained by summing over all resonance levels of the ion may be more than an order of magnitude more than that due to the radiative recombination.

VI. MOLECULAR PROCESSES

6.1. Theory of Electron–Molecule Scattering

The most important simplification in the theory of molecular structure and electron molecule collisions is the Born–Oppenheimer separation of the electronic and nuclear motion. This is based upon the fact that, owing to the slow motion of the nuclei compared with the electronic motion we can to a first approximation neglect the nuclear motion in calculating the electronic motion. The nuclei then move in a continually adjusting electronic field.

The Schrödinger equation describing the complete system is

$$\left(-\sum_i \frac{1}{2M_i} \nabla^2_{R_i} - \sum_j \tfrac{1}{2}\nabla_{r_j}^2 + V(\mathbf{R}_i, \mathbf{r}_j) - E\right)\Psi(\mathbf{R}_i, \mathbf{r}_j) = 0 \quad (6.1)$$

where $\mathbf{R}_i$ are the nuclear coordinates, $\mathbf{r}_j$ are the electronic coordinates, M_i are the masses of the nuclei in units of the electron mass and $V(\mathbf{R}_i, \mathbf{r}_j)$ is the total electrostatic interaction of the electrons and nuclei. The Born–Oppenheimer approximation is obtained by first solving (6.1) omitting the nuclear kinetic energy terms which are expected to be small because they involve M_i^{-1}. This gives

$$\left(-\sum_j \tfrac{1}{2}\nabla^2_{r_j} + V(\mathbf{R}_i, \mathbf{r}_j) - E_n(\mathbf{R}_i)\right)\phi_n(\mathbf{R}_i, \mathbf{r}_j) = 0 \quad (6.2)$$

where the eigenfunctions now depend parametrically on the nuclear coordinates $\mathbf{R}_i$. The energy function, $E_n(\mathbf{R}_i)$ which also depends on the nuclear coordinates, is now used as a potential function for the nuclear motion which satisfies

$$\left(-\sum_i \frac{1}{2M_i} \nabla^2_{R_i} + E_n(\mathbf{R}_i) - E_{n\gamma}\right)\chi_{n\gamma}(\mathbf{R}_i) = 0 \quad (6.3)$$

where the quantum number γ refers to the nuclear rotational–vibrational state and $E_{n\gamma}$ is the total energy of the system. Equations (6.2) and (6.3) correspond to approximating the total wave function by

$$\Psi \approx \phi_n(\mathbf{R}_i, \mathbf{r}_j)\chi_{n\gamma}(\mathbf{R}_i). \quad (6.4)$$

In considering the scattering of electrons by molecules it is convenient to expand the total wave function in terms of these Born–Oppenheimer states

$$\Psi = \mathscr{A} \sum_{n\gamma} \phi_n(\mathbf{R}_i, \mathbf{r}_j)\chi_{n\gamma}(\mathbf{R}_i)F_{n\gamma}(\mathbf{r}_{N+1}). \quad (6.5)$$

The function $F_{n\gamma}(\mathbf{r}_{N+1})$ which depends on the $\mathbf{R}_i$ describes the motion of the scattered electron, the operator $\mathscr{A}$ antisymmetrizes the whole expression consistent with the Pauli exclusion principle and the summation $n\gamma$ goes over the complete set of states. We have not made explicit reference to the electron spin in this discussion but this must be included in a way similar to that used in our discussion of electron–atom collisions in Section II.

One of the most important processes at low incident electron energies is rotational excitation. Following Arthurs and Dalgarno[121] it is convenient, in the case of linear diatomic molecules in a Σ state, to rewrite (6.5) in the form

$$\Psi = \mathscr{A} \sum_{njlv} \phi_n(\mathbf{R}, \mathbf{r}_j)x_{nv}(R)r_{N+1}^{-1} f^J_{jlv}(r_{N+1})y^{M_J}_{Jjl}(\hat{\mathbf{R}}, \hat{\mathbf{r}}_{N+1}) \quad (6.6)$$

where x_{nv} is the vibrational wave function and

$$y_{Jjl}^{M_J}(\hat{\mathbf{R}}, \hat{\mathbf{r}}) = \sum_{m_l m_j} c_{jl}(J M_J; m_j m_l) Y_j^{m_j}(\hat{\mathbf{R}}) Y_l^{m_l}(\hat{\mathbf{r}}) \tag{6.7}$$

where $Y_j^{m_j}(\hat{\mathbf{R}})$ is the angular part of the molecular wave function. We can obtain coupled integro-differential equations for the radial functions $f_{jlv}^J(r)$ by substituting this expression for Ψ into the Schrödinger equation for the electron–molecule system, premultiplying by $\phi_n^*(\mathbf{R}, \mathbf{r}_j) x_{nv}^*(\mathbf{R}) y_{Jjl}^{M_J*}(\hat{\mathbf{R}}, \hat{\mathbf{r}}_{N+1})$, and integrating over all coordinates except the radial coordinate of the incident electron. This gives

$$\left(\frac{d^2}{dr^2} - \frac{l(l+1)}{r^2} + k_j^2 \right) f_{jl}^J(r) = 2 \sum_{j'l'} V_{jl,j'l'}^J(r) f_{j'l'}^J(r) \tag{6.8}$$

where we have restricted our consideration to the ground electronic state and we have assumed the molecule is rigidly held at its equilibrium nuclear separation. The subscript v has thus been omitted.

As in the case of electron–atom collisions V contains direct and exchange terms. In many problems of interest the direct potential terms dominate the scattering. It is convenient to write it in the form

$$V_{jl,j'l'}^J(r) = \langle y_{Jjl}^{M_J}(\hat{\mathbf{R}}, \hat{\mathbf{r}}), V(\hat{\mathbf{R}}, \mathbf{r}) y_{Jj'l'}^{M_J}(\hat{\mathbf{R}}, \hat{\mathbf{r}}) \rangle \tag{6.9}$$

where the integration is carried out over the angular coordinates $\hat{\mathbf{R}}$ and $\hat{\mathbf{r}}$ and the potential $V(\hat{\mathbf{R}}, \mathbf{r})$ is the expectation value of the potential energy in the molecular electronic ground state. The potential can generally be expanded

$$V(\hat{\mathbf{R}}, \mathbf{r}) = \sum_\lambda V_\lambda(r) P_\lambda(\cos\theta) \tag{6.10}$$

where $\cos\theta = \hat{\mathbf{R}} \cdot \hat{\mathbf{r}}$. The asymptotic form which dominates the collision at low energies is

$$V(\hat{\mathbf{R}}, \mathbf{r}) \underset{r \to \infty}{\sim} -\frac{\alpha}{2r^4} - \frac{\mu}{r^2} P_1(\cos\theta) - \left(\frac{\alpha'}{2r^4} + \frac{Q}{r^3} \right) P_2(\cos\theta) + \ldots \tag{6.11}$$

where μ is the electric dipole moment which is zero for homonuclear diatomics, Q is the electric quadrupole moment and α, α' are related to the polarizabilities along and perpendicular to the nuclear axis by

$$\begin{aligned} \alpha &= \tfrac{1}{3}(\alpha_\| + 2\alpha_\perp) \\ \alpha' &= \tfrac{2}{3}(\alpha_\| - \alpha_\perp). \end{aligned} \tag{6.12}$$

For the important gases H_2 and N_2 we have the following values where B determines the rotational energy splitting through the relation $E_j = Bj(j+1)$.

From this table we see that the polarizability and the quadrupole moment tend to enhance for H_2 and to cancel for N_2.

Table III

	H_2	N_2
α	5·328	12·0
α'	1·250	4·2
Q	0·464	−1·1
B	$2·70 \times 10^{-4}$	$9·11 \times 10^{-6}$

A considerable amount of work has been carried out solving (6.8) with just a few rotational states retained. The cross-section can be determined from the asymptotic form of the functions $f_{jl}(r)$ in a similar way to our

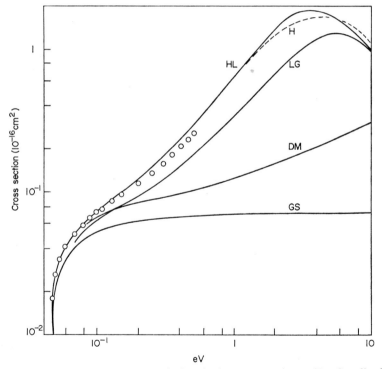

FIG. 21. The experimental and theoretical excitation cross-sections $\sigma(j = 0 \to 2)$ of H_2. HL: Henry and Lane.[123] H: Hara.[124] LG: Lane and Geltmen.[125] DM: Dalgarno and Moffett.[126] GS: Gerjuoy and Stein.[127] The circles are the experimental data of Crompton[128] (After Takayangi and Itikawa[122]).

discussion in Section II. A review of most recent work has been given by Takayanagi and Itikawa.[122] Figure 21 summarizes results obtained for the $j = 0 \to 2$ transition in $e^- - H_2$ scattering. The recent close coupling calculations of Henry and Lane,[123] Hara[124] and Lane and Geltman[125] all give a broad peak in the 2–3 eV range which is due to a $^2\Sigma_u^+$ resonance state of H_2^-. Recent studies also include $e^- - N_2$ (Burke and Sin Fai Lam[129]), $e^- - CO$ (Crawford and Dalgarno[130]) and $e^- - NH_3$ (Itikawa[131]). The work on N_2 included exchange but not polarization and showed the presence of a $^2\Pi_g$ resonance at 2·4 eV seen earlier in experiments on vibrational excitation by Schulz.[132]

The work on vibrational excitation has been far less complete; however a start in this direction has been made by Henry[133] who has close coupled the $v = 0$ and 1 and $j = 1$ and 3 states including both exchange and polarization effects for $e^- - H_2$ scattering. The results for $2 < E < 5$ eV for $\sigma(v = 0 \to 1, \Delta j = 0)$ and $\sigma(v = 0 \to 1, j = 1 \to 3)$ are 50% larger than the experimental values of Ehrhardt et al.[134] Further work on vibrational excitation has been carried out within the framework of various compound state models and will be discussed in the next Section.

The situation with respect to electronic excitation is even less clear than vibrational excitation although in principle the close coupling should have the same range of applicability as for atoms. We will therefore not discuss this further here.

6.2. Resonances

Resonances are found to play a very important role in many collision processes involving molecules where they occur as intermediate states (Bardsley and Mandl[135]). Typical processes are

(a) elastic and inelastic scattering

$$e^- + AB \to AB^- \to e^- + AB^* \tag{6.13}$$

(b) dissociative attachment or recombination

$$e^- + AB \to AB^- \to A + B^-$$
$$e^- + AB^+ \to AB \to A + B \tag{6.14}$$

(c) atom–negative ion collisions involving detachment

$$A + B^- \to AB^- \to AB + e^-$$
$$A + B^- \to AB^- \to A + B + e^- \tag{6.15}$$

The motion of the nuclei during the life-time of the resonance introduces new features which do not arise in electron resonance scattering. For example the nuclei in (6.14) dissociate during the life-time of the resonance leading to a stable negative ion state. The final state can be regarded as an alternative decay channel for the resonance which is in competition with the usual process of auto-ionization.

The role resonances play can be seen more clearly by reference to the example shown in Fig. 22. This gives the lowest potential energy curves for the H_2 and H_2^- systems. The potential energy curves were calculated by Bardsley et al.[136-137] and discussed by Burke.[138] The $^2\Sigma_u^+$ state of H_2^- corresponds to a shape resonance in the scattering of electrons by H_2 in its ground state. It is this resonance that gives rise to the broad peak in the rotational excitation cross-section in Fig. 21. It is seen that the $^2\Sigma_u^+$ state of H_2^- crosses the $^1\Sigma_g^+$ state at about 3 a.u. For values of R less than this crossing point the resonance can autoionize into the H_2 ground state with the

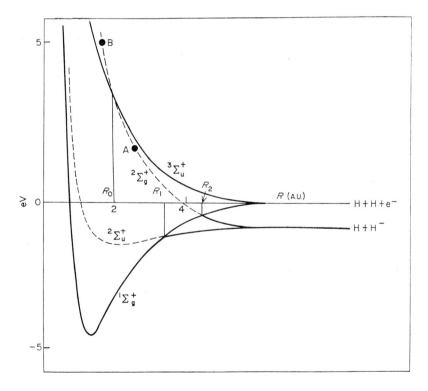

FIG. 22. The potential energy curves for the $^1\Sigma_g^+$ and $^3\Sigma_u^+$ states of H_2 and the real parts of the potential energy curves of the $^2\Sigma^+$ and $^2\Sigma_g^+$ states of H_2^-.

emission of an electron. In the Born–Oppenheimer approximation, corresponding to (6.5), we can write the wave function for the electron molecule system as

$$\Psi = \mathcal{A}\phi_0(\mathbf{R}, \mathbf{r}_j)\chi_{0\gamma}(\mathbf{R})F(\mathbf{r}_{N+1}) \qquad (6.16)$$

where we have retained just the target ground state ϕ_0 in the close coupling expansion. Now the resonance state corresponds to the outgoing wave boundary condition

$$F(\mathbf{r}) \underset{r \to \infty}{\sim} e^{ikr} \qquad (6.17)$$

where k^2 is the complex energy of the resonance

$$\tfrac{1}{2}k^2 = E_r - \tfrac{1}{2}i\Gamma \qquad (6.18)$$

where the resonance position E_r and the resonance width Γ both depend upon R. This definition of a resonance state was first introduced by Siegert.[139] It can be understood physically by noting that the time dependence of the wave function contains a term

$$\Psi \propto \exp\left(-\tfrac{1}{2}ik^2 \frac{t}{\hbar}\right). \qquad (6.19)$$

It follows that

$$|\Psi|^2 \propto \exp\left(-\Gamma \frac{t}{\hbar}\right) \qquad (6.20)$$

which means that the state decays with a life-time $\tau = \hbar/\Gamma$. It follows immediately, in a similar way to the derivation of (6.7), that the nuclear motion experiences an effective potential

$$W(R) = \tfrac{1}{2}k^2 + E_0(R) \qquad (6.21)$$

where $E_0(R)$ is the ground state potential energy curve. In the region of R where the resonance state can autoionize Γ is greater than zero and consequently $W(R)$ is complex.

Returning to Fig. 22 we see that the $^2\Sigma_g^+$ potential curve is expected to be complex for R less than the crossing point with the $^1\Sigma_g^+$ state, but to be real for R greater than this value. Calculations for the negative imaginary part of the potential curve have been carried out by Bardsley et al.[136–137]. Also shown in Fig. 22 is the $^2\Sigma_g^+$ state of H_2^-. This lies close to the $^3\Sigma_u^+$ state of H_2 and may indeed cross it close to the Franck–Condon region (Eliezer et al.[140]) only giving appreciable autoionization for smaller values state of H_2 and may indeed cross it close to the Franck–Condon region of R.

One of the most interesting resonances is the $^2\Pi_g$ state of N_2^-, which lies about 2·4 eV above the ground state of N_2 and has a width of about 0·4 eV.

It has been found by Schulz[132] to play an important role in the vibrational excitation cross-sections of N_2 as shown in Fig. 23. A resonance model fit carried out by Herzenberg and Mandl[141] (the full curves in the figure), showed that the non-resonant contribution to the excitation was very small. It is seen that structure in the resonance state is well developed indicating that the resonance life-time is not much smaller than the vibrational time.

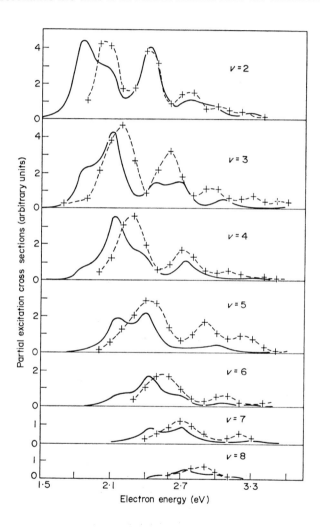

FIG. 23. Energy dependence of the cross-sections for excitation of the second to eighth vibrational levels of N_2 by electron impact. Crosses and broken curves are the measurements of Schulz.[132] and the full curves the calculations of Herzenberg and Mandl.[141] (After Herzenberg and Mandl[141]).

Recently resonances have been found in a number of other molecular systems including CO, O_2, NO, CO_2, C_2H_4 and we refer to Bardsley and Mandl[135] for a detailed discussion.

6.3. Dissociative Attachment and Recombination

Closely related to the resonance theory discussed in the previous Section is resonant dissociative attachment or recombination illustrated in the examples

$$e^- + H_2 \rightarrow H + H^-$$
$$e^- + O_2 \rightarrow O + O^- \qquad (6.22)$$
$$e^- + NO^+ \rightarrow N + O.$$

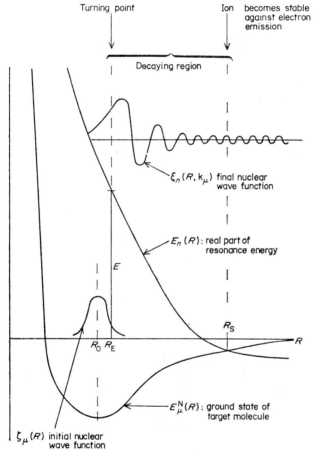

FIG. 24. Potential energy curves and nuclear wave functions in dissociative attachment. (After Bardsley et al.[137])

The basic mechanism for resonant dissociative attachment is shown in Fig. 24. Neglecting the rotational motion of the molecule, which is often slow compared with the times of interest, the molecule is initially assumed to be vibrating in the ground state potential $E_u^N(R)$. After collision, the electron is captured into the negative molecular ion state and the nuclei start to move in the potential $W_n(R)$ which is now complex. As the nuclei move apart, the compound state can relax at any value of R where $W_n(R)$ is complex, with the ejection of the electron, leaving the molecule in its ground state or in a vibrationally excited state. It follows that the amplitude of the nuclear motion in the potential $W_n(R)$ is damped until the crossing point at $R = R_S$ is reached where the system is stable against electron decay. The nuclear motion can be described by the equation

$$\left(-\frac{1}{2M}\frac{d^2}{dR^2} + W_n(R) - E\right)\chi_n(R) = J_n(R) \tag{6.23}$$

where $J_n(R)$ is a source term which is proportional to the amplitudes of the incident beam and of the initial molecular state and is responsible for the generation of the resonance. In practice one will have to solve this equation numerically but the physical features are well displayed by an approximate solution obtained by Bardsley et al.[137] using the W.K.B. approximation

$$\sigma_{DA} = \sigma_{cap} \exp\left\{-\int_{R_E}^{R_S} \frac{\Gamma_n(R)}{\hbar}\frac{dR}{v(R)}\right\} \tag{6.24}$$

where σ_{cap} is the captive cross-section for the formation of the resonant state and the exponential factor called the survival factor, represents the probability that the negative ion survives from its formation at R_E until it becomes stabilized at R_S. We note that $\Gamma_n(R)/\hbar$ is the decay rate and $dR/v(R)$ is the time taken for the nuclei to separate from R to $R + dR$.

The dissociative attachment cross-section for H_2, HD, and D_2 has been measured by Schulz and Asundi[142-143] and is shown in Fig. 25. The two peaks correspond to the $^2\Sigma_u^+$ and the $^2\Sigma_g^+$ resonances respectively (see Fig. 22). The peak at 3 eV is quite small since the survival factor in (6.24) contains a large value of Γ. The isotope effect giving rise to difficult dissociative attachment cross-sections for deuterium substituted molecules arises from the smaller velocity term in the survival factor.

In concluding this discussion of dissociative recombination we remark that the calculation of the recombination coefficient requires accurate knowledge of the potential energy curves of excited molecules and the strength of configuration interaction in the molecule. For many molecules of interest, such as N_2, NO, O_2 these calculations are difficult to perform *ab initio*. The temperature dependence of the recombination rate of course depends on the details of these potential energy curves. In the case of H_2 the rate rises to a maximum before decreasing. If on the other hand the resonant potential energy curve crosses the molecular ground state curve in the neighbourhood of its minimum then the temperature dependence is $\sim T^{-\frac{1}{2}}$ and the recombination coefficient may be as large as 10^{-7} cm^3/sec (Bates[145]).

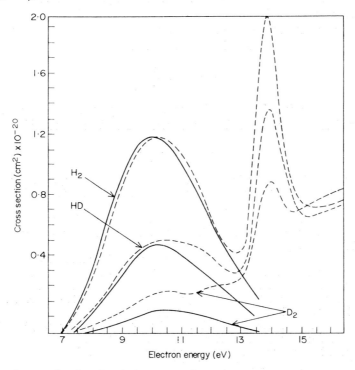

FIG. 25. Cross-section for dissociative capture in H_2, HD and D_2. Full curves are the calculations of Bardsley *et al.*[137] and the broken curves the results of Rapp *et al.*[144] (After Bardsley *et al.*[137])

6.4. Ion–Atom and Ion–Molecule Reactions

We have space here to do no more than mention the importance of ion–atom and ion–molecule reactions in the upper atmosphere and in astrophysics.

A reaction of considerable concern to astrophysicists

$$H + H^- \rightarrow H_2 + e^- \tag{6.25}$$

has been studied by Dalgarno and Browne.[146] In this work they solved Eqn (6.23), without the source term, using the potential energy curves of Taylor and Harris[147] and Bardsley et al.[136] The loss of flux into the associative detachment reaction (6.25) is taken care of by the complex potential term and they obtained a rate which varied from 1.9×10^{-9} cm^3/sec at 100°K to 1.2×10^{-9} cm^3/sec at 32,000°K.

The effective temperature independence of many ion–molecule reactions arises from the fact that they are dominated by the long range polarization potential $\alpha/2r^4$ where α is the dipole polarizability. For impact parameters below a critical value b_0 this gives rise to Langevin spiralling. If it is assumed that all collisions less than the critical radius lead to reactions then this gives a cross-section for an ion–molecule reaction of

$$\sigma = \pi b_0^2 = \frac{\pi}{v}\frac{4\alpha}{\mu} \qquad (6.26)$$

where μ is the reduced mass of the ion molecule pair and v is their relative velocity. This was discussed first by Gioumousis and Stevenson.[148]

REFERENCES

1. "Atomic and Molecular Processes", (ed. D. R. Bates), Academic Press (1962).
2. Mott, N. F. and Massey, H. S. W. "The Theory of Atomic Collisions", (3rd Edition), Oxford U.P. (1965).
3. "Lectures in Theoretical Physics", XI–C, Atomic Collision Processes", (eds. S. Geltman, K. T. Mahanthappa and W. E. Brittin), Gordon and Breach (1969).
4. Bransden, B. H., "Atomic Collision Processes", Benjamin Inc. (1970).
5. Moisewitch, B. L. and Smith, S. J., *Rev. Mod. Phys.* **40**, 238 (1968).
6. Burke, P. G., "Advances in Atomic and Molecular Physics", **4**, 173 (1968).
7. Bardsley, J. N. and Mandl, F., *Reps. on Prog. in Phys.* **31**, 471 (1968).
8. Rudge, M. R. H., *Rev. Mod. Phys.* **40**, 564 (1968)
9. Bely, O. and van Regemorter H., *Ann. Rev. Astron. Astrophys.* **8**, 329 (1970).
10. Klein, A. and Zemach, C., *Ann. Phys. (N.Y.)* **7**, 440 (1959).
11. Blatt, J. M. and Jackson, J. D., *Phys. Rev.* **76**, 18 (1949).
12. Bethe, H. A., *Phys. Rev.* **76**, 38 (1949).
13. Levy, B. R. and Keller, J. B., *J. Math. Phys.* **4**, 54 (1963).
14. O'Malley, T. F., Spruch, L. and Rosenberg, L., *J. Math. Phys.* **2**, 491 (1961).
15. Seaton, M. J., *Comp. Rend.* **240**, 1317 (1955).
16. Seaton, M. J., *M.N.R.A.S.* **118**, 504 (1958).
17. Glauber, R. J., in "Lectures in Theoretical Physics" Vol. 1, (eds. W. E. Brittin and L. G. Dunham), Wiley (Interscience) N.Y. (1959).
18. Kohn, W., *Phys. Rev.* **74**, 1763 (1948).
19. Seaton, M. J., *Proc. Phys. Soc.* **77**, 174 (1961).
20. Bely, O., *Il Nuovo Cimento* **49**, 66 (1967).
21. Ochkur, V. I., *Sov. Phys. J.E.T.P.* **18**, 503 (1964).
22. Lane, A. M. and Thomas, R. G., *Revs. Mod. Phys.* **30**, 257 (1958).
23. Ross, M. H. and Shaw, G. L., *Ann. Phys. (N.Y.)* **13**, 147 (1961).

24. Gailitis, M., *Sov. Phys. J.E.T.P.* **17,** 1328 (1964).
25. Seaton, M. J., *Proc. Phys. Soc.* **88,** 801 (1966).
26. Temkin, A., *Phys. Rev.* **116,** 358 (1959).
27. Schwartz, C., *Phys. Rev.* **124,** 1468 (1961).
28. Swan, P., *Proc Roy. Soc. A* **228,** 10 (1954).
29. Hahn, Y., O'Malley, T. F. and Spruch, L., *Phys Rev.* **134,** B397 (1964).
30. Hahn, Y., O'Malley, T. F. and Spruch, L., *Phys. Rev.* **134,** B911 (1964).
31. McKinley, W. A. and Macek, J. H., *Phys. Lett.* **10,** 210 (1964).
32. Sugar, R. and Blakenbecler, R., *Phys. Rev.* **136,** B472 (1965).
33. Feshbach, H., *Ann. Phys. (N.Y.)* **19,** 287 (1962).
34. Spruch, L., in "Lectures in Theoretical Physics". Vol. 4, (ed. W. E. Brittin) Wiley (Interscience) N.Y. (1963).
35. Burke, P. G., Schey, H. M. and Smith, K., *Phys. Rev.* **129,** 1258 (1963).
36. Burke, P. G., Ormonde, S. and Whitaker, W., *Proc. Phys. Soc.* **92,** 319 (1967).
37. Taylor, A. J. and Burke, P. G., *Proc. Phys. Soc.* **92,** 336 (1967).
38. Kauppila, W. E., Ott, W. R. and Fite, W. L., *Phys. Rev.* **A1,** 1099 (1970).
39. Burke, P. G. and Webb, T. G., *J. Phys.* **B3,** L131 (1970).
40. Tai, H., Bassel, R. H., Gerjouy, E. and Franco, V., *Phys. Rev.* **A1,** 1819 (1970).
41. Seaton, M. J., *Proc. Phys. Soc.* **79,** 1105 (1962).
42. Saraph, H. E., *Proc. Phys. Soc.* **83,** 763 (1964).
43. Gryzinsky, M. *Phys. Rev.* **115,** 374 (1959).
44. Percival, I. C. and Richards, D., *J. Phys.* **B3,** 1035 (1970).
45. Burke, P. G., McVicar, D. D. and Smith, K., *Proc. Phys. Soc.* **83,** 397 (1964).
46. Dance, D. F., Harrison, M. F. A. and Smith, A. C. H., *Proc. Roy. Soc.* **A290,** 74 (1966).
47. Burgess, A., Hummer, D. G. and Tully, J. A., *Phil. Trans. Roy. Soc.* **A266,** 255 (1963).
48. Burke, P. G., Cooper, J. W. and Ormonde, S., *Phys. Rev.* **183,** 245 (1969).
49. Schultz, G. J. and Fox, R. E., *Phys. Rev.* **106,** 1179 (1957).
50. Holt, H. K. and Krotkov, R., *Phys. Rev.* **144,** 82 (1966).
51. Andrick, D. and Ehrhardt, H., *Z. fur Physik* **192,** 99 (1966).
52. Vainshtein, L. A. and Vinogradov, A. V., *J. Phys.* **B3,** 1090 (1970).
53. Kunze, H. J., Gabriel, A. H. and Griem, H. R., *Phys. Rev.* **165,** 267 (1968).
54. Elton, R. C. and Koppendorfer, W. W., *Phys. Rev.* **160,** 194 (1967).
55. Bely, O., *Phys. Lett.* **26A,** 408 (1968).
56. Karule, E. M., in "Atomic Collisions" Vol. III, (ed. V. Ia. Veldre) Riga (1965). (Transl. Rept. No. 3, JILA, Boulder, Colorado, p.29, 1966).
57. Karule, E. M. and Peterkop, R. K., as ref. 56, p.1.
58. Burke, P. G. and Taylor, A. J., *J. Phys. B* **2,** 869 (1969).
59. Vo Ky Lan, *J. Phys.* **B4,** 658 (1971).
60. Perel, J., Englander, P. and Bedeson B., *Phys. Rev.* **128,** 1148 (1962).
61. Heroux, L., *Nature* **198,** 1291 (1963).
62. Heroux, L., *Proc. Phys. Soc.* **83,** 121 (1964).
63. Burke, P. G., Tait, J. H. and Lewis, B. A., *Proc. Phys. Soc.* **87,** 209 (1966).
64. Seaton, M. J., p.374 of ref. 1.
65. Boland, B. C., Jahoda, F. C., Jones, T. J. L. and McWhirter, R. W. P., *J. Phys.* **B3,** 1134 (1970).
66. Bely, O., *Proc. Phys. Soc.* **88,** 587 (1966).

67. Collins, R. E., Bederson, B. and Goldstein, M., *Phys. Rev.* **A3**, 1976 (1971).
68. Rubin, K., Bederson, B., Goldstein, M., and Collins, R. E., *Phys. Rev.* **182**, 201 (1969).
69. Burke, P. G. and Moores, D. L., *J. Phys.* **B1**, 575 (1968).
70. Newsom, G. H., *Proc. Phys. Soc.* **87**, 975 (1966).
71. Bely, O., Tully, J. and van Regemorter, H., *Ann. Phys.* (Paris) **8**, 303 (1963).
72. Osterbrock, D. E., *J. Phys.* **B3**, 149 (1969).
73. Beigman, I. L. and Vainshtein, L. A., *Sov. Phys. J.E.T.P.* **25**, 119 (1967).
74. Smith, K., Henry, R. J. W. and Burke, P. G., *Phys. Rev.* **147**, 21 (1966).
75. Henry, R. J. W., Burke, P. G. and SinFaiLam, A. L., *Phys. Rev.* **178**, 218 (1969).
76. Saraph, H. E., Seaton, M. J. and Shemming, J., *Phil. Trans. Roy. Soc.* **A264**, 77 (1969).
77. Seaton, M. J., *Phil. Trans. Roy. Soc.* **A245**, 469 (1953).
78. Smith, K. and Morgan, L. A., *Phys. Rev.* **165**, 110 (1968).
79. Burke, P. G., Hibbert A. and Robb, D. W., *J. Phys.* **B4**, 153, (1971).
80. Eissner, W., Nussbaumer, H. Saraph, H. E. and Seaton, M. J., *J. Phys.* **B2**, 341 (1969).
81. Layzer, D., *M.N.R.A.S.* **114**, 692 (1954).
82. Saraph, H. E. and Seaton, M. J., *M.N.R.A.S.* **148**, 367 (1970).
83. Edmonds, A. R., "Angular Momentum in Quantum Mechanics", Princeton U.P. (1957).
84. Peterkop, R. K., *Izv. Akad. Nauk. Latv. SSR* **9**, 79 (1960).
85. Peterkop, R. K., *Opt. Spectrosc.* **13**, 87 (1962).
86. Peterkop, R. K., *Sov. Phys. J.E.T.P.* **16**, 442 (1963).
87. Peterkop, R. K., *Sov. Phys. Dokl.* **27**, 987 (1963).
88. Rudge, M. R. H. and Seaton, M. J., *Proc. Roy. Soc.* **A283**, 262, (1965).
89. Geltman, S., Rudge, M. R. H. and Seaton, M. J., *Proc. Phys. Soc.* **81**, 375 (1963).
90. Wannier, G. M., *Phys. Rev.* **90**, 817 (1953).
91. Peterkop, R. K., *J. Phys.* **B4**, 513 (1971).
92. Rau, A. R. P., *Phys. Rev.* **A4**, 207 (1971).
93. Bethe, H. A. and Salpeter, E. E., in "Encyl. of Phys". Vol. XXXV, Springer, Berlin (1957).
94. Henry, R. J. W. and Lipsky, L., *Phys. Rev.* **153**, 51 (1967).
95. Seaton, M. J., *Rep. on Prog. in Phys.* **23**, 313 (1960).
96. Menzel, D. H. and Pekeris, C. L., *M.N.R.A.S.* **96**, 77 (1935).
97. Chandrasekhar, S., *Ap. J.* **102**, 223 (1945).
98. Doughty, N. A., Fraser, P. A. and McEachran, R. P., *M.N.R.A.S.* **192**, 255 (1966).
99. Smith, S. J. and Burch, D. S., *Phys. Rev. Lett.* **2**, 165 (1959).
100. Smith, S. J. and Burch, D. S., *Phys. Rev.* **116**, 1125 (1959).
101. Macek, J. H., *Proc. Phys Soc.* **92**, 365 (1967).
102. John, T. L., *M.N.R.A.S.* **128**, 93 (1964).
103. Madden, R. P. and Codling, K., *Ap. J.* **141**, 364 (1965).
104. Cooper, J. W., Fano, V. and Prats, F., *Phys. Rev. Lett.* **10**, 518 (1963).
105. Burke, P. G. and McVicar, D. D., *Proc. Phys. Soc.*, **86**, 989 (1965).
106. Fano, U., *Phys. Rev.* **124**, 1866 (1961).
107. Samson, J. A. R., *Phys. Rev.* **132**, 2122 (1963).
108. Schnopper, K., *Phys. Rev.* **131**, 2558 (1963).

109. Fano, U. and Cooper, J. W., *Phys. Rev.* **137**, A1364 (1965).
110. Jacobs, V., *Phys. Rev.* **A3**, 289 (1971).
111. Norcross, D. W., *J. Phys.* **B4**, 652 (1971).
112. Henry, R. J. W., *Planet. Space Sci.* **16**, 1503 (1968).
113. Conneely, M. J., Smith, K. and Lipsky, L., *J. Phys.* **B3**, 493 (1970).
114. Rich, J. C., *Ap. J.* **148**, 275 (1967).
115. Milne, E. A., *Phil. Mag.* **47**, 209 (1924).
116. Bates, D. R. and Dalgarno A., p. 245 of ref. 1.
117. Bates, D. R., Kingston, A. E. and McWhirter, R. W. P., *Proc. Roy. Soc.* **A267**, 297 (1962).
118. Bates, D. R., Kingston, A. E. and McWhirter, R. W. P., *Proc. Roy. Soc.* **A270**, 155 (1962).
119. Bates, D. R. and Kingston, A. E., *Proc. Phys. Soc.* **83**, 43 (1964).
120. Burgess, A., *Ap. J.* **139**, 776 (1964).
121. Arthurs, A. M. and Dalgarno A., *Proc. Roy. Soc.* **A256**, 540 (1960).
122. Takayanagi, K. and Itikawa, Y., "Advances in Atomic and Molecular Physics" **6**, 105 (1970).
123. Henry, R. J. W. and Lane, N. F., *Phys. Rev.* **183**, 221 (1969).
124. Hara, S., *J. Phys. Soc. Japan* **27**, 1592 (1969).
125. Lane, N. F. and Geltman, S., *Phys. Rev.* **160**, 53 (1967).
126. Dalgarno, A. and Moffett, R. J., *Proc., Nat. Acad. Sci. India* **A 33**, 511, (1963).
127. Gerjuoy, E. and Stein S., *Phys. Rev.* **97**, 1671 (1955).
128. Crompton, R. W., Gibson, D. K. and McIntosh, A. I., *Austral. J. Phys.* **22**, 715 (1969).
129. Burke, P. G. and SinFaiLam, A. L., *J. Phys.* **B3**, 641 (1970).
130. Crawford, O. H. and Dalgarno, A., *J. Phys.* **B4**, 494 (1971).
131 Itikawa, Y , *J. Phys. Soc. Japan*, **30**, 835 (1971).
132. Schulz, G. J., *Phys. Rev.* **125**, 229 (1962).
133. Henry, R. J. W., *Phys. Rev.* **A2**, 1349 (1970).
134. Ehrhardt, H., Langhans, L., Linder, F. and Taylor, H. S., *Phys. Rev.* **173**, 222 (1968).
135. Bardsley, J. N. and Mandl, F., *Rep. Prog. in Phys.* **31**, 471 (1968).
136. Bardsley, J. H., Herzenberg, A. and Mandl, F., *Proc. Phys. Soc.* **89**, 305 (1966).
137. Bardsley, J. N. Herzenberg, A. and Mandl, F., *Proc. Phys. Soc.*, **89**, 321 (1966).
138. Burke, P. G., *J. Phys.* **B1**, 586 (1968).
139. Siegert, A. J. F., *Phys. Rev.* **56**, 750 (1939).
140. Eliezer, I., Taylor, H. S. and Williams, J. K., *J. Chem. Phys.*, **47**, 2165 (1967).
141. Herzenberg, A. and Mandl, F., *Proc. Roy. Soc.* **A270**, 48 (1962).
142. Schulz, G. J. and Asundi, R. K., *Phys. Rev. Lett.* **15**, 946 (1965).
143. Schulz, G. J. and Asundi, R. K., *Phys. Rev.* **158**, 25 (1967).
144. Rapp, D., Sharp, T. E. and Briglia, D. D., *Phys. Rev. Lett.* **14**, 533 (1965).
145. Bates, D. R., *Phys. Rev.* **78**, 492 (1950).
146. Dalgarno, A. and Browne, J. C., *Ap. J.* **149**, 231 (1967).
147. Taylor, H. S. and Harris, F. E., *J. Chem. Phys.* **39**, 1012 (1963).
148. Giomousis, G. and Stevenson, D. P., *J. Chem. Phys.* **29**, 294 (1958).

Highly Excited Atoms

IAN C. PERCIVAL

University of Stirling, Stirling, Scotland

I. INTRODUCTION

Properties of highly excited atoms are required for a detailed understanding of recombination and ionization balance in stellar atmospheres.

They are also involved directly in the interstellar radio recombination lines as described in the lectures by Seaton, and I will concentrate on the properties needed to understand these lines. The hydrogen atom H ($n = 110$) will be used as a test case, as it is popular with radio astronomers.

II. ELEMENTARY PROPERTIES

For highly excited atoms the classical model of the atom can be taken seriously. We may use the Bohr model of a classical electron moving in a circular orbit around a proton to represent the ground state $n = 1$ of the hydrogen atom, where the model is obviously very crude, This is shown in column II of Table I. The geometric cross-section gives a good idea of the size of the atom, the R.M.S. velocity is exactly right, but the classical frequency v, has no meaning. The classical properties of a highly excited state can be obtained from those of the ground state, as shown in columns III and IV.

Starting with the quantal expression for the energy we find that H (110) atoms are so very weakly bound that particles at room temperature would have enough energy to destroy them, let alone the $10^{4\circ}$K protons and electrons of the H II regions. Only the low density saves them.

According to Coulomb the orbital radius of a charged particle bound to another one varies inversely with the energy and so the mean Bohr radius of an H (n) atom varies as n^2. The size of an H (110) atom is typical of a small biological organism, and is as big compared to a typical atom or small molecule as a typical atom is, when compared to its nucleus. By analogy the structure of the ionic core of the highly excited atom can usually be neglected; the core is distinguished only by its charge and mass. The weak binding and large size make highly excited atoms very fragile and difficult to study in the laboratory.

Using Kepler's Law the period of the electron in its orbit scales as the 3/2 power of its mean radius and from this we obtain the fundamental classical frequency, which corresponds to the 110α transition in our test case. For circular orbits, with maximum l, only this frequency $\nu_{\text{classical}} = \nu_c$ is emitted. For elliptic orbits with lower l, there are Fourier components of the motion at the harmonic frequencies $2\nu_c$, $3\nu_c$ etc., corresponding to $111 \to 109 (109\beta)$, $112 \to 109 (109\gamma)$ transitions, etc. The fundamental classical frequency is in reasonably good agreement with the frequency of the corresponding quantal transition and is even better if we take the mean of initial and final quantum numbers. The agreement gets worse for larger

Table I

Elementary properties of highly excited states.

I	II	III	IV
Property	$n = 1$	Arbitrary n	$n = 110$
Binding energy U_n	1 Ryd = 13·6 eV	Ryd$/n^2$	$1\cdot1 \times 10^{-3}$ eV
Radius a_n of Bohr orbit	$a_0 = a_1$ $= 5\cdot3 \times 10^{-9}$ cm	$n^2 a_0$	$6\cdot4 \times 10^{-5}$ cm
Geometric cross-section πa_n^2	$8\cdot8 \times 10^{-17}$ cm^2	$n^4 \pi a_0^2$	$1\cdot3 \times 10^{-8}$ cm
R.M.S. velocity of electron v_n	$c\alpha = v_1 = 2\cdot2 \times 10^8$ cm sec^{-1}	v_1/n	$2\cdot0 \times 10^6$ cm sec^{-1}
Period T_n	$1\cdot5 \times 10^{-16}$ sec	$n^3 T_1$	$2\cdot0 \times 10^{-10}$ sec
Fundamental classical frequency ν_c	$\nu_1 = 6\cdot576 \times 10^{15}$ Hz	$n^{-3} \nu_1$ $[(n-\tfrac{1}{2})^{-3}\nu_1]$	$4\cdot941 \times 10^9$ Hz $[5\cdot009 \times 10^9$ Hz$]$
Classical wave number λ_c^{-1}	$\lambda_1^{-1} = 2R_H$ $= 2\cdot194 \times 10^5$ cm^{-1}	$2n^{-3} R_H$ $[2(n-\tfrac{1}{2})^{-3} R_H]$	$0\cdot1648$ cm^{-1} $[0\cdot1671$ cm$^{-1}]$
Quantal $n\alpha$ frequency	None	$\dfrac{n - \tfrac{1}{2}}{[n(n-1)]^2} \nu_1$	5,009 MHz

Δn, but is generally not too bad when

$$\Delta n \ll n \tag{2.1}$$

From classical frequencies, following Bohr, we can obtain an estimate of the energy difference between initial and final states from the Planck relation

$$\Delta E = h\nu = \hbar\omega \tag{2.2}$$

This is simply an application of Bohr's correspondence principle. Clearly classical mechanics and the correspondence principle are much better for $n = 110$ than for $n = 1$, as we should expect.

In quantum or wave mechanics the wave functions of highly excited states have many maxima and minima, so traditional quantal methods, though still valid, are much more difficult to apply, unless we use rough approximations or simplifications which are equivalent to the application of the correspondence principle to each individual case.

The H II regions

Properties of a typical H II region are summarised in Table II. Here the free electrons are much faster and more energetic than the highly excited bound electrons. The free protons move with a velocity comparable to those bound electrons although in each case there is a considerable spread in the velocity.

Table II

H II regions: typical values

Property	Value
Density of electrons and protons	$10^4 \, \text{cm}^{-3}$
Temperature	$10^4 \, °\text{K}$
Mean K.E.	$2·07 \times 10^{-12} \, \text{erg} = 1·29 \, \text{eV}$
R.M.S. electron velocity	$6·75 \times 10^7 \, \text{cm sec}^{-1}$
R.M.S. proton velocity	$1·57 \times 10^6 \, \text{cm sec}^{-1}$
Thermal Doppler width/frequency $\Delta\nu/\nu$	$3·03 \times 10^{-5}$
Thermal Doppler width H109α	$0·16 \, \text{MHz}$

The distance between a highly excited atom and its nearest neighbouring charge is usually less than 1000 × atomic radius. As a result the electric potential energy on one side of the atom differs from that on the other side, and a Stark shift is produced. This effectively destroys the angular momentum l as a good quantum number. At one time it was thought that the Stark shift would be so great that the lines would be unobservable, but in practice initial and final states of a line are shifted by similar amounts in the same direction and the resultant Stark shift of the line is negligible.

These hydrogen lines are observed, together with weaker helium lines, shifted by 2·1 MHz to higher frequencies for 109α and also what are believed to be carbon lines which would be shifted a further 0·7 MHz. An infinite mass ionic core would hardly be distinguishable from carbon at these temperatures.

The spontaneous emission by the atoms is usually negligible. The emission is almost entirely induced by background continuum radiation by maser action. In addition the atoms are subject to collisions with electrons and protons: these collisions tend to bring the populations of the excited states towards thermal equilibrium and thus reduce the intensity of the maser action. They also interrupt the emission of radiation by producing collisional transitions. The resultant cut-off in the radiation produces an impact broadening of the lines. For the line broadening we need to know the total inelastic cross-section σ^{in}. A strong collision producing a large energy transfer is no more important than a weak inelastic collision producing a transition to a neighbouring level. These σ^{in} are relatively easy to obtain.

For the population problem the strong collisions are more important then the weak ones and individual inelastic cross-sections are required. This is more difficult.

III. CORRESPONDENCE PRINCIPLE FOR INTENSITY

Bohr was able to predict the approximate intensity of spontaneously emitted radiation for a transition between excited states by supposing that the mean power W emitted in the transition $n \to n'$ tends to coincide with the mean power emitted at the corresponding classical frequency. If ΔE is the energy difference and A the transition probability then

$$A \Delta E = W. \qquad (3.1)$$

Mezger has recently applied this to atoms, and by a similar correspondence principle method we obtain the oscillator strength

$$f(n', n) = \frac{8}{3} \frac{J_s(s) J_s'(s)}{s^2} \frac{n'^2}{n + n'} \qquad (n' < n), \qquad (3.2)$$

where $s = n - n'$, $J_s(x)$ is a Bessel function, and $J_s'(x)$ its first derivative. For $s = 2$, $n = 5$ the error is 10%, for $s = 1$, $n = 10$ the error is 0·5%, and for the radio recombination lines it is negligible. Encouraged by this agreement we can continue in the spirit of Copenhagen and Göttingen of the 1920's and apply the correspondence principle to collisions.

IV. CLASSICAL COLLISIONS

The line spectra of the radio-recombination lines provide clear evidence that the highly excited states are quantized and that classical mechanics on its own is not enough. We would like to use the correspondence principle to obtain cross-sections for collisions between the quantized states, but we cannot do this effectively until we know what happens to the corresponding classical model of the collision.

Suppose an electron collides with an H atom with an infinite mass nucleus, that U is the binding energy of the atom, that E is the initial energy of the incident electron, and that

$$E \gg U. \tag{4.1}$$

For ionization to take place, the bound electron must gain enough energy to leave the proton, but the incident electron must not loose so much that its energy becomes negative, leaving it bound to the proton.

We can classify collisions according to the range of values of the energy transfer

$$\Delta E = E - E' \tag{4.2}$$

from the incident electron to the initially bound electron, where E' is the final energy of incident electron. We have the possibilities:

$$\Delta E < U \quad \text{direct scattering} \tag{4.3a}$$
$$\text{(excitation } or \text{ de-excitation)}$$

$$U < \Delta E < E \quad \text{ionization} \tag{4.3b}$$

$$\Delta E > E \quad \text{classical rearrangement} \tag{4.3c}$$
$$\text{(exchange)}$$

Electron exchange is caused by more violent collisions than ionization, but is very rare when

$$E \gg U.$$

The classical ionization cross-section is given by

$$\sigma_I^c = \sigma^c(U < \Delta E < E). \tag{4.4}$$

The quantal cross-section is given to a good approximation by the same quantity

$$\sigma_I^q \approx \sigma_I^c. \tag{4.5}$$

The validity of this result will be discussed in the next Section.

V. DISCRETE STATES

Excitation is more difficult to understand. In classical mechanics every dynamical variable has a continuous range of possible values, including binding energies, angular momenta and other physically quantized variables. Classical cross-sections for these variables are differential cross-sections. In particular, we define

$$\partial\sigma/\partial\Delta E \tag{5.1}$$

to be the differential cross-section with respect to energy transfer.

This differential cross-section must be related to excitation or de-excitation cross-sections. If $\partial\sigma/\partial\Delta E$ varies by only a small fraction over values of ΔE corresponding to several neighbouring final excited states n', then there is little difficulty in obtaining from it unambiguous excitation cross-sections, by assuming a smooth distribution among the final states n' as shown in Fig. 1, and as follows:

$$\sigma(n \to n') \approx \left|\frac{\partial \Delta E}{\partial n'}\right| \partial\sigma/\partial\Delta E. \tag{5.2}$$

where ΔE is now the energy required to excite the atom from state n to state n', and

$$|\partial \Delta E/\partial n'| = 2Ry/n'^3 \tag{5.3}$$

is a mean energy range between states labelled by n', or the inverse of the density of states per unit energy range. The practical condition for the

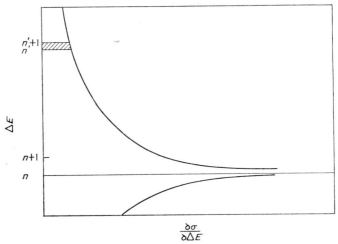

FIG. 1. Density of states correspondence for $\sigma(n \to n')$ excitation cross-sections. The shaded area gives the approximate cross section.

validity of this naive correspondence is that initial and final quantum numbers and all charges in quantum number should be large, that is

$$n, n', |\Delta n| \gg 1 \tag{5.4}$$

From Fig. 1 there is ambiguity when Δn is small because of the rapid change in the differential cross-section as a function of ΔE.

The above naive correspondence based on density of states enables us to apply the laws of classical mechanics to such problems as the motion of the planets, which is done without hesitation and without invoking any correspondence principle explicitly. Notice that the condition on $|\Delta n|$ in (5.4) is in the opposite sense to that of Bohr's correspondence principle.

A more solid theoretical foundation for the above naive correspondence is the inequality

$$S_{\text{classical}} \gg \hbar \tag{5.5}$$

where $S_{\text{classical}}$ represents relevant classical action functions for the scattering problem, including those of the atom, the incident electron and their interaction. When (5.5) is satisfied we can make quantal wave-packets which are much smaller than the dimensions of the system and which do not spread or diffract significantly. Ehrenfest showed that their mean positions follow the laws of classical mechanics. In this way we can justify both (4.5) for ionization and (5.2) for excitation and de-excitation. In practice, for the important range of incident energy, we do not need to verify (5.3). Condition (5.4) is enough.

VI. SCALING LAWS

Systems of particles interacting through Coulomb forces satisfy simple scaling laws. If old trajectories $R_i(t)$ for 3 particles $i = 1, 2, 3$ satisfy the equations of motion, then new trajectories $R_i'(t')$ also satisfy the equations of motion provided that for some θ

$$R_i(t) = \theta R_i'(t')$$
$$t = \theta^{3/2} t'. \tag{6.1}$$

This may be verified by substitution in the equations of motion. All positions, lengths and times must be similarly scaled, Velocity V, energy E and cross-section σ are scaled according to the transformations

$$V = \theta^{-1/2} V', \quad E = \theta^{-1} E', \quad \sigma = \theta^2 \sigma' \tag{6.2}$$

Consequently, if our H atom has binding energy U, the quantity

$$\frac{\sigma(E/U)}{\sigma^g} \tag{6.3}$$

is a universal function of E/U, independent of U, where $\sigma^g = n^4 \pi a_0^2$ is the geometrical cross-section and where the initial conditions are similar. As a result a classical cross-section need be obtained as a function of energy for *one* highly excited level only, and (6.3) may then be used to obtain the cross-section for other levels.

VII. INITIAL CONDITIONS

These clearly depend on the distribution of atoms among the sub-states (nlm) or Stark states ($n_1 n_2 m$) of a highly excited level n. In the H II regions, where $n \to n'$ collisions are important, these sub-states are nearly always almost equally populated.

The corresponding classical distribution consists of an ensemble of atoms with electrons moving in elliptical orbits with:

$$\text{binding energy } U_n = Ry/n^2 \tag{7.1a}$$

$$\varepsilon^2 \text{ uniformly distributed between 0 and 1} \tag{7.1b}$$

$$\text{arbitrary orientation, spherically symmetric distribution.} \tag{7.1c}$$

The eccentricity ε varies continuously from $\varepsilon = 1$ for lowest angular momentum (straight line orbit) to $\varepsilon = 0$ for maximum angular momentum (circular orbit). This distribution is similar for all n, so only one classical cross-section need be obtained. For all excited levels the appropriate cross-sections can then be obtained by scaling. The resulting momentum distribution for the bound electron, which has a critical effect on the collision cross-section, is given by

$$p(\mathbf{p}) = \frac{8 p_n^5}{\pi^2} \frac{1}{(\mathbf{p}^2 + p_n^2)^4} \tag{7.2}$$

where

$$p_n = m v_n = (2 m U_n)^{\frac{1}{2}}. \tag{7.3}$$

This is *exactly* the same as the quantal distribution for all n (Fock,[1] Norcliffe and Percival[2]).

VIII. BINARY ENCOUNTER OR CLASSICAL IMPULSE APPROXIMATION

When the velocity V of the incident charged particle is large compared to v_n, and the collision is fairly violent, we can use the binary encounter approximation:

$$V \gg v_n \tag{8.1}$$

The approximation has a long and tortuous history (see Burgess and Percival[3]) and is based on the assumption that the distribution of energy ΔE transferred from an incident electron of energy E to a bound electron is the same as would be obtained if it passed through a gas of electrons with the same velocity distribution as the bound electrons.

Collisions between incident and bound electrons are supposed to occur according to the Rutherford scattering law. For high n the Mott correction for exchange interference is negligible. The effect of the nuclear proton is neglected during this binary encounter. For ionization we obtain

$$\frac{\sigma^i}{\sigma^g} = \frac{4}{\bar{E}}\left(1+\frac{1}{2\bar{E}}\right)^{-1}\left(\frac{5}{3}-\frac{1}{\bar{E}}-\frac{2}{3}\frac{1}{\bar{E}^2}\right), \quad \bar{E} = E/U \gg 1 \qquad (8.2)$$

where σ^g is the geometrical cross-section and $\bar{E} = E/U$ is the incident electron energy in units of the ioniaztion energy. The use of the velocity distribution with fixed magnitude v_n for $\mathbf{v}$ does not significantly affect this result in the high energy region where the binary encounter theory is valid. For incident protons

$$\frac{\sigma^i}{\sigma^g} = \frac{4}{\bar{E}}\left(\frac{5}{3}-\frac{1}{4(\bar{E}-1)}\right), \quad \bar{E} = \frac{m_e E}{M_p U} \gg 1. \qquad (8.3)$$

This cross-section is valid for protons with *velocity* much higher than that of the bound electrons, and has the same high-velocity limit as the cross-section for electrons of the same velocity. The quantum mechanical correction to these cross-sections is completely negligible for highly excited states.

For excitation the theory is only valid when the collision is sufficiently strong. For weak collisions, specifically, when, for electrons

$$E\Delta E \lesssim U^2 \qquad (8.4)$$

adiabatic effects become significant. We then have for electron and proton binary encounter excitation cross-sections

$$\frac{\sigma(n \to n')}{\sigma^g(n)} = \frac{2n^2}{n'^3}\frac{4}{\bar{E}+2}\left(\frac{1}{(\Delta\bar{E})^2}+\frac{4}{3}\frac{1}{(\Delta\bar{E})^3}\right) \qquad (8.5)$$

(electrons, $\Delta n \gg 1$, $\bar{E}\Delta\bar{E} \gg 1$, $\bar{E} = E/U$, $\Delta\bar{E} = \Delta E/U$)

and

$$\frac{\sigma(n \to n')}{\sigma^g(n)} = \frac{2n^2}{n'^3}\frac{4}{\bar{E}}\left(\frac{1}{(\Delta\bar{E})^2}+\frac{4}{3}\frac{1}{(\Delta\bar{E})^3}\right) \qquad (8.6)$$

(protons, $\Delta n \gg 1$, $\bar{E}\Delta\bar{E} \gg 1$, $\bar{E} = m_e E/M_p U$, $\Delta\bar{E} = \Delta E/U$).

The adiabatic theory for smaller energy transfers has been treated by Percival and Richards.[4]

Flannery[5] (and papers cited there) has applied binary encounter theory to collisions between normal neutral systems with geometric cross-section of order πa_0^2 and highly excited atoms. The forces are short range, so the cross-sections add to no more than the total cross-section between the orbiting electron and the small neutral. The resultant factor n^{-4} makes these cross-sections astrophysically insignificant for high n, in comparison with the cross-sections for charged particles.

IX. ORBIT INTEGRATION AND MONTE CARLO METHOD

For low velocity incident charged particles, such as protons in H II regions, no simplifications can be made. The classical problem must be solved directly by repeated numerical integration of the classical equations of motion, usually chosen by appropriate random sampling, forming in effect a computer model of the scattering process. The only effective errors in this "Monte Carlo" procedure (apart from the classical approximation) are the statistical errors analogous to those of laboratory experiments, with counts of the order of 1,000 or 2,000 events (Abrines and Percival[6]). The computing time involved is large but not prohibitive.

Results are presented in Fig. 2 for ionization of hydrogen by protons in a classical model for $n = 1$. Cross-sections for other levels may be obtained

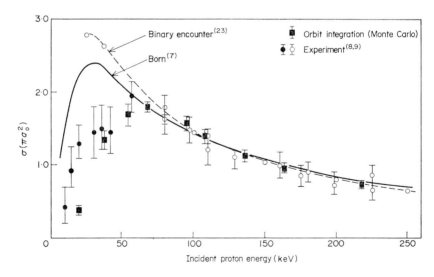

FIG. 2. p–H ionization cross-section for initial $n = 1$.

directly by using the scale-independent form σ^i/σ^g as a function of $\bar{E}$. Cross-sections for electrons were also obtained (Abrines, Percival and Valentine[10]). Excitation cross-section calculations are in progress.

In conclusion it should be mentioned that much of the classical theory of collisions was developed for atoms of low quantum number, where its validity is in greater doubt than for highly excited levels. Nevertheless comparison with experiment even for low levels is not too bad.

X. CORRESPONDENCE PRINCIPLES

The classical theory and correspondence principle for density of states is inadequate for small Δn. Not only is the assignment of the energy range ambiguous, but also the form of the cross-section is wrong.

For sufficiently high incident particle energies the Bethe–Born approximation is valid and provides a quantal cross-section which has the form

$$\sigma^q(n \to n') = \frac{A \ln E}{E} + \frac{B}{E} \qquad (E \text{ large}). \tag{10.1}$$

For large n and large Δn and values of E of physical interest the logarithmic term is insignificant.

The classical theory above provides a cross-section which has the wrong form

$$\sigma^c(n \to n') = B'/E \qquad (E \text{ large}, \Delta n \text{ small}). \tag{10.2}$$

The additional logarithmic term in (10.1) comes from weak collisions with large impact parameters, where the classical energy transferred to the bound particle is less than the energy splitting between the states; the neighbouring energy states are thus classical inaccessible. Two clues enable us to overcome this problem.

Williams[11] pointed out that classical theory can give a correct value for energy loss of fast particles, even for these weak collisions, so that the classical energy transfer behaves as if it were shared among the quantum mechanical states to produce the same mean energy transfer.

Bohr's correspondence principle for radiation deals with a similar situation, for a classical radiating electron never gives up enough energy in a single orbit to reach a neighbouring quantal state. The energy transfer to the radiation field in this case is made up of a series of Fourier components

$$\Delta E = \sum_s \Delta E_s \qquad (s = n - n'). \tag{10.3}$$

Bohr's correspondence principle relates each Fourier component to a quantal transition.

The classical theory of perturbed orbits due to Lagrange can be used to express the energy transferred in a weak collision as a Fourier sum of the form (10.3) (Born,[12] Percival and Richards[4]). The analogy between the radiating system and the weak collision allows us to start a correspondence principle (analogous to Bohr's) for collisions:

> The mean net energy transferred to an atom by a passing charged particle due to upward and downward transitions $n \leftrightarrow n'$ tends to coincide with the mean net energy transferred according to classical theory from Fourier components of order s and $-s$ *provided* the perturbation of the classical orbit is small *and* the probability of a transition from the initial state is small.

Classical mechanics used with this correspondence principle provides cross-sections with the correct logarithmic form (10.1) for transitions with Δn comparable to 1. For intermediate values of Δn it agrees with the density-of-states theory. It is not, however, the best way to obtain cross-sections.

XI. HEISENBERG'S FORM

The above adaptation of Bohr's correspondence principle lacks generality. In founding the matrix formulation of quantum mechanics, Heisenberg obtained a more general form of correspondence principle relating matrix elements for explicit functions of position to Fourier components. Suppose

$$x(t) = \sum_{\text{all } s} x_s \exp(-is\omega t) \qquad (11.1)$$

is the position of a particle moving periodically in one dimension, for example an anharmonic oscillator. Then any classical function $F^c(x)$ of position may also be expressed as a Fourier series in time:

$$F^c(x) = \sum_s F_s^c \exp(-is\omega t). \qquad (11.2)$$

Then Heisenberg's form of the correspondence principle states that the quantal matrix elements between states n and n' of the dynamical variable F tend to coincide with the classical Fourier components

$$\langle n' | F | n \rangle \approx F_s^c \qquad (s = n - n'; s \ll n, n') \qquad (11.3)$$

where the classical orbit is taken near the mean energy of initial and final states.

Using this correspondence principle we can find matrix elements of first-order perturbation theory (Which gives us the same result as the previous section) or, for example, for the sudden approximation (Alder and Winther,[13] Kramer and Bernstein[14]).

XII. STRONG COUPLING

Heisenberg's principle has been extended by Percival and Richards[15] to provide a correspondence principle for strongly coupled states. The conditions on the validity of this principle are that n should be large and Δn small. No assumptions are made about the validity of quantum perturbation theory: collisions can be so strong that the probability of loss from the initial state n approaches unity. For highly excited states this is quite compatible with the assumption that the quantum numbers change relatively little.

The transition amplitudes $S(n', n)$ for a one-dimensional system, according to the three approximations discussed, are

$$S_{\text{weak}}(n', n) = \frac{\omega}{2\pi} \int_0^{2\pi/\omega} d\tau \exp(is\omega\tau) \left[-\frac{i}{\hbar} \int_{-\infty}^{\infty} dt V^c(x(t + \tau), t) \right],$$

$$(|s| \ll n, n'; \sum_{n'} |S(n', n)|^2 \ll 1) \quad (12.1)$$

$$S_{\text{sudden}}(n', n) = \frac{\omega}{2\pi} \int_0^{2\pi/\omega} d\tau \exp \left[is\omega\tau - \frac{i}{\hbar} \int_{-\infty}^{\infty} dt V^c(x(\tau), t) \right],$$

$$(|s| \ll n, n'; \omega \times (\text{Time of collision}) \ll 1) \quad (12.2)$$

$$S_{\text{strong}}(n', n) = \frac{\omega}{2\pi} \int_0^{2\pi/\omega} d\tau \exp \left[is\omega\tau - \frac{i}{\hbar} \int_{-\infty}^{\infty} dt V^c(x(t + \tau), t) \right],$$

$$(|s| \ll n, n'). \quad (12.3)$$

The function $V^c(x(t + \tau), t)$ is the value of the classical interation potential at time t. We suppose that at time $t = 0$ the bound particle is at a point on the orbit reached after time τ, from some initial point. The transition amplitude appears as a coherent sum over the amplitudes for classical orbits with different starting points. The many dimensional-theory for separable systems is more complicated, but differs in no essentials.

XIII. RANGES OF VALIDITY

These are illustrated in Fig, 3. in a b–V plot of impact parameter and incident particle velocity. Incident electrons in H II regions are usually in region $\mathcal{B}$ where there is a considerable range of impact parameter for which both weak coupling and sudden approximations are valid. Incident protons in H II regions are usually in region $\mathcal{D}$ where orbit integration is necessary.

The Born approximation is not valid for incident electron energies below 1 Ry, independently of initial quantum number change of quantum number and even nuclear charge. It therefore should not be used in H II regions. Note that the Born theory here breaks down below about 10^6 times threshold energy!

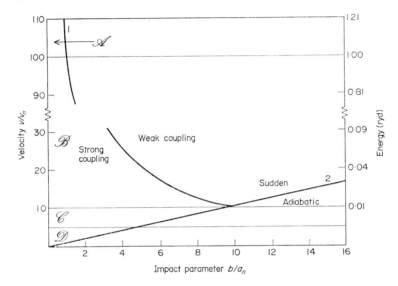

FIG. 3. b–V plot for $n = 100$.

We now look at these regions in more detail.

Region $\mathscr{A}$. Quantal perturbation theory is valid in this region, so that

$$\sum_{\text{all } n' \neq n} p(n \to n') \ll 1. \tag{13.1}$$

The probabilities can be obtained from the correspondence principle. If this is to be satisfied for all impact parameters greater than the atomic radius, we find that

$$\left(\frac{V}{v_n}\right)^2 > n^2 \quad \text{or} \quad E > 1 \, Ry \quad \text{(electrons)} \tag{13.2}$$

for first-order perturbation theory or the Born approximation to be valid.

Region $\mathscr{B}$. At lower energies we enter region $\mathscr{B}$ where first-order theory is valid for a limited range of impact parameters

$$b > b_w = \frac{na_n v_n}{V}. \tag{13.3}$$

In this region the nature of the collision depends strongly on the "adiabaticity parameter"

$$\frac{\omega b}{V} = \frac{b v_n}{a_n V} \tag{13.4}$$

which is the ratio of the collision time to the fundamental period. When the impact parameter is so large that the collision time is large compared to the period, the collision is adiabatic, and the probability of transitions is small. Conversely, for smaller impact parameters the collision is "sudden" and the sudden approximation is valid. If we write

$$b_{\text{ad}} = \frac{V}{\omega} = \frac{v_n}{a_n V} \tag{13.5}$$

then in region $\mathscr{B}$

$$b_{\text{ad}} > b_w \tag{13.6}$$

so that there is an overlap between the ranges of validity of sudden and weak coupling theories. Equality of the two impact parameters gives the lower energy boundary of this region, so that region $\mathscr{B}$ is defined by

$$\frac{2Ry}{n} < E < 1\,Ry. \tag{13.7}$$

This is one of the main energy ranges of interest for the radio recombination lines.

A remarkable property of this region is that the logarithmic term is $2A \ln E/E$, which is exactly double that for the Born region. Thus the slope of the Bethe plot is doubled in this region, compared with the slope in the Born region. Region $\mathscr{C}$ where $b_w < b_{\text{ad}}$ has not been studied in detail. In region $\mathscr{D}$, where the velocity of the incident particle is comparable to that of the bound electron, Monte Carlo methods are appropriate.

XIV. CROSS-SECTIONS (QUANTAL)

Quantal cross-sections for highly excited states with $\Delta n = \pm 1$ have been obtained by Seaton[16] and by Saraph[17] using the impact parameter method and the dipole approximation. The incident particle was assumed to follow a classical path and the dipole approximation was used. The energy transfer ΔE is so small that these approximations are certainly valid over incident energy ranges of interest. In addition a weak coupling approximation, first-order quantal perturbation theory, was used for large impact parameters b; and for small b, where the first order probability exceeded $\frac{1}{2}$, a rough estimate of the probability was made. The cut-off on b is essentially the Weisskopf radius b_w, so the cross-section is given by

$$\sigma = 2\pi \left[\int_0^{b_w} \tfrac{1}{2} b \, db + \int_{b_w}^{\infty} P_{\text{weak}}(b) \, b \, db \right] \quad (14.1)$$

$$= \frac{\pi}{2} b_w^2 + 2\pi \int_{b_w}^{\infty} P_{\text{weak}}(b) \, b \, db.$$

For higher energies they obtained, in effect, an approximation to Born. They found a kink in the cross-section, which represents the transition between region $\mathscr{B}$ and region $\mathscr{A}$.

Kingston and Lauer[18–19] used the Born approximation and extrapolation from lower n to obtain cross-sections for $\Delta n = \pm 1, \pm 2$. This included non-dipole terms and is valid in region $\mathscr{A}$, with incident electron energy above 1 Ry, although it is used by those authors also for much lower energies where it should not be used.

XV. CROSS-SECTIONS (CORRESPONDENCE)

The correspondence principle methods we have described were applied by Percival and Richards[15, 20–22] to the general problem of direct collisions of particles of mass M and charge $Z'e$ ($Z' = -1$ for electrons) with highly excited hydrogen-like ions of nuclear charge Ze ($Z = 1$ for neutral H atoms) in region $\mathscr{B}$, which for these more general collisions is given by

$$\frac{2M}{m_e} \frac{|ZZ'|}{n} < \frac{E}{Ry} < \frac{M}{m_e} Z'^2 \quad (15.1)$$

The classical path and dipole approximations were made for all orbits. The weak coupling and sudden approximations were made in their respective regions of validity but the combination of the two was checked against the full strong coupling correspondence principle result for circular orbits. We believe the result to be valid within about 20%. For intermediate $s = \Delta n$ the results agree with the binary encounter theory. A binary encounter correction term was inserted which is insignificant for $s \approx 1$, but necessary for $s \approx n$.

For energies in the range

$$5s^2 \frac{|ZZ'|}{n} \frac{M}{m_e} < \frac{E}{Ry} < \frac{M}{m_e} Z'^2 \quad (15.2)$$

the cross-section is given by

$$\frac{m_e}{M}\left(\frac{Z}{Z'}\right)^2 \frac{\sigma(n \to n')}{\sigma^g(n)} = \frac{Ry}{(EE_>)^{\frac{1}{2}}} \left\{\frac{16}{3}\left(\frac{n'}{n}\right)^3 I_0(s) \ln\left[0.8 \frac{m_e}{M}\frac{(nn'EE')^{\frac{1}{2}}}{|ZZ'|Ry}\right]\right.$$

$$\left. + \frac{8n'}{(n'^2 - n^2)^2}\left(\frac{n_<}{n}\right)^5 \left[\frac{8n^2 n'^2 I_1(s)}{n_<^2(n+n')} + 1\right]\right\} \quad (15.3)$$

$(n, n' \gg 1; \sigma^g(n) = \pi n^4 a_0^2; E_> =$ greatest of $E, E'; n_< =$ least of $n, n')$

or for electrons on hydrogen

$$\frac{\sigma(n \to n')}{\sigma^g(n)} = \frac{Ry}{(EE_>)^{\frac{1}{2}}} \left\{\frac{16}{3}\left(\frac{n'}{n}\right)^3 I_0(s) \ln\left[\frac{0.8(nn'EE')^{\frac{1}{2}}}{Ry}\right]\right.$$

$$\left. + \frac{8n'}{(n'^2 - n^2)^2}\left(\frac{n_<}{n}\right)^5 \left[\frac{8n^2 n'^2 I_1(s)}{n_<^2(n+n')} + 1\right]\right\} \quad (15.4)$$

The functions $I_0(s)$ and $I_1(s)$ are given by

$$I_0(s) = I_0(-s) = J_s(s)J_s'(s)/s^3 \approx 0.184 s^{-4}$$
$$I_0(1) = 0.143$$
$$I_0(2) = 0.988 \times 10^{-2} \quad (15.5)$$
$$I_0(3) = 0.203 \times 10^{-3}$$
$$I_1(1) = 0.077$$

$$I_1(s) \approx \frac{1}{6|s|} - \frac{0.0933}{|s|^{1.41}} \quad (15.6)$$

The range of the calculations was extended by a further factor of $5s^2$ to lower energies by a more complicated formula involving modified Bessel functions. This shows deviations from the simple $2A \ln E/E + B'/E$ form above. Comparison with the work of Kingston and Lauer and of Saraph is made in Fig. 4.

XVI. LITERATURE

There appears to be no general review of highly excited atoms. Purely classical methods and in particular binary encounter methods have been reviewed by Burgess and Percival[3] and by Vriens.[23] In each case the authors were more concerned with problems that arise in the application of the theory to low states of atoms. Abrines and Percival[6, 24-25] obtained proton–hydrogen ionization cross-sections using Monte Carlo methods and discussed the naive density-of-states correspondence principle. The Monte-Carlo electron–hydrogen ionization cross-sections were calculated by Abrines, Percival and Valentine[10] and verified in later calculations by Brattsev and Ochkur[26] and by Mansbach and Keck[27] (region $\mathscr{D}$). Excitation cross-sections for regions $\mathscr{C}$ and $\mathscr{D}$ are not yet available. They might be useful for cool interstellar regions.

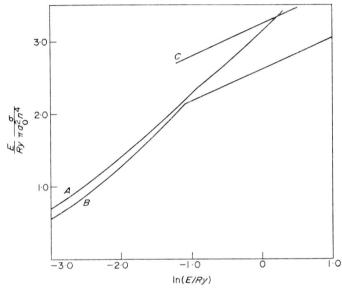

Fig. 4. The Bethe plot of the 20 → 21 transition: A is the correspondence cross-section, B is that of Saraph[17] and C is that of Kingston and Lauer.[19]

The correspondence principle methods were developed by Percival and Richards and by Abrines and Percival in the papers cited. The background classical theory is best presented by Born[12] in his book on "Mechanics of the Atom". Most modern classical mechanics texts have an inadequate treatment of Hamilton–Jacobi theory.

A semi-classical theory has been presented by Beigman et al.,[28] which Richards has shown to be equivalent to the strong coupling correspondence principle in one dimension. However, in applying their theory Beigman et al. make such rough approximations that its advantages are lost. Richards has also shown that the theory of Presnyakov and Urnov[29] which is a quantum-mechanical theory for states of equally spaced energy levels and special coupling terms, reduces to the strong coupling correspondence principle. They have just obtained their first results (August 1971), and comparison with the correspondence cross-sections is awaited.

Further theory of highly excited states is given by Minaeva et al.[30] and calculations on states with $n \approx 5$ are presented by McCoyd and Milford[31] and by Omidvar.[32] Highly excited states ($n \approx 15$) have been produced in hydrogen in beams by those concerned with the production of controlled thermonuclear reactions (see, for example, Berkner et al.[33]), and have been observed in barium up to $n \approx 50$.

REFERENCES

1. Fock, V., *Z. für Phys.* **98**, 145 (1935).
2. Norcliffe, A. and Percival, I. C., *J. Phys.* **B1**, 784 (1968).
3. Burgess, A. and Percival, I. C., *Adv. in At. and Mol. Phys.* **4**, 109 (1968).
4. Percival, I. C. and Richards, D., *Proc. Phys. Soc.* **92**, 311 (1967).
5. Flannery, M. R., *J. Phys. B* **4**, 892 (1971).
6. Abrines, R. and Percival, I. C., *Proc. Phys. Soc.* **88**, 873 (1966).
7. Bates, D. R. and Griffing, G. W., *Proc. Phys. Soc.* **A66**, 961 (1953).
8. Gilbody, H. B. and Ireland, J. V., *Proc. Phys. Soc.* **277**, 137 (1963).
9. Fite, W. L., Stebbings, R. F., Hammer, D. G. and Brackmann, R. T., *Phys. Rev.* **119**, 663 (1960).
10. Abrines, R., Percival, I. C. and Valentine, N. A., *Proc. Phys. Soc.* **89**, 515 (1966).
11. Williams, E. J., *Proc. Roy. Soc.* **A130**, 328 (1931).
12. Born, M., "Mechanics of the Atom", F. Ungar Publ. Co., New York (1960).
13. Alder, K. and Winther, A., *Math-fys. Meddr.* **32**, (1960).
14. Kramer, K. H. and Bernstein, R. B., *J. Chem. Phys.* **40**, 200 (1964).
15. Percival, I. C. and Richards, D., *J. Phys.* **B3**, 1035 (1970).
16. Seaton, M. J., *Proc. Phys. Soc.* **79**, 1105 (1962).
17. Saraph, H. E., *Proc. Phys. Soc.* **83**, 763 (1964).
18. Kingston, A. E. and Lauer, J. E., *Proc. Phys. Soc.* **87**, 399 (1966).
19. Kingston, A. E. and Lauer, J. E., *Proc. Phys. Soc.* **88**, 597 (1966).
20. Percival, I. C. and Richards, D., *Astrophys. Lett.* **4**, 235 (1970).
21. Percival, I. C. and Richards, D., *J. Phys.* **B4**, 918 (1971).
22. Percival, I. C. and Richards, D., *J. Phys.* **B4**, 932 (1971).
23. Vriens, L., *In* "Case Studies in Atomic Collision Physics". No. 1. North Holland (1969).
24. Abrines, R. and Percival, I. C., *Phys. Lett.* **13**, 216 (1964).
25. Abrines, R. and Percival, I. C., *Proc. Phys. Soc.* **88**, 861 (1966).
26. Brattsev, V. F. and Ochkur, V. I., *Soviet Physics J.E.T.P.* **25**, 631 (1967).
27. Mansbach, P. and Keck, J., *Phys. Rev.* **181**, 275 (1969).
28. Beigmann, I. L., Vainshtein, L. A. and Sobel'man, I. I., *Sov. Phys. J.E.T.P.* **30**, 920 (1970).
29. Presnyakov, L. P. and Urnov, A. N., *J. Phys.* **B3**, 1267 (1970).
30. Minaeva, L. A., Sobel'man, I. I. and Sorochenko, R. L., *Astron. Zh. (U.S.S.R.)* **44**, 995 (1967).
31. McCoyd, G. C. and Milford, S. N., *Phys. Rev.* **130**, 206 (1963).
32. Omidvar, K., *Phys. Rev.* **140A**, 38 (1965).
33. Berkner, K. H., King, I. I. and Riviere, A. C., *Nuclear Fusion* **7**, 115 (1967).

Spectral Line Broadening

H. VAN REGEMORTER

*Dept. Astrophysique Fondamentale,
Observatoire de Paris, 92-Meudon*

INTRODUCTION

The detailed study of the line profiles emitted by laboratory and astrophysical plasmas can produce a large volume of information. The line broadening theory has been considerably stimulated both by the needs of plasma phycisists and astrophysicists and is a field which is progressing rapidly.

It is not our purpose to give a complete review of the subject: that would require a book, not a few lectures. Many review articles deal with the theory of pressure broadening, by Unsold,[1] Chen and Takeo,[2] Breene,[3] Bohm,[4] Traving,[5] Margenau and Lewis,[6] Mazing,[7] Breene,[8] Baranger,[9] Breene,[10] Griem,[11] Sobelman,[12] Wiese,[13] van Regemorter,[14] Cooper[15] and Traving.[16]

We shall only consider the pressure broadening of spectral lines, due to the interactions of the radiating atom with surrounding particles. Of course, thermal motion of the emitting atom—Doppler broadening—has also to be considered as well as natural broadening and interaction of the atom with the radiating field. We ought to be able to combine these different kinds of broadening or to consider the case when the pressure broadening is more important than the others (i.e. when the density of the plasma is high enough, the radiation field is weak). We have to bear in mind that, in some astrophysical situations, the profile of a line may be more sensitive to the model of the emitting region (variation of the temperature or of the density with the optical depth) or to the mode of scattering of a quantum from one atom to another, or to the turbulence of the emitting region, than to pressure broadening. But when pressure broadening is important it can give good estimations of the temperature and of the density of the plasma.

In these lectures, we shall concentrate mainly on the basic principles and basic methods in line broadening theory and not on detailed calculations or comparisons with experiments. Instead of using a difficult mathematical apparatus, full of complicated operators originating from the many body problems—which is simplified anyway for solving practical problems—we shall try to understand the real nature of the interactions and how they are giving rise to the broadening of a line.

I. BASIC PRINCIPLES

1.1. General consideration

The theory of pressure broadening involves two distinct problems. The first is the study of the interaction between the radiating atom and one perturber. This is a *quantum mechanical problem* which is not peculiar to pressure broadening. For instance for interaction with neutral atoms at very low temperature the main interaction may be the Van Der Waals interaction proportional to r^{-6}. When the perturbers are charged particles (ions and electrons) the interaction is via the Coulomb interaction $V = -\mathbf{d} \cdot \mathbf{E}$ where $\mathbf{d}$ is the electric dipole moment of the radiator and $\mathbf{E}$ is the electric field produced by perturber. For a large distance between perturber and radiator this interaction scales as r^{-2}.

When the perturbers may induce inelastic collision in the radiating atom, the interaction becomes much more complicated and we must solve a difficult quantum problem. Fortunately, in many cases the solution can be obtained using a quasi-classical approximation, which simplifies the problem and makes it more descriptive.

The second problem is a *statistical problem* of combining the effect of a large number of perturbers and of averaging over all possible motions of these. If the quantum mechanical problem is not easy, the statistical problem is typical to pressure broadening and is what makes it difficult and more difficult than the problem of the collision between two particles. For example, if we know how one perturber interacts with the atom, do we know how many perturbers interact? Can we neglect the interactions between the perturbers and assume that they move independently of each other?

1.2. The line shape

The absorption coefficient within a spectral line is

$$K(\Delta\omega) = \frac{\pi e^2}{mc} f N_i \frac{F(\Delta\omega)}{2\pi}$$

where $\Delta\omega$ is the angular frequency from the line centre, N_i is the population of level i and f the oscillator strength when stimulated emission is neglected. $F(\Delta\omega)$ is the normalized profile of the line.

$$\int_{-\infty}^{+\infty} F(\Delta\omega)\, d(\Delta\omega) = 1.$$

For spontaneous emission, the power radiated per unit frequency is

$$P(\Delta\omega) = \frac{4\omega^4}{3c^3} F(\Delta\omega).$$

From the theory of the interaction of a quantum system with electromagnetic radiation, we know that for dipolar transitions $i \to f$

$$F(\omega) = \lim_{T \to \infty} \frac{1}{2\pi T} \left| \int_0^T dt\, e^{i\omega t} \langle \Psi_f(t) |\mathbf{d}| \Psi_i(t) \rangle \right|^2 \tag{1}$$

where $\mathbf{d}$ is the electrical dipole of the system. Ψ_f and Ψ_i are the initial and final total wave function of the radiating system. The exact nature of the system will be explained later on. In the semi-classcial approach it is simply the radiating atom. In the pure quantum approach it is a composite system of a cell with the radiating atom and many perturbers.

It is more convenient to write the line shape as a function of its Fourier transform,

$$\Phi(s) = \int_{-\infty}^{+\infty} e^{-i\Delta\omega s} F(\Delta\omega)\, d(\Delta\omega)$$

to calculate $\Phi(s)$ first, and then come back to $F(\Delta\omega)$

$$F(\Delta\omega) = \frac{1}{\pi} \mathcal{R} \int_0^{\infty} e^{i\Delta\omega s} \Phi(s)\, ds.$$

This is because one has to solve a statistical problem. When this statistical problem can be avoided (far in the wings of the line) a straightforward calculation of $F(\Delta\omega)$ can be done very easily.

1.3. The statistical problem

The statistical problem can only be solved in very simple cases. In fact, it is necessary to assume that the perturbers contribute to the line shape independently. Suppose we have an atom and N perturbers. Assuming scalar additivity of the interactions between perturbers and atom the total line shape $F(\omega)$ is the distribution of the variable

$$\omega = \omega_1 + \omega_2 + \ldots + \omega_N$$

where each ω_i is distributed according to some distribution $f(\omega_i)$. Also assuming no interactions between the perturbers themselves, so that they may be regarded as completely independent, each probability distribution $f(\omega_i)$ is the same.

In this case if the Fourier transform $\phi(s)$ of $f(\omega)$ is

$$\phi(s) = \int_{-\infty}^{+\infty} e^{-i\omega s} f(\omega)\, d\omega.$$

The total line shape $F(\omega)$ is given by

$$F(\omega) = \frac{1}{2\pi} \int_{-\infty}^{+\infty} e^{i\omega s}\, \tilde{\Phi}(s)\, ds$$

with

$$\tilde{\Phi}(s) = [\phi(s)]^N. \qquad (2)$$

This is the fundamental theorem of the statistical problem in line broadening theory which underlines the limitation of the actual theories.

In fact there are situations in which

—the statistical problem does not arise: the perturber is only given by the nearest neighbour. This the one perturber approximation valid in the line wing in which the total system radiating is the quantum system atom + 1 perturber. At the static limit this gives the so-called "nearest neighbour approximation".

—the statistical problem can be solved using the theorem above, because the real superposition in time of interactions with different perturbers does not arise. We can predict that the superposition problem will be avoided if the moving perturbers act one after another in the time sequence of their closest approach to the radiating atom. This must be true for fast particles at low densities, at the "impact" limit where each interaction is in fact a collision. On the contrary, for slow particles, for which one uses static theories, the superposition problem arises, and consequently the problem of additivity of interactions.

1.4. Characteristic times

The line shape can be rewritten

$$F(\Delta\omega) = \frac{1}{2\pi} \int_{-\infty}^{+\infty} e^{i\Delta\omega s}\, \Phi(s)\, ds$$

with $\Delta\omega = \omega - \omega_0$, ω_0 being the undisturbed frequency, in order to factor out the high frequency part of $\tilde{\Phi}(s) = \Phi(s)\,e^{-i\omega_0 s}$. $\Phi(s)$ is then a slow function of s and the time interval which is important for calculating the profile at $\Delta\omega$ is of the order of

$$\tau = \frac{1}{\Delta\omega}$$

which is the time of interest.

Assuming a straight line classical path for the perturber, if ρ is the impact parameter or the distance of closest approach, then

$$\tau_c = \frac{\rho}{v}$$

is the collision time, where v is the relative velocity of the perturber and the atom.

If $\tau \gg \tau_c$ the interaction is completed in a very short time compared to the time of interest τ. The interaction can then be called a collision and the line broadening problem is just an application of collision theory.

At the same time ρ has a clear meaning (not always clear in the literature): ρ has nothing to do with the mean distance between particles ($n^{-1/3}$, if n is the density) but is the mean impact parameter for one type of interaction. If Q is the cross section, ρ is defined after the calculation of Q by

$$Q = \pi\rho_{\text{eff}}^2$$

ρ_{eff} can be estimated by using the Weisskopf radius, or the strong collision cutoff as defined later on. The relation

$$\Delta\omega \ll \frac{v}{\rho_{\text{eff}}} \tag{3}$$

means that "the collisions are completed" which is one of the conditions of the usual impact theory.

As soon as one can think in terms of collisions, one can also define the time between two collisions. This average time is of the order of

$$\tau_i = \frac{1}{\gamma}$$

because $\gamma = n\langle vQ\rangle$ is the probability that the atom experiences a collision if n is the particle density. As we shall see γ is in fact the half width.

Not too far in the wings of the line $\Delta\omega \sim \gamma$, and condition (3) means that

$$\gamma \ll \frac{v}{\rho_{\text{eff}}} \qquad (4)$$

which is equivalent to $\rho_{\text{eff}} < n^{-1/3}$.

1.5. The impact approximation

If $1/\gamma \gg \rho_{\text{eff}}/v$, the collision time is very short compared to the time between two collisions. It is the same to say that the collision volume ρ_{eff}^3 is very small compared to the volume per perturber n^{-1}. This means that each collision acts individually, one after another. There is no superposition problem. We can solve the statistical problem using formula (2).

In the usual impact approximation, there are in fact two different approximations: first, the "binary interaction approximation" (each collision acts individually), second, the collision approximation (each interaction is completed during the time of interest).

In the line core $\Delta\omega$ is of the order of γ and a few collisions can influence the profile, but if the impact approximation is valid one can solve the statistical problem using (2).

On the contrary in the line wings $\Delta\omega \gg \gamma$, and we can predict that only the closest particle will affect the profile. In this case, we as shall see later on, the exact expression of the profile can be obtained using the "one perturber approximation". The statistical problem does not arise. With the one perturber approximation, condition (3) may or may not be satisfied.

1.6. The static limit

The condition of validity of the impact approximation (4) which can be written

$$n \rho_{\text{eff}}^3 \ll 1$$

imposes an upper limit on the perturber density. At high densities we have to find a different formulation of the problem to take account of the simultaneous interaction of many particles.

If for one preturber the collision time is very long compared to the time of interest $\Delta\omega^{-1}$, i.e.

$$\tau_c \gg \Delta\omega^{-1}$$

it is a reasonable approximation to suppose that the perturber does not move and that the intensity emitted by the gas as a whole is given directly from a knowledge of the probability distribution for stationary perturbers. Holtsmark derived the field distribution for charged perturbers. His theory and the different improvements have been reviewed in the literature. Only the physical principles are given here.

Because we have to solve the statistical problem, we first have to show that we can again use the fundamental theorem (2).

First we consider neutral perturbers for which the long range interactions (Van der Waals potentials) add scalarly. The frequency shift for each perturber, $\Delta\omega$, has the same probability distribution $f(\Delta\omega)$ given by

$$f(\Delta\omega)\,d(\Delta\omega) = \frac{d^3x}{V}$$

if we assume that the perturber is distributed uniformly through volume V and if x is the 3 space coordinate. Since the conditions of theorem (2) are satisfied:

For one perturber

$$\phi(s) = \int_V e^{-i\Delta\omega(x)s}\,\frac{d^3x}{V}$$

$$= 1 - \frac{1}{V}\int_0^R [1 - e^{-i\Delta\omega(r)s}]4\pi r^2\,dr.$$

For $N = nV$ perturbers

$$\Phi(s) = \exp\left\{-4\pi n \int_0^\infty [1 - e^{-i\Delta\omega(r)s}]r^2\,dr\right\}. \tag{7}$$

When perturbers are charged particles, if we can assume that the perturbers are independent of each other, the treatment is the same but all variables are vectors and all integrals are triple. This has been done by Holtsmark neglecting all correlation between perturbers (protons and electrons). These effects of correlation have been taken into account by using Debye screened fields instead of Coulomb fields, in order to be able to solve the statistical problem for "independent" screened particles.

1.7. The nearest neighbour approximation

In the wing of a line, when the static approximation is valid, the probability of having a particle very near the atom (at the distance corresponding to $c_r/r^r = \Delta\omega$) is very small and it is a reasonable approximation to consider only the interaction with the closest particle. This is the so called "nearest neighbour approximation" which clearly corresponds to the "one perturber approximation" of the impact approach.

The probability of finding one particle in the interval dr and no particle closer than r is

$$W(r)\,dr = \exp\left(-\tfrac{4}{3}\pi r^3 n\right) n 4\pi r^2\,dr. \tag{8}$$

With $\Delta\omega = c_r/r^r$ and $\Delta\omega_0$ the normal shift corresponding to the mean distance $n^{-1/3}$, from the probability distribution $W(r)$ we can deduce the profile

$$F(\Delta\omega) = \frac{1}{r}\left(\frac{\Delta\omega_0}{\Delta\omega}\right)^{(r+3)/r} \exp\left[-\left(\frac{\Delta\omega_0}{\Delta\omega}\right)^{3/r}\right]\frac{1}{\Delta\omega_0}. \tag{9}$$

This formula is generally correct to obtain the far wings, but is not accurate in the line core.

1.8. Relationship between impact and static theories

We are considering a plasma at temperature T where the perturber density is n. kT gives the average speed of the particles. Of course because of the actual distribution of velocities, according to the value of the temperature the range of velocities can be large (high temperature) or small (low temperature).

For a given temperature, heavy particles like protons, or atoms, are much slower than light particles. The impact approximation must be valid for the latter, the static approximation tends to be valid for the former. But this statement must be considered with caution, because if the speed of the particle is one factor which is important in the validity criteria, it has to be compared with other factors, like the time of interest $\Delta\omega^{-1}$ or the range of the interaction potential.

For the given value of T, a proton and an H atom have the same speed, but for the case of broadening of a hydrogenic line by a proton the interaction potential is of the form Cr^{-2}, whereas in a collision between two unlike atoms (A + H) this potential is of much shorter range Cr^{-6} and the average impact parameter is very small. In fact in the first case the static approximation can be used and in the second (broadening by neutral H) the impact approximation can be valid.

The static limit is nothing but a low temperature limit. In each case the quantity $\bar{v}\Delta\omega^{-1}$, the actual distance traversed during the time of interest, must be compared to the mean distance between perturbers (the statistical problem) and to the range of the interaction potential (is the collision completed or not).

In the case of electrons, a high energy approximation can be valid to solve the "collision problem". For example, we can use the semi-classical approach.

Most of the perturbation is given by distant collisions ($l = mv\rho/\hbar > 1$). But at the same time, the time during which the atom is perturbed can be very short in the line wings (of the order of $\Delta\omega^{-1}$) and the static limit can be valid, as a limit of a semi-classical approach valid at high velocities.

In fact for isolated lines (not hydrogenic lines) for which the long range adiabatic potential is of the form Cr^{-4}, the usual impact approximation is always valid for electrons. In the case of hydrogenic lines because of the level degeneracy the field is of the form Cr^{-2}. The effective ρ—let us say, the Weisskopf radius—is very big and can become of the order of $n^{-1/3}$, the time of interest $\Delta\omega^{-1}$ can be very short and the particles act as if they are stationary.

In both extreme situations, the complete collision approximation on one hand and the static approximation on the other, there is a common feature: the particles act individually even when they act simultaneously in time. In both situations Eqn (2) can be used

In the far wings in both situations, only one particle is really perturbing the atom, and one can guess that it is possible to go smoothly from one to another. In all present "unified theories" valid from the impact regime to the static regime, the statistical relation (2) is assumed.

THE SEMI-CLASSICAL APPROACH

2.1. A simple classical model

We shall first use a very simple semi-classical approach in which we can represent the radiating atom by an oscillator which emits the frequency ω_0.

Let us make also the following assumptions

—the perturbing particles are following classical paths (straight lines)

—the perturbation is adiabatic, i.e. does not induce transitions between different states of the atom.

The general expression of the profile

$$F(\omega) = \lim_{T \to \infty} \left| \int_{-T/2}^{+T/2} dt \, e^{-i\omega t} \langle \Psi_f(t) | \mathbf{d} | \Psi_i(t) \rangle \right|^2 \frac{1}{2\pi T} \quad (10)$$

can be written in the form

$$F(\omega) \approx \frac{1}{2\pi T} \left| \int_{-T/2}^{+T/2} e^{-i\omega t} f(t) \, dt \right|^2 \quad (11)$$

$F(\omega)$ being proportional to the oscillation of the atomic oscillator

$$f(t) = \exp\left(i\omega_0 t - i \int_{-\infty}^{t} \omega(t') \, dt'\right). \tag{12}$$

The interaction with the perturbers changes the phase of the oscillator. Because of the adiabatic approximation, the amplitude of the oscillator does not change

$$\eta(t) = \int_{-\infty}^{t} \omega(t') \, dt'$$

is the sum of the phase shifts in the interval t. For simplicity, let us also suppose that only the upper level i is perturbed. If, for example, the interaction is of the long range form $C_r/r^r = V_i$ when the perturbers are at a long distance from the atom ($r \gg n^2 a_0$ the mean radius)

$$\omega(t') = \frac{1}{\hbar} C_r \sum_j \{\rho_j^2 + v^2(t' - t_j)^2\}^{-n/2} \tag{13}$$

for a large number of collisions with parameters ρ_j and t_j. It is convenient for practical calculations to convert formula (11) into a Fourier integral.

Using new variables $t_2 = t$ and $t_1 - t_2 = s$

$$F(\omega) = \lim_{T \to \infty} \frac{1}{2\pi T} \int_{-T/2}^{+T/2} \int_{-T/2}^{+T/2} f(t_1) f^*(t_2) \exp[-i\omega(t_1 - t_2)] \, dt_1 \, dt_2$$

$$= \lim_{T \to \infty} \frac{1}{2\pi T} \int_{-T}^{+T} e^{i\omega s} \, ds \int_{-T/2}^{+T/2} f^*(t) f(t + s) \, dt \tag{14}$$

$$= \frac{1}{2\pi} \int_{-\infty}^{+\infty} e^{i\omega s} \tilde{\Phi}(s) \, ds \tag{15}$$

where the autocorrelation function $\Phi(s)$ is the time average or can be replaced by an average over the ensemble

$$\tilde{\Phi}(s) = \langle f^*(t) f(t + s) \rangle_{\text{AV}}. \tag{16}$$

With all our previous approximations (classical path, adiabatic approximation)

$$f(t) = \exp[i\omega_0 t - i\eta(t)]$$

$$F(\omega) = \frac{1}{2\pi} \int_{-\infty}^{+\infty} \exp[i(\omega - \omega_0)s] \Phi(s) \, ds$$

with

$$\Phi(s) = \langle \exp\{-i[\eta(t + s) - \eta(t)]\} \rangle_{\text{AV}}. \tag{17}$$

For one perturber

$$\phi_j(s) = \exp\{-i\eta_j(s)\} \quad \text{with} \quad \eta_j(s) = \frac{1}{\hbar}\int_{-\infty}^{s} V_j\, dt \qquad (18)$$

where V_j is the potential of particle j.

2.2. The ensemble average. The impact approximation

To perform the average over the ensemble, let us assume the impact approximation is valid when $s \sim 1/\Delta\omega$ is very large compared to ρ/v the collision time. When the impact approximation is valid (for large values of s and for short range potentials V_i)

$$\eta_j(s \to \infty) = \frac{1}{\hbar}\int_{-\infty}^{+\infty} V_j\, dt. \qquad (19)$$

This corresponds to the "complete collision" approximation.

$$S = \exp[-i\eta_j(\rho)] = \exp\left(-\frac{i}{\hbar}\int_{-\infty}^{+\infty} V_j\, dt\right) \qquad (20)$$

is in fact the S matrix in the very simple case examined here.

For one particle in a very large volume V, the average value of $\phi(s)$ is

$$\phi(s) = \frac{1}{V}\int_V e^{-i\eta}\, dV = 1 + \frac{1}{V}\int_V [e^{-i\eta} - 1]\, dV \qquad (21)$$

because the mean effect of one particle is weak.

For N particles in this volume interacting individually with the atom (time between collision much bigger than the collision time)

$$\Phi(s) = [\phi(s)]^N = [\phi(s)]^{nV} \qquad (22)$$

where n is the particle density.

For an impact parameter ρ during the time s if $\bar{v}$ is the mean perturber velocity. $V = \pi\rho^2 \bar{v} s$ and $dV = 2\pi\rho\bar{v}\, d\rho s$

$$\Phi(s) = \exp\left(n\bar{v}\int_0^{\infty} 2\pi\rho\, d\rho\, [e^{-i\eta(\rho)} - 1]s\right). \qquad (23)$$

For simplicity we use $\bar{v}$ but in fact one has to integrate over the Maxwellian distribution of velocities.

Performing the Fourier integral the line has a Lorentz shape

$$F(\omega) = \frac{\gamma/2\pi}{(\Delta\omega - d)^2 + (\gamma/2)^2} \tag{24}$$

where the half-width γ and the shift d are associated to the real and to the imaginary part of the exponent

$$\gamma = n\bar{v}Q \quad \text{with} \quad Q = 2\int_0^\infty [1 - \cos\eta(\rho)] \, 2\pi\rho \, d\rho$$

which is in fact the collision cross section of elastic scattering by level i (we have assumed that only the upper atomic level is perturbed).

One finds

$$d = n\bar{v}\int_0^\infty \sin\eta(\rho) \, 2\pi\rho \, d\rho.$$

This is the Lindholm adiabatic approximation, which has been extensively used in particular with simple potentials of the form $V = C_n r^{-n}$, and using a straight trajectory classical path for computing the phase shifts

$$\eta(\rho) = \frac{1}{\hbar}\int_{-\infty}^{+\infty} V_j \, dt = \frac{C_n}{\hbar} \frac{1}{v\rho^{n-1}}. \tag{25}$$

We shall not give here the details of the Lindholm theory when applied to different types of interactions. This is explained in length in the literature—see for example reference 16. Let us say that it is an interesting exercise to rediscover the Lindholm approximation, starting from the exact expression of the half-width within the impact approximation, and enumerating all the successive approximations, which give at the end the Lindholm results.

The linear dependence in s of the exponent in $\Phi(s)$ is typical of the impact approximation. One still obtains a Lorentz profile when the non-essential approximations just described are dropped out:

(i) when both the interactions with the upper and lower atomic level are taken into account. In this case the phase shift in (20) is replaced by the difference of two phase shifts

$$\exp\{-i[\eta_i(\rho) - \eta_f(\rho)]\} = S_i \, S_f^*$$

(ii) when inelastic collisions are taken into account. From the classical point of view we have now a damped oscillator whose amplitude varies

$$f(t) = \exp\left(i\omega_0 t - i\eta(t) - \frac{\Gamma}{2}t\right)$$

where Γ is associated to the collisional decay of the atomic level, due to inelastic collisions.

2.3. Modern semi-classical approach. Impact broadening by electrons

In 1958, Baranger[9,17] has given a very complete quantum impact theory (see later on) which has been translated into semi-classical terms by him and by Vainshstein and Sobelman.[18]

The half-width γ and the shift d for an isolated line $i \to f$ (non hydrogenic line not sensitive to the linear Stark effect) broadened by electrons are given by

$$\frac{\gamma}{2} = N_e \bar{v} \int 2\pi\rho \, d\rho \, \mathcal{R}[1 - S_{ii} S_{ff}^*]$$

$$d = - N_e \bar{v} \int 2\pi\rho \, d\rho \, \mathcal{I}[1 - S_{ii} S_{ff}^*] \tag{26}$$

where S_{ii} and S_{ff} are the diagonal elements of the scattering matrix, and N_e is the electron density.

Using the conservation conditions well known in collision theory (exact S is unitary and symmetric) γ can be given in terms of elastic and inelastic processes,

$$\gamma = N_e \bar{v} \left[\sum_j Q_{ij} + \sum_j Q_{fj} + \int \sin\theta \, d\theta \, d\phi \, |f_i(\theta, \phi) - f_f(\theta, \phi)|^2 \right] \tag{27}$$

where the Q_{ij} and Q_{fj} are the inelastic cross-sections to any level j of the atom which shorten the life time of each level i and f, and where an interference term contains the scattering amplitude f_i and f_f of the scattering process: atom in level i (or in level j) + electron.

For simplicity we neglect eventual angular average appearing in the results and we use the mean velocity $\bar{v}$. $N_e \bar{v} \Sigma Q_{ij}$ and $N_e \bar{v} \Sigma Q_{fj}$ are related to the collisional life time of the two atomic levels i and f. The last interference term is the adiabatic term, the only one retained in the Lindholm theory.

In the semi-classical approach, the atom experiences an external time dependent field. The Schrödinger equation which is satisfied by the atomic wave function ψ of Eqn. (1) is solved using the second order non stationary perturbation theory.

Because we are solving the problem in the frame of the impact approximation we can first solve the problem of the perturbation by one electron and solve the statistical problem separately.

For one perturber we solve

$$i\hbar \frac{\partial \psi}{\partial t} = [H_0 + V(t)]\psi \qquad (28)$$

using

$$\psi_i(t) = \sum_j a_{ij}(t)\, \psi_j(0) \exp\left(-i\frac{E_j}{\hbar} t\right). \qquad (29)$$

With the complete collision approximation, corresponding to Eqns (19) and (20) (the time of interest $\Delta\omega^{-1}$ is very large compared to the collision time ρ/v where ρ is the mean range of potential V) the scattering matrix is simply the matrix a for $t = \infty$

$$S_{ij} = a_{ij}(\rho, v, t = \infty). \qquad (30)$$

2.4. The use of pertubation theory

The Schrödinger equation can be solved exactly, but the calculations are difficult and not always meaningful because the semi-classical model may break down when strong close collisions are important. Then it is usual to use second order perturbation theory

$$T_{ii} = 1 - S_{ii} = \frac{i}{\hbar}\int_{-\infty}^{+\infty} V_{ii}(t)\, dt + \frac{1}{\hbar^2} \sum_j \int_{-\infty}^{+\infty} V_{ij}(t) \exp(i\omega_{ij}t)$$

$$\int_{-\infty}^{t} V_{ji}(t') \exp(i\omega_{ji}t')\, dt'\, dt \qquad (31)$$

$$2\mathcal{R} T_{ii} = \frac{1}{\hbar^2} \sum_j \left| \int_{-\infty}^{+\infty} V_{ij}(t) \exp(i\omega_{ij}t)\, dt \right|^2 = \sum_j P_{ij}(\rho) \qquad (32)$$

$$\mathcal{I}\, T_{ii} = \frac{1}{\hbar}\int_{-\infty}^{+\infty} V_{ii}(t)\, dt + \frac{1}{\hbar^2} \mathcal{I}\left[\sum_j \int_{-\infty}^{+\infty} V_{ij}(t) \exp(i\omega_{ij}t)\right.$$

$$\left. \times \int_{-\infty}^{t} V_{ji}(t') \exp(i\omega_{ji}t')\, dt'\, dt \right]. \qquad (33)$$

In the imaginary part of $T_{ii} = 1 - S_{ii}$, the first term cannot be a dipolar term for an isolated line (no dipole term between identical levels with the same orbital quantum numbers), and corresponds to a quadrupole contribution. The second term is the polarization term which scaler like αr^{-4}. This is the long range term corresponding to the induced dipole by the electron, in fact the only one retained in the adiabatic calculation of Lindholm.

The phase shifts are related to T_{ii} by

$$2 \tan \varphi_i = - \mathscr{I} T_{ii} \simeq - \eta(\rho) \tag{34}$$

and the cross-sections are related to the probability given in formula (32) by the integration over the impact parameter ρ

$$Q_{ij} = \int_0^\infty 2\pi\rho \, d\rho \, P_{ij}(\rho) \tag{35}$$

γ and d can be written as

$$\gamma = N_e \bar{v} \int 2\pi\rho \, d\rho \left[\sum_j P_{ij}(\rho) + \sum_j P_{fj}(\rho) + 4 \sin^2(\varphi_f - \varphi_i) \right] \tag{36}$$

$$d = N_e \bar{v} \int 2\pi\rho \, d\rho \, [\sin 2(\varphi_f - \varphi_i)]. \tag{37}$$

For weak interactions for which the perturbation theory is valid φ_f and φ_i are small compared with unity.

In fact this semi-classical approach is equivalent to the Born approximation and more precisely to the second order Born approximation for elastic terms because of the inclusion of the polarization term. The approximation is even worse, because the concept of a straight classical path is meaningless where the trajectory penetrates the atom. The real interaction potential has the form

$$V = \frac{e^2}{|\mathbf{r}_a - \mathbf{r}|} - \frac{e^2}{r} \tag{38}$$

where r_a is the atomic coordinate. When the perturber is outside the target, $r > r_a$ and we have

$$|\mathbf{r}_a - \mathbf{r}|^{-1} = \sum_\lambda P_\lambda(\hat{\mathbf{r}}_a \cdot \hat{\mathbf{r}}) \frac{r_a^\lambda}{r^{\lambda+1}} \tag{39}$$

where, for dipolar interaction, $\lambda = 1$. The use of this approximation is equivalent to the well known Bethe approximation in collision theory.

2.5. The role of the strong collisions

For close collisions both this last approximation and perturbation theory break down. If ρ_1 is the impact parameter below which a collision is strong (the phase shift is of the order of unity or, for an inelastic term the probability $P_{ij}(\rho_1)$ is of the order of unity) one uses ρ_1 as a lower cutoff in the integration on ρ. Usually this ρ_1 is bigger than the mean radius of the atom $n^2 a_0$. In the adiabatic approximation this ρ_1 is usually called the Weisskopf radius ($\eta(\rho) = 1$ in Eqn (19)). Each cross-section in formula (27) is calculated using

$$Q = \frac{\pi}{2}\rho_1^2 + \int_{\rho_1}^{\rho_D} P(\rho)\, 2\pi\rho\, d\rho \tag{40}$$

with the cutoff ρ_1 and eventually for long range fields an upper cutoff ρ_D which is the Debye radius.

Only for $\rho > \rho_1$ can the perturbation theory be used. Strong collisions are just estimated with the condition $P(\rho) \simeq \frac{1}{2}$. Therefore the semi-classical results will be reliable only if the contribution of the strong collisions is small compared to the contribution of the weak collisions (a condition which can only be satisfied a posteriori). In fact the condition is not probably so drastic, because what really matters in the expression of the half-width are infinite sums of cross-sections

$$\sum_j Q_{ij} \quad \text{over all atomic levels.}$$

For each value of ρ the total probability is $\Sigma_j P_{ij}(\rho) = 1$. This is probably the reason why even when strong collisions are important (short range potential for isolated lines, inelastic collisions important) the semi-classical results can be reliable in many cases. Everything happens as if all the strong collisions can be treated as a mean collision against a sphere the radius of which is a mean Weisskopf radius ρ_1.

When inelastic terms are important, the semi-classical approximation is again not reliable because it neglects exchange of energy between perturbers and the target (what is called back reaction in the literature on line broadening). Consequently the reciprocity condition

$$\omega_i P_{ij}(\rho) = \omega_j P_{ji}(\rho) \tag{41}$$

where the ω are the statistical weights, is not satisfied. When the kinetic energy of the perturber is not much bigger than ΔE_{ij} one has to use symmetrized expressions.[19]

In many papers on line broadening theory it is explicitly assumed that the usual semi-classical theory is correct as soon as the impact parameter is bigger than the de Broglie length $\lambda = \hbar(mv)^{-1}$
This means:

$$l = \frac{mv\rho}{\hbar} > 1.$$

In fact, as we have just seen: ρ must be $> n^2 a_0$ the mean radius of the atom, ρ must be $> \rho_1$ the Weisskopf radius and $\pi\rho_1^2$ the strong collision contribution must be small compared to the weak collision contribution for which perturbation theory is valid.

2.6. An alternative treatment for weak interactions

In order to avoid the use of lower cutoff ρ_1, different authors have used other approximate expressions for the S matrix.

When we solve the Schrödinger equation (28) in the interaction picture we can define the S matrix in terms of a matrix more commonly known as the evolution operator U_1 for one perturber. Thus, putting

$$\psi'(t) = U_1(t, t_0) \psi'(t_0) \quad (42)$$

where

$$\psi'(t) = \psi(t) \exp(iH_0 t/\hbar)$$

the S matrix becomes

$$S = U_1(+\infty, -\infty) \quad (43)$$

or can be written in the very compact form

$$S = \theta \exp\left[-\frac{i}{\hbar} \int_{-\infty}^{+\infty} V'(t) \, dt\right] \quad (44)$$

where the time ordering operator θ gives the order of integration in the power series. Instead of using the second order development (31) one can simply neglect the time ordering operator θ.

Defining

$$R = -\frac{1}{2\hbar} \int_{-\infty}^{+\infty} V'(t) \, dt$$

S can be written as

$$S = \exp(2iR) \quad (45)$$

which is in fact the approximate value of the S matrix used in many recent papers on line broadening theory.[20] This form of S is very useful because it retains its unitarity even for small ρ for which strong collisions are important. As it is well known in collision theory this form is only correct for weak interactions when it gives the same result as the second order development. In fact, it is a way to avoid the use of a lower cutoff and in some cases to avoid divergence for small ρ, but, in practical calculations, the electronic computer is choosing a cutoff ρ, which is roughly the Weisskopf radius below which the approximate R matrix does not have any meaning.

To avoid any confusion for plasma physicists who are more familiar with the problem of the time evolution operators in the many-body problem, what we have just said concerns the importance of the coupling of different induced interactions $i \to j \to i$ in the collision problem with one particle, and does not concern the eventual superposition effects of many particles.

III. QUANTUM THEORY OF LINE BROADENING

3.1. The time independent point of view

The general expression of the line shape is given by formula (1)

$$F(\omega) = \lim_{T \to \infty} \frac{1}{2\pi T} \left| \int_0^T dt \exp(i\omega t) \langle \Psi_f(t) | \mathbf{d} | \Psi_i(t) \rangle \right|^2. \tag{1}$$

From the quantum mechanical point of view it is natural to consider the quantum system of a cell consisting of the radiating atom and all relevant perturbers interacting with it included in the volume V of the cell. We consider that the perturbers are electrons and that the electromagnetic field does not perturb the wave functions. The total Hamiltonian H of the system is then time independent (neglecting the interaction between the cells) with $\Psi_f(t)$ and $\Psi_i(t)$ as the final and initial wave functions of this total radiating system, in the Schrödinger picture.

If we introduce the Schrödinger evolution operator $T(t)$ which transforms a state at time o into a state at time t

$$\Psi(t) = T(t) \Psi(0) \quad \text{with} \quad T(t) = \exp\left(-i \frac{H}{\hbar} t\right).$$

With the semi-classical approach, the system is simply the radiating atom subjected to a time dependent interaction with all the perturbers. The Hamiltonian is then time dependent $H = H_0$ (unperturbed atom) $+ H_1(t)$. Consequently we have to use the time dependent way to formulate the problem.

From the quantum standpoint, $F(\omega)$ can be written as

$$F(\omega) = \frac{1}{2\pi} |\langle \Psi_f(0) | \mathbf{d} | \Psi_i(0) \rangle|^2 \, \delta(\omega - \omega_{if}) \tag{46}$$

where $\omega_{if} = (E_i - E_f)\hbar^{-1}$ is the frequency, and E is the total energy of the system. Hence

$$T(t)\Psi(o) = \exp\left(-i\frac{E}{\hbar}t\right)\Psi(0). \tag{47}$$

During radiation of light quantum $\hbar\omega_{if}$ the whole system changes from E_i to E_f, in accordance with the law of conservation of energy.

3.2. The quantum expression of the line profile

Following Baranger,[17] the initial wave function for the whole system is

$$\Psi_i = \varphi_i(\mathbf{r}_a)\,\psi_{ik_1}(\mathbf{r}_1)\,\psi_{ik_2}(\mathbf{r}_2)\ldots\psi_{ik_N}(\mathbf{r}_N) \tag{48}$$

where $\mathbf{r}_a$ represents the atomic, and $\mathbf{r}_j$ ($j = 1, 2\ldots N$) the perturber, coordinates. The conditions of validity of this simple form of the wave function given in the literature are not always very clear. This form implies that adiabatic approximation and is only valid when $\mathbf{r}_1, \mathbf{r}_2\ldots\mathbf{r}_n$ are large compared to the atomic dimensions. Correlations between the perturbers are also neglected. This form can be used if the basic contribution in the integral is given by large values of r_N (all $r_N \gg n^2 a_0$), a condition which is coherent with the validity condition of the impact approximation (see later on).

We are interested in a small frequency interval around the unperturbed frequency of the atom, and not in the continuum radiation. Therefore we can take $\mathbf{d} = e\mathbf{r}_a$ taking only account of the atomic dipole moment.

Then if the initial atomic state is a and b the final state, for the atomic transition $a \to b$ (in fact a stands for the quantum numbers nLJ) the line shape is proportional to a constant factor

$$|\langle\varphi_b|\mathbf{d}|\varphi_a\rangle|^2$$

the matrix element of the atomic dipole. The line shape is proportional to

$$|\langle\psi_{fk_1'}|\psi_{ik_1}\rangle|^2\,|\langle\psi_{fk_2'}|\psi_{ik_2}\rangle|^2\ldots. \tag{49}$$

The wave functions ψ_i and ψ_f correspond to different potentials $V_a(r)$ and $V_b(r)$ of the atom respectively in state a and state b and therefore are not orthogonal. Consequently the perturber wave function integral is different from zero

$$\int \psi_{fk'}^* \psi_{ik}\,d\mathbf{r} \neq 0$$

for any value of k and k' if $V_a \neq V_b$.

In fact this notation is not clear because it does not include properly all the quantum numbers which characterise the atom in a degenerate state a and the perturbers each one being characterised by the numbers $k\,l\,m\,s\,m_s$. Neglecting the spin, assuming spherical atomic states, assuming adiabaticity —no exchange of energy, no exchange of angular momentum—the simple notation of Baranger can be used.

The total energy being $\varepsilon_a + \varepsilon_1 + \varepsilon_2 + \ldots \varepsilon_N$ the frequency of the light emitted, counted from the unperturbed line frequency $\omega_{ab} = (\varepsilon_a - \varepsilon_b)/\hbar$ is

$$(\varepsilon_1 - \varepsilon_1') + (\varepsilon_2 - \varepsilon_2') + \ldots + (\varepsilon_N - \varepsilon_N').$$

The important fact here is that the N variables $(\varepsilon_1 - \varepsilon_1') \ldots (\varepsilon_N - \varepsilon_N')$ are independent, because the total probability (49) is a product of N factors. Therefore we can apply the general statistical equation (2), calculate $\phi(s)$ for one perturber, and take the N power of $\phi(s)$ to get the total $\Phi(s)$. where

$$\phi(s) = \sum_{kk'} \rho_k |\langle \psi_{fk'}|\psi_{ik}\rangle|^2 \exp\left[-i(\varepsilon - \varepsilon')s/\hbar\right] \qquad (50)$$

averaging over initial states and summing over final states.

3.3. The one electron approximation

In the line wings the one electron approximation becomes valid. The quantum system consisting of the atom + one electron is isolated. If it is not, the amplitude of the wave function is damped with the factor $\exp(-\gamma t/2)$ where γ is precisely the half-width. In the wings $\Delta\omega \gg \gamma$ and the system just defined is isolated.

In this case one can show[21] that the line shape is given by

$$F(\omega) = |\langle \varphi_b|\mathbf{d}|\varphi_a\rangle|^2 \sum_{a'} \iiint \frac{1}{4\pi} |\langle \chi_b|\psi_{a'a}\rangle|^2 \, d\hat{\mathbf{k}}_a \, d\hat{\mathbf{k}}_{a'}$$

$$\rho(\varepsilon_{a'})f(\varepsilon_{a'})\,n\,\hbar^{-1}\,d\varepsilon_{a'} \qquad (51)$$

where we have assumed that only the upper state a is perturbed: χ_b represents a plane wave, a' stands for the initial channel in the collision problem and n is the electron density.

$$\Psi_{a'}(\mathbf{r}_1, \mathbf{r}_2) = \sum_{a''} \psi_{a'a''}(\mathbf{r}_2)\,\varphi_{a''}(\mathbf{r}_1). \qquad (52)$$

But only $a'' = a$ is included in the expression (51) because we are interested in the line $a - b$.

On the other hand the sign $\Sigma_{a'}$, the integration over $d\hat{k}_a, d\hat{k}_{a'}, d\varepsilon_{a'}$, the term $\rho(\varepsilon_{a'})$—the density of initial states—are included for the necessary sums and averages (over velocities and direction of perturbers).

Finally, to obtain the wave function $\psi_{a'a}$ correctly, for the collision problem, it is of course necessary to solve the coupled problem properly

$$\left[\frac{\hbar^2}{2m}\nabla^2 + \varepsilon_a\right]\psi_{a'a} = \sum_{a''} V_{aa''}\psi_{a'a''} \tag{53}$$

where the first index a' stands only for the initial channel.

Solving these coupled equations, we can obtain the wave functions and their asymptotic forms, consequently the terms of the S matrix. But as is well known, it is necessary first to solve the coupled system even if one needs the asymptotic forms only.

The value of r which is important is the overlap integrals is $r \approx \Delta k^{-1}$. If ρ is the mean range of the potentials V appearing in (53), and if $\rho \ll \Delta k^{-1}$, only the asymptotic forms of the wave functions are important in the overlapping integrals, which can be expressed simply in terms of S matrix, But $\rho \Delta k \ll 1$ is equivalent to

$$\frac{\rho}{v} \ll \Delta\omega^{-1}$$

which is precisely the validity condition for the impact approximation. In this case only the profile can be expressed in terms of the S matrix and cross-sections. The line broadening is simply an application of collision theory.

For the wings, one finds

$$F(\Delta\omega) = \frac{\gamma}{2\pi}\frac{1}{(\Delta\omega)^2} \quad \text{with} \quad \gamma = n\bar{v}\sum_{a''} Q_{aa''}.$$

From what we have said we can guess the possibility to go beyond the impact approximation in the region $\Delta\omega \sim v/\rho$ or $> v/\rho$. One can simply use the exact wave functions instead of their asymptotic forms. As predicted in Section I.3 the one-electron approximation must converge towards the nearest-neighbour approximation at the static limit.

3.4. A few remarks on the quantum mechanical problem

Within the impact approximation, the line broadening problem is just an application of collision theory. We only need the asymptotic forms of the perturber wave functions. Beyond this approximation we need also these wave functions.

Therefore we have in any case to solve correctly the coupled problem (53) adding exchange terms eventually, because we need the wave functions at very low energies. These exact wave functions can be found, and this means in particular that one allows for exchange of energy between the atom and the perturber (what is generally called "back reaction" in the literature on line broadening) which is neglected in the semi-classical approach (it is only empirically taken into account when symmetrised expressions are used to satisfy Eqn (52)).

It is necessary to note that the "adiabatic" approximation in the sense used in some papers on line broadening theory[17] with a total wave function of the type (48) does not have much meaning. At low energies it is always necessary to solve the coupled system (53) taking account of the closed channels[22] and eventually of the resonances.[23,24] One can show how the coupling with the closed channel gives rise to a polarization potential.[22]

For excited non-spherical states ($L \neq 0$) there appears a quadrupole potential of the form βr^{-3} which has to be added to the αr^{-4} potential for non-hydrogenic lines and to the dipole field Cr^{-2} for hydrogenic lines.

All these questions have been clarified by those who have used the quantum formulation of the collision problem to improve the quasi-classical calculations[22,25,26] and by those who have used accurate close coupling calculations for electron scattering by H,[27] Ca^+,[24] Mg^+.[23]

3.5. Quantum system with many perturbers

We come back to the many-electron case for which we cannot avoid the calculation of $\Phi(s)$, and to the impact approximation in the sense that the "collisions are completed": $\Delta\omega < v/\rho$. For such collisions the time between collisions γ^{-1} is much bigger that the collision time ρ/v and we can follow the arguments given in 2.2. (all demonstrations follow the same type of arguments).

In such a case

$$\Phi(s) = [\phi(s)]^N \qquad (2)$$

where $N = nV$ is the total number of perturbers. It is the same to say that the time evolution operator for N perturbers is a simple product of N operators for one particle, or to say that the perturbers (even if they interact strongly with the radiator, e.g. if they induce inelastic effects and suffer from "back reaction") act individually one at a time (see paragraph 3.5).

The mean value of $\phi(s)$ has a form analogous to (50) with good perturber wave functions of 3.3. Avoiding here all the problems of angular average and of angular algebra, one obtains

$$\phi(s) = \frac{1}{V}\frac{\pi}{k^2}\sum_l (2l + 1) \int_0^\infty dk' \exp\left[\frac{i}{\hbar}(\varepsilon - \varepsilon')s\right] \left|\int F_{ik} F_{fk} \cdot dr\right|^2 \quad (54)$$

after a partial wave analysis. F are the radial perturber wave functions. In the impact approximation the F can be replaced by their asymptotic forms of the type

$$\mathscr{E}^- \text{ (incoming wave)} - \mathscr{E}^+ \text{ (outgoing wave)} S_{ij}$$

in terms of the S matrix. We have also

$$\varepsilon = \frac{\hbar^2}{2m}k^2, \quad \frac{1}{\hbar}(\varepsilon - \varepsilon') = v[k - k']$$

and l is the angular momentum quantum number.

After a few transformations analogous to the ones used with the classical model in 2.1. and 2.2., $\langle\phi(s)\rangle$ takes a form analogous to (21)

$$\phi(s) = 1 - \frac{\bar{v}s}{V}\frac{\pi}{k^2}\sum_l (2l + 1)[1 - S_{ii} S_{ff}^*]. \quad (55)$$

Raising to the power N

$$\Phi(s) = \exp\left[n\bar{v}s\frac{\pi}{k^2}\sum_l (2l + 1)[1 - S_{ii} S_{ff}^*]\right] \quad (56)$$

the typical exponential form for the impact approximation, giving rise to a Lorentz profile with

$$\frac{\gamma}{2} = n\bar{v}\frac{\pi}{k^2}\sum_l (2l + 1)\,\mathscr{R}\,[1 - S_{ii} S_{ff}^*] \quad (57)$$

$$d = -n\bar{v}\frac{\pi}{k^2}\sum_l (2l + 1)\,\mathscr{I}\,[1 - S_{ii} S_{ff}^*] \quad (58)$$

the quantum analogues of formula (26).

In fact the levels a and b can be degenerate and, for simplicity, we have neglected here all kinds of degeneracy. The general case is often not clear in the literature on line broadening. It has been clarified recently.[23,24,26]

3.6. The Fourier transform of the line shape

Using (1) the absolute square is written as the product of the two complex conjugate quantities

$$F(\omega) = \frac{1}{2\pi T} \int_0^T dt \int_0^T dt' \exp[i\omega(t-t')] \langle \Psi_i(0)|T^*(t')\,\mathbf{d}T(t')|\Psi_f(0)\rangle$$
$$\times \langle \Psi_f(0)|T^*(t)\,\mathbf{d}T(t)\,\Psi_i(0)\rangle. \quad (59)$$

By summing over all final states (dropping the factor $|\Psi_f(0)\rangle \langle \Psi_f(0)|$) and averaging over initial states (introducing the density matrix $\rho(0)$)

$$F(\omega) = \frac{1}{2\pi T} \int_0^T dt \int_0^T dt' \exp[i\omega(t-t')] \operatorname{Tr}[\rho(0)\,T^*(t')\,\mathbf{d}T(t')T^*(t)\,\mathbf{d}T(t)]. \quad (60)$$

If we define $T(t-t')$ as the operator that transfroms the state at time t' into the state at time t, and putting $t-t' = s$, $F(\omega)$ can be written as the Fourier transform of the autocorrelation function $\Phi(s)$

$$F(\omega) = \frac{1}{2\pi} \int_{-\infty}^{+\infty} \exp(i\omega s)\,\Phi(s)\,ds \quad (61)$$

with

$$\Phi(s) = \operatorname{Tr}[T(s)\,\rho(0)\,\mathbf{d}T^*(s)\,\mathbf{d}]. \quad (62)$$

Starting with this general formula, Baranger[17] has found formula (56) following a pure quantum mechanical approach and using the same usual approximations of the impact theory as above.

Defining the total time evolution operator in the interaction representation

$$U(t,o) = \exp\left[\frac{i}{\hbar}(H_A + K)t\right] \exp\left[-\frac{i}{\hbar}Ht\right]$$

with total Hamiltonian $H = H_A + K + V$ where K is the sum of the kinetic energies of the perturbers, and $V = \Sigma_N V_j$ one assumes

$$U_{AV}(t+s,t) = \langle U_j(t+s,t)\rangle^N. \quad (63)$$

The demonstration is too long to be given here. The real impact approximation is made in fact when, for computing the matrix element,

$$\langle \chi_{fk'}|V_j|\Psi_{ik}\rangle \quad (64)$$

with energy difference $k^2 - k'^2$ very small ($\varepsilon - \varepsilon'$ of the order of $\hbar\Delta\omega$) one assumes that the two energies are the same.

This approximation is valid when $\Delta k \rho \ll 1$ if ρ is the mean range of potential V_j. This condition is equivalent to

$$\frac{\rho}{v} \ll \Delta\omega^{-1}$$

the well known condition of validity of the impact approximation.

As soon as this is true, the above matrix element (64) can be written as a scattering amplitude. The results can be given in terms of cross sections and of scattering matrices as in (56).

3.7. On the actual unified theories

When the impact approximation is not valid for computing all the profile (electron in the case of H lines, broadening by neutral atoms at high densities) it is important to have a theory valid from the impact regime to the static regime. This has been first done by Anderson and Talman[28] in 1955 and in fact modern theories are generalisations of this work.

The principle of the method is very simple. Coming back to Section II.1 and to the Lindholm's formulation of the problem, one drops the "complete collision assumption". Instead of using (19) in formula (23) one uses the true value (18) for all values of s.

This has been done by Anderson and Talman[28] and by others since, using the adiabatic approximation for calculating the broadening by neutral particles assuming that the effective potential is of the form $\Sigma_r c_r r^{-r}$.

In the more general case, still assuming the semi-classical approach, one uses the true value of the time evolution operator U_1 for one perturber as defined by Eqn (42). Instead of (44)

$$U_1(s, 0) = \theta \exp\left[-\frac{i}{\hbar}\int_0^s V' \, dt\right]. \tag{65}$$

In practical calculations, one drops the operator θ, and as we saw in 2.6 the treatment is not correct for strong interactions. Thanks to the exponential form of U_1, one can go smoothly from the impact approximation to the static regime (s very small).[20]

When the one-electron approximation is valid, we have seen that dropping the impact approximation, the results must converge to the nearest-neighbour static approximation. In the general case, the unified theories assume that the general statistical approximation (2) can be applied. This is valid when strong collisions are separated in time, not when strong interactions are going on at the same time with different perturbers.

Near the static limit, the time of interest $s = \Delta\omega^{-1}$ is very short, and the exponant in (65) can be small even if the interaction V is strong. Second order terms in V are small and formula (2) is valid. In between it is better to check that the product of the interaction by the characteristic time is small compared to $\hbar$, when the complete collision approximation breaks down.

In actual unified theories, each interaction acts individually, if strong. The superposition problem arises only for weak interactions but it does not matter really and the general statistical approximation (2) can still be assumed.

IV. APPLICATIONS TO A FEW CASES

We are considering three important applications:

(i) The broadening of non-hydrogenic lines by charged particles
(ii) The broadening of hydrogenic lines by charged particles
(iii) The broadening due to collisions with neutral particles.

4.1. The impact approximation for electrons in the case of non-hydrogenic lines

Until recently all the calculations have been done using the semi-classical approach just described. Most of the broadening is due to electrons compared to the proton contribution.

Assuming a classical path, as soon as the trajectory $r(t)$ is given it is possible from formula (31) to obtain T_{ii} as a function of the impact parameter ρ and of v. From the real and the imaginary parts of T_{ii} one obtains

$$\tfrac{1}{2}P_{ij} + 2i\,\varphi_{ij} = \frac{e^4}{mv^2} \frac{f_{ij}}{\Delta E_{ij}} \frac{1}{\rho^2} [A(\beta) + iB(\beta)] \tag{66}$$

where f_{ij} is the oscillator strength of transition $i \to j$, and

$$\beta = \frac{\rho \Delta \omega_{ij}}{v}$$

for dipolar interaction with a neutral atom.

When the radiating atom is ionized, the classical path cannot be a straight line. It is necessary to use a hyperbolic trajectory and expression (66) is replaced by a more complicated expression in terms of the eccentricity ε of the trajectory and of two different functions

$$A(\xi, \varepsilon)$$

and

$$B(\xi, \varepsilon)$$

with

$$\xi = \frac{'a\omega_{ij}}{v}, \quad a = \frac{z}{mv^2}.$$

The functions A are known from the semi-classical calculation of inelastic scattering.[19,25,26] Accurate calculations of the function B have been made.[25,29,30]

It is interesting to note that when all $\rho\omega_{ij}/v \gg 1$ for $\beta \to \infty$, $A(\beta) \to 0$ and we come back to the adiabatic approximation already discussed. For dipolar terms, $B(\beta) \to \pi/4\beta$ which gives in fact the polarization term, the only one included in the Lindholm approximation for the phase shift φ. Incidently, although it has been used for both neutrals and positive ions, the usual Lindholm theory breaks down for positive ions. As is well known in collision theory, for neutrals the phase goes to zero when $k^2 \to 0$.

With plane waves and a pure polarization potential αr^{-4}

$$\tan \varphi = \frac{\pi \alpha}{8} [(l + \tfrac{3}{2})(l + \tfrac{1}{2})(l - \tfrac{1}{2})]^{-1} k^2$$

with $k = mv/\hbar$. Using $l = \rho k$, we obtain in the limit $l \gg 1$

$$\tan \varphi = \frac{\pi \alpha}{8} \frac{\hbar}{m} \frac{1}{v\rho^3}$$

which is equivalent to the Lindholm result (see formula (25)).

For positive ions, to Coulomb waves there correspond hyperbolic paths, and because of the ions field, phase shifts and cross sections are finite at threshold. The half-width is much bigger at low energies.

All semi-classical calculations using a straight path for positive ions, as for example in Griem's book "Plasma Spectroscopy" have yielded Stark widths which are too low by factors between 2 and 10. Recent calculations have been done properly[22][25] by many different authors. This improvement is also implicit in a semi-empirical approximation of Griem[31] using an effective Gaunt factor[32] and allowing for elastic collisions.

Many differences still exist between the different semi-classical calculations (in particular the way of defining the proper cutoffs in formula (40) because of the use of perturbation theory to obtain (66)). They have been discussed in many recent papers[26,33,35] in connection with comparison with the few experiments available.[13]

Let us simply stress that comparison between experiments and semi-classical calculations, on the one hand, and between semi-classical calculations and two recent quantum calculations on the other, shows that the electron broadening of an atomic line can be represented apparently using the semi-classical approach. Accurate measurements and semi-classical calculations, have been made for the resonance lines of Mg^+ and Ca^+ by Chapelle and Sahal[36] as well as quantum calculations by Bely and Griem[23] and Barnes and Peach.[24] For reasons already discussed in Section II.5, semi-classical calculations seem to be accurate even when the contribution of strong collisions is important. For complex atoms the accuracy is not so good, since other uncertainties (on atomic levels and on wave functions) are added to the uncertainties of the broadening problem itself.

4.2. Hydrogenic lines

Hydrogenic lines are those which are subject to the linear Stark effects, i.e., lines of hydrogenic ions, lines involving very excited states for which the distance of two interacting levels i, j of different orbital quantum numbers is not large compared to the width of the line $i \to f$.

For isolated lines, the quadratic Stark splitting given by protons is very small and in fact because the polarization field is "short" range αr^{-4} the impact condition is more or less satisfied. The adiabatic impact approximation is commonly used to calculate the small contribution of the protons (less than 20%) to the broadening.

The linear Stark effect is nothing but a small correction and, relative to this interaction, the proton, because of its small velocity, has to be treated using the static approximation. Therefore for hydrogenic lines, the usual model is the following: the hydrogenic levels are split by the quasi-static field of the protons; the lines which arise from these sublevels are then broadened by electron collisions which induce phase shifts and transitions between the nearly degenerate sublevels of both the initial and final levels.

The impact approximation has been used first for treating the electrons, by Kolb (who has used an abiadatic approximation of Lindholm neglecting the induced transition between the sublevels) and then by Griem[37] using the so called "overlapping line version of the impact approximation" (overlapping because two interacting levels with $L \neq L'$ are very close). In fact, because of the ion field, $L\ J$ are not good quantum numbers, and one has to calculate correctly the phase shifts and the induced transitions between Stark sublevels characterized by "parabolic" numbers.

The model just described implies that electron collisions are statistically independent of the momentary value of the proton field F. In this case for one value of F, one component of the hydrogenic line is displaced by

$$\Delta\omega_k = c_k F.$$

If F_0 is the mean field of a proton at distance $n_p^{-1/3}$ (n_p is the proton density) the probability of having the k component displaced by $\Delta\omega_k$ is a given function W and the total profile of one component is of the form

$$F_k(\Delta\omega) = \int_0^\infty W\left(\frac{\Delta\omega_k}{c_k F_0}\right) \frac{\gamma/2\pi}{(\Delta\omega - d - \Delta\omega_k)^2 + (\gamma/2)^2} d\left(\frac{\Delta\omega_k}{c_k F_0}\right) \quad (67)$$

where one sees the Lorentz profile due to electron impacts. The total line profile is then

$$F(\Delta\omega) = \sum_k I_k F_k(\Delta\omega) \quad (68)$$

where I_k is the relative intensity of component k ($\sum_k I_k = 1$).

For uncorrelated perturbers W is the Holtsmark distribution.[38] In the core of the line this distribution of the field (of the distant perturbers) has to be corrected by taking account of correlation between protons and of shielding by the electron cloud surrounding them.[39] Electron collisions with the atom are not entirely statistically independent of the proton field, and because of the long range nature Cr^{-2} of both the electron and the proton field, the broadening is in fact given by many perturbers in the line core.

4.3. The electron contribution for hydrogenic lines

Many calculations have been done using the semi-classical approach of the impact approximation and assuming complete degeneracy of the hydrogenic levels (neglecting the Stark splitting at the stage of the calculation of the electron contribution).

As is well known because of the long range nature of the field $V = \alpha/r^2$ which gives for the transition probability for dipolar excitation of degenerate levels

$$P(\rho) = \frac{\alpha^2}{\hbar^2 \rho^2 v^2}. \quad (69)$$

The integral giving the cross-section

$$Q = \pi \rho_1^2 + \int_{\rho_1} 2\pi\rho \, d\rho \, P(\rho) \quad (70)$$

is divergent for all values of v, when $\rho \to \infty$. As before ρ_1 is the Weisskopf radius of the order of $\rho_1 = \alpha/\hbar v$.

The first idea is to use ρ_D the Debye radius as an upper cutoff ignoring the collision with particles beyond the Debye sphere, the Debye length being the distance beyond which the charge of a perturber is neutralized by the charges of other particles.

This gives the expression used by Griem et al.[37] for γ

$$\gamma = n_e \bar{v} \left[\pi \rho_1^2 + 2\pi \rho_1^2 \ln \frac{\rho_D}{\rho_1} \right]. \tag{71}$$

But still assuming the impact approximation (instantaneous collisions) two other effects can remove the divergence.

First, particularly for high lines (large values of n) one has to take proper account of the Stark splitting of the order of $\Delta E = n^2 e^2 a_0 n_p^{+2/3}$. Secondly, it is necessary to take account of the real life time of the atomic level (natural life time, radiative life time, collisional life time). If $1/\Gamma$ is this life time, this introduces an upper cutoff of the order of v/Γ which may be shorter than ρ_D.

But as discussed at length before, the time of interest which really matters at the distance $\Delta\omega$ from the line centre (in fact $\Delta\omega - \Delta\omega_k$ for a given component) is of the order of $1/\Delta\omega$. The collision is not complete any more. This fact has been first interpreted by Lewis[40] by introducing an upper cutoff of the order of $v/\Delta\omega$: distant collisions are too weak to induce any perturbation.

4.4. Modified impact approximation

For a given line, for

$$\Delta\omega < \Delta\omega_L = \frac{v}{\rho_D} = \left(\frac{4\pi n_e e^2}{m} \right)^{1/2}$$

the usual impact approximation can be used. Beyond this frequency in (71) the term $\ln \rho_D/\rho_1$ can be replaced by $\ln v/\rho_1 \Delta\omega$. Modified formulae for the profile and for the line wings have been given by Lewis[40] and Griem.[41] This correction is particularly important for some high lines for which Ferguson and Schlüter[42] have found good agreement between measurement and calculation taking account of the Lewis correction.

The Lewis correction reduces the contribution of weak collisions. This is not in favour of the validity of the use of the perturbation semi-classical theory ($\pi \rho_1^2$ must be small compared to the weak collision contribution). In fact for dipolar excitation of degenerate levels, in many physical situations

$$\rho_1 = n^2 \frac{\hbar}{mv} \gg n^2 a_0$$

the perturber is outside the radius ρ_1 with correspondingly high values of the angular momentum. A correct semi-classical approach can be used, solving the Schrödinger equation for the wave function (29) developed at least on the basis of the different Stark levels.[43] A dipolar approximation is not satisfactory. Higher order multipoles have to be taken into account. In this case, a pure quantum approach is not more complicated than a semi-classical approach. Comparison by Griem[27] of close-coupling calculations with semi-classical results gives some evidence for our last comment.

Many earlier results for Balmer lines have been recalculated[44] using simply the Lewis correction and many other improvements (lower state interaction, Stark splitting of the level, accurate proton fields, quadrupole interaction) and good agreement has been found over a large domain of temperature and of electron density. For these lines good agreement is also found for the line wings using interpolation formulae of Griem[45] between the impact regime and the static regime (these formulae are given in the literature in terms of the Lewis cutoff).

But one has to have in mind that comparisons between theory and experimental profile are generally made for rather high densities. In astrophysical conditions, the electron density is very low, ρ_D is very large and the frequency $\Delta\omega_L$ beyond which the usual impact approximation breaks down can be very small. The theory as corrected by Lewis[40] can be used but it is better to use an improved impact theory, the so-called unified theory.

4.5. Lines involving high excited states

For high Balmer lines[42] the Lewis correction is also very important and in fact the $\Delta\omega$ beyond which the static theory is valid for electrons is very small. Experimental profiles have roughly a static shape.[46] This suggests that the old Inglis and Teller[47] formula, which gives the principal quantum number n_m of n for which merging of high Balmer lines occurs, should be correct when used with $N = N_e + N_i$ (where here N_e and N_i are the electron and proton density) instead of N_i

$$\log 2N = 23\cdot 26 - 7\cdot 5 \log n_m. \tag{72}$$

The total intensity in the line wing is $I = 2I_i$ where I_i is the usual Holtsmark profile due to the proton field which scales as $\Delta\lambda^{-5/2}$.

This discussion reveals that the impact approximation modified in an empirical way with the Lewis cutoff (still assuming that each collision is complete in the sense that the time evolution operator U_1 of formula (43) is simply the corresponding S matrix) is not satisfactory. It is intrinsically not able to describe the transition region between the impact regime and the static regime. The wing formulae of Griem[45] have merely the character of an interpolation. This is why recent attempts[20,48,49] have been made to go beyond the impact approximation dropping the complete collision assumption as it is explained in 3.6. Another approach consists of using the solution of a simple model of the microfield of the electrons, its probability distribution and the covariance of its fluctuations.[50]

Finally, calculations have been done by Griem[51] for the broadening of $n \to n+1$ radio frequency transitions (n very large) which have yielded Stark width too large to explain the observed intensities of these radio lines in H II regions of the interstellar medium. More recent calculations[52] have been done using purely classical methods for this specific case.

4.6. Pressure broadening due to neutral particles

We are not considering here the resonance broadening, i.e., the resonance interaction of the radiating atom with identical atoms of the plasma. When an atom in an excited state is brought near an atom of the same kind in its ground state, a very strong interaction which scales as r^{-3} is dominating the broadening. We are simply examining, briefly the broadening by unlike atoms, a problem which has been thoroughly discussed in two fundamental papers[53,54] and in all review articles.

The width due to neutral atoms is generally estimated using the Van der Waals force law: the field of the perturbing atom (a dipole) polarizes the radiating atom, producing in it a dipole whose moment is of the force r^{-3}. The potential energy of interactions between these two dipoles is of the form Cr^{-6}. Using this long range potential together with the Lindholm semi-classical adiabatic approximation yields the well known formulae for the width and the shift of the line, and the width to shift ratio which is a constant for all lines

$$\frac{2\gamma}{d} = 2 \cdot 76.$$

This is the impact result, i.e., the width is proportional to n, the perturber density. When the density is very high the impact approximation breaks down and one can use an Anderson and Talman[28] "unified" theory to explain the transition region towards the static regime.

But in many laboratory or astrophysical situations the impact approximation is valid (see Section I.8) because of the r^{-6} law, so that the field of a neutral atom decreases very quickly when r increases. In fact for heavy perturbers with high polarizabilities the simple formula using the Van der Waals law gives good results, but there is now much experimental evidence that it is not the case for light perturbers like He[55] for which the width to shift ratio differs very much from the theoretical value 2·76. In order to explain the experiments, Hindmarsh[56] and others [57,58] have to use an empirical Lennard–Jones potential of the form

$$\frac{c_{12}}{r^{12}} - \frac{c_6}{r^6}$$

and have found that the repulsive part of the potential plays an important role in the broadening processes.

This repulsive potential is much more difficult to calculate. *A priori* calculations using molecular wave functions[59] or a simple model valid in the intermediate range (r bigger than the mean radius of the atoms)[60] can be used to derive the different potential curves of the quasi molecule (radiating atom + perturber).

For the broadening by neutral helium or neutral H (important in the solar atmosphere) theoretical evidence has here been given recently that the real interaction is not the Van der Waals potential but a much shorter range interaction. It is true at room temperature and *a fortiori* at the temperature of the sun.[61] The same is true, for the same reasons, for calculating the relaxation of excited states due to the transfer of excitation between fine structure levels (given by fluorescence experiments).

In these problems of interaction between atoms the "adiabatic" approximation of Lindholm may break down. One cannot always neglect the rotation of the molecular axis and the potential of exchange between the perturber and the valence electron of the radiating atom. Within the semiclassical approximation, which is good because the perturber is heavy, it is necessary to solve coupled equations analogous to those of Section II.3, in order to find the different terms of the scattering matrix S taking account eventually of the fine structure of the atomic levels.

These simple remarks are simply a warning against the use of the Van der Waals law (still used by astrophysicists with many disappointments). The broadening by neutral atoms is a very difficult and exciting problem like all the problems related to the interaction between atoms at low energies.

REFERENCES

1. Unsöld, A., "Physik der Sternatmosphären" (Springer–Verlag, Berlin, 1955).
2. Chen, S. and Takeo, M., *Rev. Mod. Phys.* **29**, 20 (1957).
3. Breene, R. G., *Rev. Mod. Phys.* **29**, 94 (1957).
4. Bohm, K. H., *In* "Stellar Atmospheres" (Greenstein, J. L., Ed., Univ. of Chicago Press, Chicago, 1960).
5. Traving, G., "Über die Theorie der Druckverbreitung von Spektrallinien" (Braun, Karlsruhe, 1960).

6. Margenau, H. and Lewis, M., *Rev. Mod. Phys.* **31,** 569 (1959).
7. Mazing, M. A., *Trans. Lebedev Inst., Moscow* (1961).
8. Breene, R. G., "The Shift and Shape of Spectral Lines" (Pargamon, London, 1961).
9. Baranger, M., *In* "Atomic and Molecular Processes" (Bates, D. R., Ed., Academic Press, New York, 1962).
10. Breene, R. G., *In* "Handbuch der Physik" XXVII (Springer-Verlag, Berlin, 1964).
11. Griem, H. R., *In* "Plasma Spectroscopy" (McGraw-Hill, 1964).
12. Sobelman, I. I., "Theory of Atomic Spectra" (in Russian)(Moscow, 1963).
13. Wiese, W. L., *In* "Plasma Diagnostic Techniques" (Academic Press, 1965).
14. Van Regemorter, H., *Annual Review of Astr. and Ap.* **3,** 71 (1965).
15. Cooper, J., *In* "Lectures in Theoretical Physics" (Boulder) **11** (1965).
16. Traving, G., *In* "Plasma Diagnostics" (North Holland, 1968).
17. Baranger, M., *Phys. Rev.* **111,** 481, 494; **112,** 855 (1958).
18. Vainshtein, L. A. and Sobelman, I., *Opt. Spektroskopiya* **6,** 440 (1959).
19. Seaton, M. J., *Proc. Phys. Soc.* **79,** 1105 (1962).
20. Smith, E. W., Cooper, J., and Vidal, C. R., *Phys. Rev.* **185,** 140 (1969).
21. Van Regemorter, H., *Phys. Letters* **30A,** 365 (1969).
22. Brechot, S. and Van Regemorter, H., *Ann. Astrophys.* **27,** 432 (1964).
23. Bely, O. and Griem, H. R., *Phys. Rev.* **1A,** 97 (1970).
24. Barnes, K. and Peach, G., *J. Phys. B* **3,** 350 (1970).
25. Sahal-Brechot, S., *Astr. and Ap.* **1,** 91 (1969).
26. Sahal-Brechot, S., *Astr. and Ap.* **2,** 322 (1969).
27. Griem, H. R., *Comments on Atom and Mol. Phys.* **2,** 52 (1970).
28. Anderson, P. W. and Talman, J. D., *Conf. Broadening of Spectral Lines, Pittsburgh* (1955).
29. Cooper, J. and Oertel, G. K., *Phys. Rev.* **180,** 286 (1969).
30. Klarsfeld, S., *Phys. Letters* **32A,** 26 (1970).
31. Griem, H. R., *Phys. Rev.* **165,** 258 (1969).
32. Van Regemorter, H., *Ap. J.* **136,** 906 (1962).
33. Roberts, D. E. and Davis, J., *J. Phys. B* **1,** 48 (1968).
34. Roberts, D. E., *Astr. and Ap.* **6,** 1 (1970).
35. Griem, H. R., *Comments in Atom. and Mol. Phys.* **1,** 27 (1969).
36. Chapelle, J. and Sahal-Brechot, S., *Astr. and Ap.* **6,** 415 (1970).
37. Griem, H. R., Kolb, A. and Shen, K. Y., *Ap. J.* **135,** 272 (1962).
38. Chandrasekhar, S., *Rev. Mod. Phys.* **15,** 1 (1943).
39. Hooper, C. F., *Phys. Rev.* **165,** 215; **169,** 193 (1968).
40. Lewis, N., *Phys. Rev.* **121,** 501 (1961).
41. Griem, H. R., *Ap. J.* **136,** 422 (1962).
42. Ferguson, E. and Schlüter, H., *Ann. Phys.* **22,** 351 (1963).
43. Pfennig, H., *Phys. Lett.* **34A,** 292 (1971).
44. Kepple, P. and Griem, H. R., *Phys. Rev.* **170.** 317 (1968).
45. Griem, H. R., *Phys. Rev. Letters* **17,** 509 (1966).
46. Schlüter, H. and Avila, C., *Ap. J.* **144,** 785 (1966).
47. Inglis, D. R. and Teller, E., *Ap. J.* **90,** 439 (1939).
48. Smith, E. W., Cooper, J. and Vidal, C. R., *JQSRT* **10,** 1011 (1970).
49. Volslamber, D., *Z. Naturforsch.* **24a,** 1458 (1969).
50. Brissaud, A. and Frish, U., *JQSRT* **11,** 1753 (1971).
51. Griem, H. R., *Ap. J.* **148,** 547 (1967).

52. Brocklehurst, M. and Leeman, S., *Astrophys. Lett.* **9,** 35 (1971).
53. Chen, S. and Takeo, M., *Rev. Mod. Phys.* **29,** 20 (1957).
54. Tsao and Curnutte, *JQRST* **2,** 41 (1962).
55. Hindmarsh, W. R., *M.N.R.A.S.* **119,** 11 (1959).
56. Hindmarsh, W. R., Petford, A. D. and Smith, G., *Proc. Roy. Soc.* **A207,** 296 (1967).
57. Behmenburg, W., *JQSRT* **4,** 177 (1964).
58. Lemaire, J. L. and Rostas, F., *J. Phys. B* **4,** 555 (1971).
59. Lewis, E. L., McNamara, L. F. and Michels, H., *Phys. Rev.* **3,** 1939 (1971).
60. Roueff, E., *Astr. and Ap.* **7,** 4 (1970).
61. Roueff, E. and Van Regemorter, H., *Astr. and Ap.* **1,** 69 (1969).

The Spectra of Gaseous Nebulae

M. J. SEATON

University College London, England

I. INTRODUCTION

Gaseous nebulae are clouds of rarefied ionized gas, characterized by bright emission lines in the visible spectrum. The purpose of these lectures is to investigate the physical and chemical properties of these objects, and to explain the physical processes involved. Special attention is given to recent work on spectrum lines observed at radio wavelengths. Some basic physics formulae are collected in the Appendix (equations (A1), (A2), etc).

The first comparisons of astronomical and laboratory spectra were made by Huggins who, over 100 years ago, was able to conclude that the sun and stars are composed of the same chemical elements as the earth. An investigation of the spectra of several diffuse objects detected telescopically revealed that while some had spectra similar to those of the stars (galaxies), many possessed spectra which consisted of very strong emission lines (gaseous nebulae). A few weaker lines correspond with solar lines, but the stronger lines could not be identified. The suggestion that they were emitted by a new element (nebulium) was refuted by Huggins. He proposed, correctly, that the emission was from a known element, but that the physical conditions in the emitting region were totally different from conditions available in laboratories.

The introduction of large optical telescopes led to the resolution of galaxies into individual stars, and to the classification of gaseous nebulae into two types. Diffuse nebulae, such as Orion, are of irregular shape, and contain a number of hot stars. They are regions of gas, dust and star formation, and are found in galactic spiral arms. Planetary nebulae are more regular, and contain a single very hot highly evolved star. The physical processes in the two types of nebula are very similar. For general accounts of the problems of gaseous nebulae we refer to review articles by Seaton,[1,2] Osterbrock[3] and Flower and Seaton,[4] and books by Wurm,[5] Dufay[6] and Aller.[7]

Radiation intensity is observed as a function of angle, wavelength and, occasionally, time. From observations we have to deduce (1) the physical conditions of the emitting region (2) the chemical composition and (3) the origin and evolution of the nebulae. The procedure is that of trial and error. It is found that hot stars (20,000°K–200,000°K) are surrounded by a low density gas ($\leqslant 10^6$ particle/cm^3), composed of 90% H, 10% He and $\approx 0.1\%$ heavy elements. It is these trace elements which produce the strong emission lines observed by Huggins.

II. PHYSICAL PROCESSES

The primary physical process is that of photoionization of ground state atoms by dilute stellar ultra-violet radiation. Most of the free electrons are produced by ionization of hydrogen

$$H_1 + h\nu \to H^+ + e. \tag{1}$$

This process has a cross-section of $a_\nu \text{cm}^2$. Electron–electron collisions are very effective in maintaining a Maxwellian distribution and the concept of electron temperature, T_e, is meaningful. Radiative capture can take place to the ground state or any excited state n

$$H^+ + e \to H_n + h\nu. \tag{2}$$

The recombination rate is $\alpha_n \text{ cm}^3 \text{ sec}^{-1}$ where

$$\alpha_n = \langle v\sigma \rangle \tag{3}$$

v **is** the velocity of the colliding electron, σ is the capture cross-section and $\langle \ \rangle$ denotes an average over a Maxwellian velocity distribution.

The observed spectrum is produced by cascade transitions

$$H_n \to H_{n'} + h\nu_{nn'} \tag{4}$$

and by collisions

$$H_n + A \to H_{n'} + A. \tag{5}$$

Because of the large transition probabilities for radiative decay, the ionization equilibrium is not affected by ionization from excited states.

III. IONIZATION EQUILIBRIUM

Consider a simple plane parallel model with a monochromatic photon flux $F \text{ cm}^{-2} \text{ sec}^{-1}$. Then, equating the number of photo-ionizations per unit time to the number of recombinations

$$Fa_v N(\mathrm{H}_1) = \alpha N_e N_+ \qquad (6)$$

where

$$\alpha = \sum_{n=1}^{\infty} \alpha_n$$

if all the radiation escapes, and

$$\alpha = \sum_{n=2}^{\infty} \alpha_n$$

if all ionizing radiation is absorbed (i.e., excluding ground state capture, which gives a photon capable of producing further ionization). As we pass through the gas, the number of ionizing photons decreases, leading to the equation of radiative transfer

$$\frac{dF}{dx} = - Fa_v N(\mathrm{H}_1). \qquad (7)$$

At this point, some assumptions have to be made about the densities involved. If we assume

$$N(\mathrm{H}_1) + N(\mathrm{H}^+) = N = \text{constant}, \qquad N(\mathrm{H}^+) = N_e \qquad (8)$$

then the Eqns (6) and (7) can be integrated to give F, $N(\mathrm{H}^+)/N$ and $N(\mathrm{H}_1)/N$ as functions of x. The results obtained are shown in Fig. 1. As x increases, F decreases. But as F decreases, $N(\mathrm{H}_1)/N$ increases which gives an increase in the rate at which F decreases. Eventually we reach a point $x = x_0$ at which $N(\mathrm{H}_1)/N = N(\mathrm{H}^+)/N = \frac{1}{2}$. For $x > x_0$, F decreases very rapidly. We may therefore say that the ionized region has a rather sharp boundary, at $x = x_0$.

The size, x_0, of the ionized region can be calculated in a simple approximate way. We make the simplifying assumption that the gas is completely ionized for $x < x_0$, completely neutral for $x > x_0$. The total number of ionizing quanta is $F(x = 0) \, \text{cm}^{-2} \, \text{sec}^{-1}$ and the total number of recombinations is $\alpha N_e N_+ x_0$. Equating the total number of ionizations to the total number of recombinations,

$$F(x = 0) = \alpha N_e N_+ x_0. \qquad (9)$$

We may now make a generalization to the case of spherical symmetry. Suppose that a star has luminosity $L_v \, dv \, \text{erg sec}^{-1}$ in the frequency range dv.

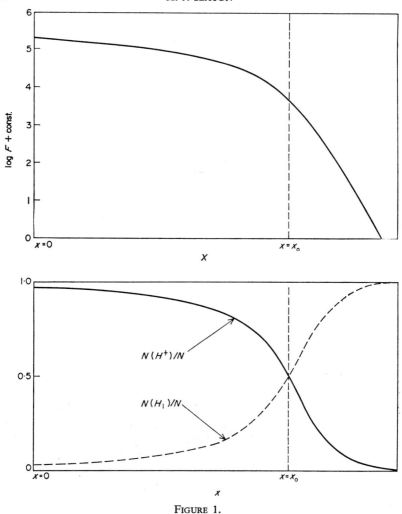

FIGURE 1.

The available number of ionising quanta is $\int_{v_1}^{\infty} (L_v/hv)\, dv$ where v_1 is the threshold frequency for the process (1). Suppose that we have an ionized gas of uniform density within a sphere of radius r_0, and neutral gas for $r > r_0$. Then

$$\int_{v_1}^{\infty} \frac{L_v}{hv}\, dv = \tfrac{4}{3}\pi r_0^3 \alpha\, N_e N_+ ; \qquad (10)$$

r_0 is known as the Strömgren radius.

IV. TEMPERATURES OF CENTRAL STARS

Since

No. of recombinations on level n/unit volume/unit time $= \alpha_n N_e N_+$

we may write

No. of quanta in line $n \to n'$/unit volume/unit time $= \alpha_{n,n'} N_e N_+$

where $\alpha_{n,n'}$ is an effective recombination coefficient for the line (the calculation of $\alpha_{n,n'}$ will be considered in detail later). Consider the H_β line ($n = 4 \to 2$). The luminosity L_β is given by

$$L_\beta = h\nu_\beta \alpha_{4,2} N_e N_+ \tfrac{4}{3}\pi r_0^3. \tag{11a}$$

From (10)

$$L_\beta = \frac{\alpha_{4,2}}{\alpha} h\nu_\beta \int_{\nu_1}^{\infty} \frac{L_\nu}{h\nu} d\nu. \tag{11b}$$

If the star is observed in the visible ($\nu = \nu_{\text{vis}}$), then the measured ratio

$$\frac{L_\beta}{L_{\nu_{\text{vis}}}} = \frac{\alpha_{4,2}}{\alpha} \nu_\beta \frac{\int_{\nu_1}^{\infty} (L_\nu/\nu) d\nu}{L_{\nu_{\text{vis}}}} \tag{12}$$

is a function of the star temperature, T_s, only. If the star acts like a black body of radius r_s and temperature T_s,

$$L_\nu = 4\pi r_s^2 (2\pi h\nu^3/c^2)(e^{h\nu/kT_s} - 1)^{-1} \tag{13}$$

and substitution of the appropriate values in (12) gives the star temperature. This idea was first put forward by Zanstra.[8]

V. ELECTRON TEMPERATURES AND FORBIDDEN LINES

To a good approximation, the free electrons form a Maxwellian distribution of energies, characterized by an electron temperature T_e (see Appendix, A6). The electron temperature is insensitive to density, and most nebulae are in the temperature range $T_e = 7\,10^3\,°K - 1\cdot4\,10^4\,°K$. At these temperatures, $kT_e \approx 1$ eV. Why is the mean electron energy about 1 eV? This question will be answered in Section 5.2.

5.1. Forbidden lines

The strong lines observed by Huggins, 4959 Å and 5007 Å, were identified by Bowen[9] as forbidden transitions in O^{2+}. Most observed lines are due to transitions in configurations p^2, p^3 and p^4. For example, we have configurations

$$1s^2\, 2s^2\, 2p^2 : O^{2+}, N^+$$
$$1s^2\, 2s^2\, 2p^3 : O^+$$
$$1s^2\, 2s^2\, 2p^4 : O^0, Ne^{++}$$

each giving three terms (see Burke's lectures, Fig. 14, for a typical level diagram). The transitions are rigorously forbidden for electric dipole and electric quadrupole radiation in LS coupling. They are allowed for magnetic dipole and electric quadrupole radiation when LS coupling breaks down. These two processes usually have comparable probabilities.

The excitation mechanism for a two-level atom is

$$\begin{aligned} A_1 + e &\to A_2 + e \\ A_2 &\to A_1 + h\nu \\ A_2 + e &\to A_1 + e \end{aligned} \quad (14)$$

where the radiative transition probability A_{21} for forbidden lines is in the range 1 to 10^{-2} sec^{-1} (cf., 10^9 sec^{-1} for allowed transitions). Let $q_{12}N_e$ and $q_{21}N_e$ be the probabilities for collisional transitions. The number ratio

$$\frac{N_2}{N_1} = \frac{q_{12}N_e}{q_{21}N_e + A_{21}} \quad (15)$$

can be considered in two limits.

(i) For $q_{21}N_e \gg A_{21}$

$$\frac{N_2}{N_1} = \frac{q_{12}}{q_{21}} \quad (16a)$$

since collisions are dominant. From (A7) we have

$$\frac{N_2}{N_1} = \frac{\omega_2}{\omega_1} \exp(-E_{21}/kT_e) \quad (16b)$$

where ω_i is the statistical weight of level i, and $E_{21} = E_2 - E_1$.

(ii) For $q_{21}N_e \ll A_{21}$

$$\frac{N_2}{N_1} = \frac{q_{12}N_e}{A_{21}}. \quad (17)$$

In this case, the number of photons/cm^3/sec $= N_2 A_{21} = N_1 q_{12} N_e$. Note that an case (ii) the number of photons depends only on the collision cross-sections, whereas in case (i) it depends on the radiative transition probability. The general case in gaseous nebulae is somewhere between the two limiting cases, so that we require both q_{12} and A_{21}.

In three-level atoms, the intensity ratio

$$\frac{I_{32}}{I_{21}} = f(T_e, N_e) \tag{18}$$

is measured. The ratio may be both temperature and density dependent, since one transition may be close to case (i) and the other close to case (ii). If both transitions are case (ii),

$$\frac{I_{32}}{I_{21}} = \frac{q_{13}}{q_{12}} \frac{A_{32}}{(A_{31} + A_{32})} = \text{constant} \frac{\Omega_{13}}{\Omega_{12}} \exp(-E_{32}/kT_e) \tag{19}$$

where Ω is the collision strength (Seaton[2]). The O^{2+} ratio $(^1S - {}^1D)/({}^1D - {}^3P)$ is very temperature sensitive, while the O^+ ratio $(^2D_{3/2} - {}^4S)/({}^2D_{5/2} - {}^4S)$ is sensitive to electron density because of the long lifetimes of the ^2D states. The ratios from fine structure levels can be accurately measured since reddening and instrumental effects cancel out.

For each observed ratio a density versus temperature curve may be constructed. Many observations of different ratios from the same object should give curves with a common intersection corresponding to the local values of T_e and N_e. In some nebulae, lack of a unique intersection indicates that N_e is not constant at all points in the nebula (Danks,[10] Osterbrock and Flather[11]).

5.2. Thermal balance

Consider again the case of a plane parallel model. A photon flux $F_\nu d\nu$ cm^{-2} sec^{-1} provides the nebula with a kinetic energy input

$$\int_{\nu_1}^{\infty} F_\nu a_\nu N(H)(h\nu - h\nu_1) d\nu \tag{20}$$

where $(h\nu - h\nu_1)$ is the K.E. of the photo-electron. Three energy loss mechanisms balance this input:

(i) Capture: $H^+ + e \to H_n + h\nu$
(ii) Free-free transitions: $H^+ + e \to H^+ + e + h\nu$
(iii) Excitation of forbidden lines.

The excitation process (iii) is the most important. Considering only (iii) for one line, the thermal balance equation is

$$N(H) \int_{\nu_1}^{\infty} F_\nu a_\nu (h\nu - h\nu_1) d\nu = N_1 q_{12} N_e E_{12} \tag{21}$$

where N_1 is the number of ground state ions of the element which gives the forbidden line. The ionization equilibrium gives

$$N(H) \int_{v_1}^{\infty} F_v a_v \, dv = N(H^+) N_e \alpha. \tag{22}$$

Combining Eqns (21) and (22), we obtain

$$\frac{\int_{v_1}^{\infty} F_v a_v (hv - hv_1) \, dv}{\int_{v_1}^{\infty} F_v a_v \, dv} = \frac{N_1}{N(H^+)} \frac{q_{12}}{\alpha} E_{12} \tag{23}$$

where $N_1/N(H^+)$ is an abundance factor, and q_{12}/α has an exponential temperature dependence. The solutions of (23) are insensitive to changes in F_v and always give $T_e \approx 10^4 \,°K$. A hotter central star would produce stronger forbidden line emission, but only a slightly larger electron temperature. This implies a mean electron K.E. of 1 eV, approximately half the excitation energy of the forbidden lines.

VI. MODEL CALCULATIONS

The basic equations are:
 (i) Transfer equation: $dF_v/dx = -F_v a_v N(H)$
 (ii) Ionization equilibrium

$$N(H) \int_{v_1}^{\infty} F_v a_v \, dv = N(H^+) N_e \alpha$$

 (iii) Thermal balance equation

$$N(H) \int_{v_1}^{\infty} F_v a_v (hv - hv_1) \, dv = N_1 q_{12} E_{12} N_e.$$

These give a first approximation to the structure. A more detailed model includes
 (iv) The ionization equilibrium for the heavier elements.
 (v) Calculations of the line emission at each point.

Iterations may be necessary since the equations are inter-related. Having constructed a model, the abundances and electron density are adjusted to get the best fit with observations (Flower[12,13,14]). The agreement is generally good, although there is a disagreement for low ionization potential lines, for which calculated intensities are too weak. If the lines are emitted in the transition region (H becoming neutral), there will be very few exciting electrons. Disagreements between theory and observation may be caused by neglecting the gas dynamics of the transition region. Such a treatment would involve the life history of the object.

6.1. Level populations for radiative recombination

If we consider only radiative capture and cascade, and neglect all non-radiative collisional processes, the equations for $N_i = N(X_i)$, the number of atoms X in level i, are

$$N_e N(X^+)\alpha_i + \sum_{i'=i+1}^{\infty} N_{i'} A_{i'i} = N_i \sum_{i'=i_0}^{i-1} A_{ii'}. \qquad (24)$$

If all radiation can escape freely, we have $i_0 = 1$ ($i_0 = 1$ is the ground state). In general N_1 is large compared with N_i for $i > 1$, and radiation in transitions $i \to 1$ is re-absorbed at a point close to where it is emitted. In this case we take $i_0 = 2$ ($i_0 = 2$ is the first excited state). The solutions of the equation for level populations are usually expressed in terms of factors b_i, giving departures from Saha equilibrium (A8),

$$N_i = b_i N_i(TE : \text{Saha}). \qquad (25)$$

In the radiative case (collisions neglected), b_i is independent of density.

The equations (24) cannot be strictly correct: we have assumed that collisions are very effective in setting up a Maxwell distribution for the free electrons; it follows that collisions must also be important in determining the populations of the highly excited states.

When collisions are included the equations become

$$N_e N(X^+)[\alpha_i + \gamma_i N_e] + \sum_{i'} N_{i'}[C_{i'i} N_e + A_{i'i}]$$
$$= N_i \left\{ \sum_{i'} [A_{ii'} + C_{ii'} N_e] + \beta_i N_e \right\} \qquad (26)$$

where $C_{ii'} = \langle vQ_{i \to i'} \rangle$, $\beta_i = \langle vQ_{i \to c} \rangle$ and γ_i is the rate coefficient for 3-body capture (inverse of electron impact ionization), $Q_{i \to i'}$ and $Q_{i \to c}$ being the excitation and ionization cross-sections.

For highly excited states ($n \gtrsim 40$) collisional redistribution of angular momentum by ions is very effective and one has

$$N_{nl} = \frac{(2l+1)}{n^2} N_n. \qquad (27)$$

Accurate solutions of the equations (26) have been obtained by Brocklehurst[15,16] for the highly excited states (Fig. 2), using the correspondence principle cross-sections of Percival and Richards,[17,18] and for the less highly excited states, not assuming (27).

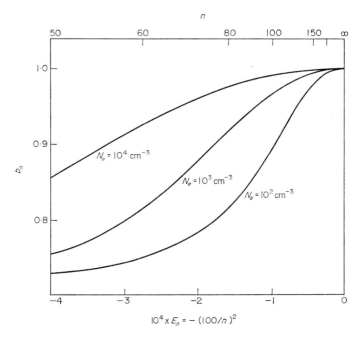

FIGURE 2.

VIII. OPTICAL OBSERVATIONS OF RECOMBINATION LINES

For a well observed bright planetary nebula, forty or more lines may be observed in the recombination spectra of H^0, He^0 and He^+. There has been considerable interest in comparing the theoretical and experimental Balmer line ratios $I_{n,2}/I_{4,2}$ ($n = 3, 4, \ldots 30$). Early observations showed that, compared with theory, the lines at larger wavelengths were too strong. Berman[19] allowed for the effects of interstellar absorption, and found that the amount of reddening was dependent on distance. A single adjustable parameter (corresponding to the amount of absorption) brought calculated and observed values into reasonable agreement.

In later years, observations of several nebulae (Aller et al.,[20] Aller and Minkowski[21]) indicated that there was good agreement for $n \leqslant 7$, but for $n \approx 20$ the observed intensity ratios were often 3 to 5 times the theoretical values. Seaton[22] suggested that the observations were in error, since the discrepancy appeared to be a function of intensity. Careful remeasurement, however, appeared to confirm the previous results (Aller et al.[23]).

Until recently, all the high Balmer lines were observed photographically. Miller[24] has observed these lines in several nebulae, using photo-electric techniques. The measured intensities are 3 to 5 times smaller than the previous photographic intensities, and in good agreement with theory (Brocklehurst[16]).

VIII. INTERPRETATION OF ABSOLUTE LINE INTENSITIES

The observed absolute intensity, I, is proportional to the number of positive ions along the line of sight through the nebula,

$$I \propto \alpha_{n',n} \int_0^\sigma N_e N(X^+) \, ds \qquad (28)$$

which is proportional to the emission measure, $E = \int_0^\sigma N_e N_+ \, ds$. For a homogeneous spherical nebula, radius r, the observed surface brightness S is given by

$$S \propto \frac{N_e N_+ 4\pi r^3/3}{4\pi r^2} = N_e N_+ \frac{r}{3}. \qquad (29)$$

If $N_e N_+$ is known (from, for example, the relative intensities of forbidden lines) this equation may be used to deduce the nebular size r and hence distance $R = r/\theta$, where θ is the angular radius. Conversely, if R and θ are known, (29) gives $N_e N_+$. Comparing the relative intensities of the spectra of H^0, He^0 and He^+ provides information on the element abundances. For most nebulae, $N(He)/N(H) \approx 0.10 \pm 0.01$ (Osterbrock[25]).

IX. THE SPECTRUM OF NEUTRAL HELIUM

The calculation of He level populations is complicated by the metastability of 2^1S and 2^3S. Most work has been concerned with the triplet lines, which are stronger than the singlet lines by a factor ≈ 3. Effects arising from the metastability of 2^3S have been discussed by Osterbrock,[3] Capriotti[26] and Robbins.[27,28]

Re-absorption of a $2^3P \to 2^3S$ photon does not affect the line intensity, since another $2^3P \to 2^3S$ photon will be emitted. Re-absorption of a $3^3P \to 2^3S$ photon will weaken the $3^3P \to 2^3S$ line and strengthen the lines $3^3P \to 3^3S$, $3^3S \to 2^3P$ and $2^3P \to 2^3S$. These effects are observed in many nebulae. It is, however, difficult to calculate the amount of reabsorption, since this depends on mass motions and resulting line profiles. In many nebulae the $2^3P \to 2^3S$ line, at 10830 Å, is very strong. This is due to collisional excitation of 2^3P from 2^3S. The population of 2^3S is determined by (a) capture and cascade (b) $2^3S \to 1^1S$ radiative transitions (c) collisional transitions to singlet states, and (d) photo-ionization by stellar quanta and by $L\alpha$ quanta produced in the nebula. Process (b) was, until recently, thought to proceed by a very slow 2-photon transition. Griem[29] and Drake,[30] however, show that the single photon magnetic dipole radiation is not forbidden and has a much larger probability of occurring.

Transitions of the type $nd \to 2p$ are insensitive to the above-mentioned problems. To a reasonable first approximation, hydrogenic calculations, with the introduction of appropriate statistical weights, may be used. The helium abundance ratio may be obtained from observations of these transitions.

X. THE OPTICAL CONTINUUM

The continuum radiation is produced by radiative recombination, free-free transitions and 2-quantum emission:

$$H^+ + e \to H_n + h\nu, \; \nu \geqslant \nu_n, \; I_n = h\nu_n$$
$$H^+ + e \to H^+ + e + h\nu \qquad (30)$$
$$H(2s) \to H(1s) + h\nu + h\nu'.$$

As the electron density increases, the two-quantum emission becomes less important because of the collisional transition

$$H(2s) + H^+ \to H(2p) + H^+ \qquad (31)$$

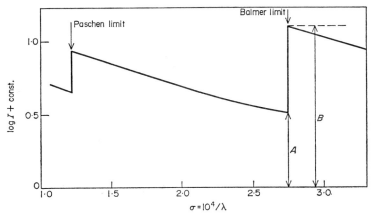

FIGURE 3.

followed by emission of a $Ly\alpha$ photon. Figure 3 shows the form of the continuous spectrum. The Balmer discontinuity, $D = (B - A)/B$ is temperature and density dependent, but independent of the absolute calibration of the observations. Balmer line blending, however, makes it difficult to observe the continuum near the long-wavelength side of the Balmer limit.

XI. THE RADIO CONTINUUM

The radio continuum is predominantly due to the free-free process

$$X^+ + e \rightleftharpoons X^+ + e + hv \tag{32}$$

where $X^+ = H^+$ or He^+. The absorption coefficient may be calculated using classical theory (quantum corrections are small). One obtains (Oster[31])

$$\kappa = N_e^2 r_e^3 \lambda^2 (8/3\sqrt{2\pi})(mc^2/kT_e)^{3/2}$$
$$\times \ln\{(2/\gamma)^{5/2} (kT_e/mc^2)^{3/2} (2\pi)^{-1} (\lambda/r_e)\} \tag{33}$$

where it is assumed that $N_e = N(H^+) + N(He^+)$, and where $r_e = e^2/(mc^2)$ is the electromagnetic radius of the electron, $\lambda = c/v$ is the wavelength and $\ln \gamma = 0.5772...$ is the Euler–Mascheroni constant. Note that

$$\kappa \propto v^{-2} T_e^{-3/2} \tag{34}$$

neglecting slowly varying logarithmic factors.

Let j be the continuum emissivity. The ratio (j/κ) depends only on the energy distribution for the free electrons. This distribution is Maxwellian and (j/κ) is therefore given by the relation (Kirchoff's law, (A10))

$$j/\kappa = B(T_e), \tag{35}$$

where B is the black-body intensity defined by (A.9). At radio frequencies we have $hv \ll kT_e$ and hence, from (A.9),

$$B(T_e) = (2v^2/c^2)kT_e. \tag{36}$$

It follows from (34), (35) and (36) that the emissivity j depends on frequency through the logarithmic factors only.

The absorption coefficient must become large at low frequencies, and it is then necessary to consider the transfer of radiation. The intensity $I(s)$ is calculated on solving the transfer equation (see Section A.2)

$$\frac{dI(s)}{ds} = \kappa I(s) - j. \tag{37}$$

Defining the optical depth at the point s,

$$t(s) = \int_0^s \kappa(s')\,ds', \qquad (38)$$

and using (35) and (38), (37) may be written

$$\frac{dI}{dt} = I - B \qquad (39)$$

with solution for the emergent intensity

$$I(0) = \int_0^\tau B(T_e)\,e^{-t}\,dt \qquad (40)$$

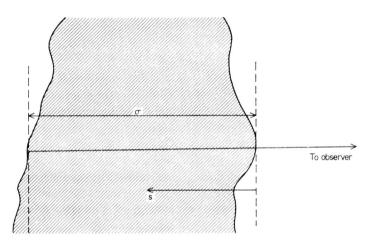

FIGURE 4.

where σ is the length of the nebula along the line of sight (see Fig. 4), and where $\tau = t(\sigma)$ is the total optical thickness of the nebula. If T_e is constant this gives

$$I(0) = B(1 - e^{-\tau}). \qquad (41)$$

Using (41) we have:

(i) For ν sufficiently small, τ is large ($\tau \gg 1$) and

$$I(0) = B = (2\nu^2/c^2)\,kT_e.$$

The source is a blackbody and observations give T_e.

(ii) For ν sufficiently large, τ is small ($\tau \ll 1$) and

$$I(0) = B\tau = B\kappa\sigma = j\sigma.$$

Note that $j\sigma$ is proportional to the emission measure, $E = N_e^2 \sigma$. The general form of the continuum intensity for a plane parallel nebula of emission measure $E = 6 \times 10^6$ cm^{-6} pc is shown in Fig. 5. One parsec (pc) corresponds to a length of 3,085 10^{18} cm.

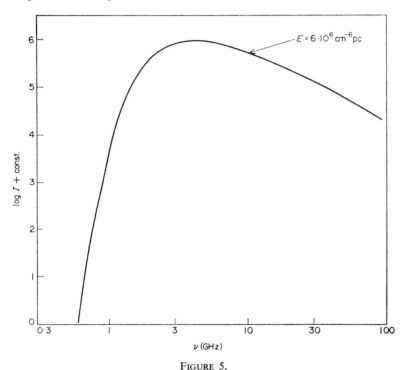

FIGURE 5.

XII. RADIO RECOMBINATION LINES

The interpretation of radio recombination line observations has led to some new astrophysical ideas. Disagreements between observations and early theory were confirmed by new and more accurate calculations of collision cross-sections, level populations and line widths. The assumptions of constant electron density and temperature were questioned, and allowance for variations of these parameters provided new radiative transfer problems. It is found that variable density models give reasonable agreement with observation.

12.1 Line frequencies

For level n of H, $E_n = -\chi_1/n^2$ where $\chi_1 = 13 \cdot 60$ eV. For the lines $n + m \to n$,

$$hv = \chi_1 \left(\frac{1}{n^2} - \frac{1}{(n+m)^2} \right). \tag{42}$$

Suppose that $n \gg 1$ and $m \ll n$. Then

$$hv \approx 2\chi_1 m/n^3 \tag{43}$$

giving

$$v \approx 6 \cdot 58 \, m(100/n)^3 \text{ GHz}. \tag{44}$$

The lines are referred to as $n\alpha, n\beta, n\gamma \ldots$ for $m = 1, 2, 3, \ldots$. Figure 6 presents some observations of recombination lines in Orion A.

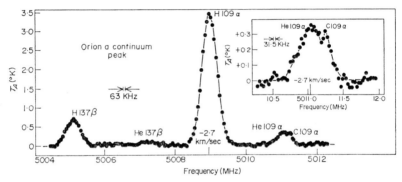

FIG. 6. Lines observed in Orion A by Churchwell and Mezger,[45] at frequencies close to 5008 MHz. The antenna temperature T_A is proportional to the intensity of the line above the intensity of the adjacent continuum. The inset figure shows further detail on the observations of He 109α and C 109α.

12.2. Emissivities and absorption coefficients

We have already seen that we may define b_n, such that

$$N_n = b_n N_n(TE). \tag{45}$$

For $n \gg 1$, $b_n \approx 1$. We may also express line emissivities and opacities, j_v^l and κ_v^l, in terms of values for thermodynamic equilibrium, $j_v^l(TE)$, $\kappa_v^l(TE)$ (Goldberg[33]).

From (A3) and (45),

$$j_v^l = b_{n+m} j_v^l(TE) \tag{46}$$

and from (A4) and (45),

$$\kappa_v^l = b_n \beta_{n,m} \kappa_v^l(TE) \tag{47}$$

where

$$\beta_{n,m} = \left(1 - \frac{b_{n+m}}{b_n} e^{-h\nu/kT_e}\right)(1 - e^{-h\nu/kT_e})^{-1}. \tag{48}$$

For the radio lines we have $(h\nu/kT_e) \ll 1$. Putting

$$e^{-h\nu/kT_e} \approx 1 - h\nu/kT_e \tag{49}$$

and

$$\frac{b_{n+m}}{b_n} \approx 1 + m \frac{d \ln b_n}{dn} \tag{50}$$

we then obtain

$$\beta_{n,m} \approx 1 - \frac{kT_e}{h\nu} m \frac{d \ln b_n}{dn} \tag{51}$$

and, using (43),

$$\beta_{n,m} \approx 1 - \frac{kT_e}{\chi_1} \frac{n^3}{2} \frac{d \ln b_n}{dn} = 1 - kT_e \frac{d \ln b_n}{dE_n}. \tag{52}$$

In this approximation, $\beta_{n,m}$ is independent of m. We may neglect differences between b_{n+m} in (46) and b_n in (47). Dropping subscripts n,

$$j_\nu^l = b j_\nu^l(TE), \qquad \kappa_\nu^l = b\beta \kappa_\nu^l(TE) \qquad \beta = 1 - kT_e \frac{d \ln b}{dE}. \tag{53}$$

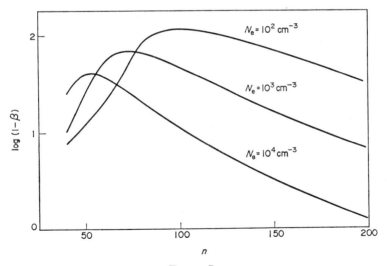

FIGURE 7.

Figure 7 shows $\log(1-\beta)$ against n for $T_e = 10^4\,°K$ and $\log N_e = 2, 3$ and 4. The quantity $(1-\beta)$ is always positive. The line intensities may therefore be increased due to stimulated emission (maser action).

Using (A5) and (A8) we obtain

$$\kappa_v^l(TE) = N_e N_+ \frac{\pi e^2}{mc}(h^2/2\pi m k T_e)^{3/2}$$
$$\times n^2 \exp(-E_n/kT_e)(1-\exp(-hv/kT_e))f\phi_v. \quad (54)$$

For $n \gg 1$ and $m \ll n$ it may be shown (Menzel[34]) that

$$hvf = \chi_1 \frac{2m}{n^2} K(m)\{1 + O(m/n)^2\} \quad (55)$$

where $K(1) = 0.1908$, $K(2) = 0.02633$, $K(3) = 0.00810, \ldots$. Using this result we obtain, after some re-arrangement of (54),

$$\kappa_v^l(TE) = N_e N_+ r_e a_0^3 \lambda \exp(-E_n/kT_e) \cdot (\chi_1/kT_e)^{5/2} 8\pi^{5/2} 2mK(m)(v\phi_v) \quad (56)$$

where a_0 is the Bohr radius. Given $\kappa_v^l(TE)$, $j_v^l(TE)$ is calculated using

$$j_v^l(TE) = B\kappa_v^l(TE). \quad (57)$$

Our convention is as follows: we put the subscript v on quantities involving the profile factor ϕ_v, which varies rapidly with frequency. This subscript is omitted for quantities integrated over frequency: thus

$$j^l = \int j_v^l\,dv, \qquad \kappa^l = \int \kappa_v^l\,dv. \quad (58)$$

The subscript v is also dropped for quantities referring to the continuum.

12.3. Line profiles for constant density models

The profile factor ϕ_v is obtained on convoluting a Gaussian profile, ϕ_v^D, for Doppler broadening with a Lorentz profile, ϕ_v^C, for collision broadening:

$$\phi_v = \int \phi_{v-v'+v_0}^C \phi_{v'}^D\,dv' \quad (59)$$

where v_0 is the frequency at the centre of the line,

$$\phi_v^D = (\alpha/\pi^{1/2} v_0)\exp\{-[\alpha(v-v_0)/v_0]^2\} \quad (60)$$

with

$$\alpha = (Mc^2/2kT_D)^{1/2}$$

$$T_D = \text{"Doppler temperature"}$$

and

$$\phi_v^C = \left(\frac{\delta}{\pi}\right)[(v - v_0)^2 + \delta^2]^{-1}. \quad (61)$$

The line broadening parameter δ has been calculated by Griem[35] and by Brocklehurst and Leeman[36] who obtain

$$\delta = 4.7 \left(\frac{n}{100}\right)^{4.4} \left(\frac{10^4}{T_e}\right)^{0.1} N_e \text{ Hz} \quad (62)$$

with T_e in °K and N_e in cm^{-3}.

Let us introduce the quantities $x = \alpha(v - v_0)/v_0$ and $f(x) = \phi_v/\phi_{v_0}$. Then $f(0) = 1$. The profile factors $f(x)$ of H $n\alpha$ lines emitted by a nebula of constant electron density $N_e = 10^4$ cm^{-3} are shown in Fig. 8. A pure Doppler profile (independent of n) is given for reference. The electron impact widths for $N_e = 10^4$ cm^{-3} are very much larger than the observed widths, especially for the high lines (Davies,[38] Pedlar and Davies,[39] Churchwell and Edrich[40]). In an attempt to overcome these difficulties, models of variable density must be considered.

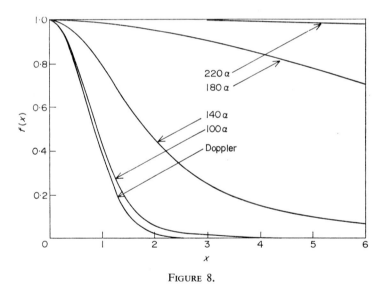

FIGURE 8.

12.4. Solution of the transfer equation (Brocklehurst and Seaton[37])

The intensity in the line, I_v^l, is calculated as

{intensity of line + continuum} − {intensity of continuum}.

For the emergent intensity we obtain

$$I_v^l = \int_0^\sigma \{(j_v^l + j)\exp[-(t_v^l + t)] - j\exp(-t)\}\,ds. \qquad (63)$$

We now make three approximations:

(i) $|t_v^l| \ll 1$, which seems to be valid for all situations of practical interest (Fig. 9).

(ii) $j_v^l \ll j$, which is good for $n > 100$, less good for $n < 100$ (Fig. 10).

(iii) $T_e = $ constant.

Using (i), $\exp(-t_v^l) = 1 - t_v^l$ and (63) reduces to

$$I_v^l = \int_0^\sigma \{j_v^l - (j_v^l + j)t_v^l\}e^{-t}\,ds \qquad (64)$$

and using (ii)

$$I_v^l = \int_0^\sigma \{j_v^l - jt_v^l\}e^{-t}\,ds. \qquad (65)$$

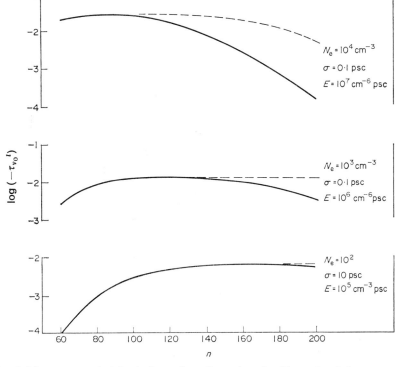

FIG. 9. Line-centre optical depths for $n\alpha$ lines, for various densities and emission measures. The effect of impact broadening is neglected for the broken lines, included for the full lines.

Using $(j/\kappa) = (j_\nu^l(TE)/\kappa_\nu^l(TE)) = B$, we have

$$j_\nu^l \, ds = B \frac{\kappa^l(TE)}{\kappa} b\phi_\nu \, dt \tag{66}$$

$$jt_\nu^l \, ds = j\left(\int_0^s \kappa_\nu^l \, ds'\right) ds = B\left\{\int_0^t \frac{\kappa^l(TE)}{\kappa} b\beta\phi_\nu \, dt'\right\} dt. \tag{67}$$

Using (iii) B is constant and $\kappa^l(TE)/\kappa$ is constant: (64) reduces to

$$I_\nu^l = B \frac{\kappa^l(TE)}{\kappa} y_\nu \tag{68}$$

where

$$y_\nu = \int_0^\tau \left\{b\phi_\nu - \int_0^t b\beta\phi_\nu \, dt'\right\} e^{-t} \, dt \tag{69}$$

which, after an integration by parts, reduces to

$$y_\nu = \int_0^\tau b\phi_\nu \{e^{-t} + (1-\beta)(e^{-t} - e^{-\tau})\} \, dt. \tag{70}$$

We may put

$$I_\nu^l = I^l \theta_\nu$$

where

$$\int \theta_\nu \, d\nu = 1$$

and obtain

$$\theta_\nu = y_\nu/y$$

where

$$y = \int_0^\tau b\{e^{-t} + (1-\beta)(e^{-t} - e^{-\tau})\} \, dt. \tag{72}$$

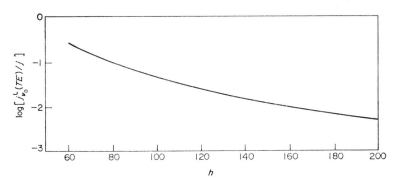

Figure 10.

The term $e^{-\tau}$ in (70) and (72) gives a reduction in line intensity due to absorption in the continuum. The term $(1-\beta)(e^{-t}-e^{-\tau})$ in (70) and (72) gives an increase in line intensity due to maser action in the line. This term is largest for small N_e and for small t. It should be noted that t is small in the outermost part of a nebula, closest to the observer: in this region N_e is usually small and maser action is very effective.

12.5. Line-to-continuum intensity ratios

The observers measure the intensities in the lines, I_v^l, relative to the intensity I in the adjacent continuum. Using Eqns (41) and (68), we obtain

$$\frac{I_v^l}{I} = \frac{\kappa^l(TE)}{\kappa} Y \theta_v \tag{73}$$

where

$$Y = y/(1-e^{-\tau}) \tag{74}$$

and where θ_v is the profile factor for the emergent intensity, defined by (71). Since $\int \theta_v \, dv = 1$, the integrated line intensity, relative to the continuum, is

$$\frac{I^l}{I} = \frac{\kappa^l(TE)}{\kappa} Y. \tag{75}$$

If the nebula is optically thin ($\tau \ll 1$ for the continuum) and in thermodynamic equilibrium ($b = \beta = 1$), we obtain $Y = 1$ and

$$\left(\frac{I^l}{I}\right)_{\tau \ll 1, TE} = \frac{\kappa^l(TE)}{\kappa}. \tag{76}$$

It should be noted that the temperature dependence of $\kappa^l(TE)/\kappa \approx \text{const.} \times T_e^{-1}$.

12.6. Solutions for models of constant density

If N_e, T_e, and T_D do not vary, ϕ_v in (70) is independent of t, and we obtain $\theta_v = \phi_v$: the profile of the emergent intensity is equal to the profile of the emission and absorption coefficients.

Evaluation of the integral in (72) gives

$$Y = b\{g(\tau) + (1-\beta)[1 - g(\tau)]\} \tag{77}$$

where

$$g(\tau) = \tau/(e^\tau - 1). \tag{78}$$

The behaviour of this function is shown in Fig. 11.

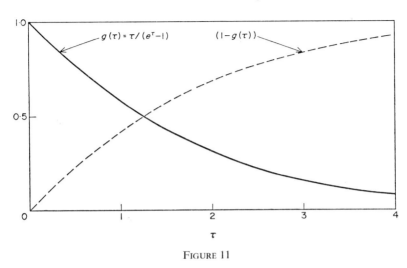

FIGURE 11

Let us consider the case of $n\alpha$ lines. The frequency is

$$v = 6\cdot58 \times (100/n)^3 \text{ GHz}.$$

Let κ be the continuum opacity at the frequency of an $n\alpha$ line: we have $\kappa \propto v^{-2}$ giving $\kappa \propto n^6$.

The continuum optical depth τ is small for small n (in practice for $n \lesssim 100$). Expanding $g(\tau)$ we obtain $g(\tau) = 1 - \tfrac{1}{2}\tau + \ldots$ and hence

$$Y \approx 1 + \tfrac{1}{2}\tau(1 - \beta) \qquad (\tau \ll 1). \tag{79}$$

We here omit the factor b, which is close to unity. It should be noted that we assume $\tau \ll 1$ but do not assume $(1 - \beta)\tau \ll 1$ since $(1 - \beta)$ may be of order 10^2 (see Fig. 7). Since $(1 - \beta)$ is positive, (79) shows that the line intensity may be increased by maser action.

The optical depth τ is large for large n (say $n \gtrsim 200$). For this case we have $g(\tau) \ll 1$ and

$$Y \approx (1 - \beta) \qquad (\tau \gg 1). \tag{80}$$

For this case, in the absence of maser action the line would not be observed, due to absorption in the continuum. If we know $(1 - \beta)$ as a function of N_e, from the observed line intensities N_e may be deduced.

12.7. Electron temperatures

If we assume the case ($\tau \ll 1, TE$), we have

$$\left(\frac{I^l}{I}\right)_{\tau \ll 1, TE} \propto T_e^{-1}. \tag{81}$$

From the observed ratios T_e may be deduced.

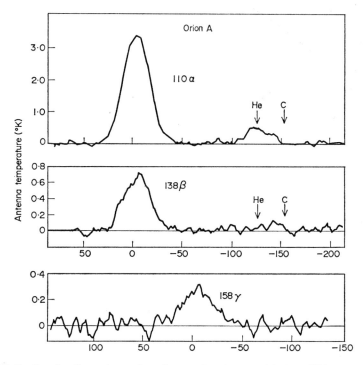

FIG. 12. Profiles for lines observed at frequencies close to 4·9 GHz.[38] Frequency is expressed in terms of velocity, in km sec^{-1}, with respect to the local standard of rest.

Suppose that we observe a number of lines, $n\alpha$, $n'\beta$, $n''\gamma$, ... all at nearly the same frequency. Thus at $v \approx 4\cdot 9$ GHz ($\lambda \approx 6$ cm) one can observe 110α, 138β, 158γ, 173δ and 186ε. Figure 12 shows a set of such observations of Orion A (Davies[38]). At this frequency, we usually have $\tau \ll 1$ and (79) may be used. The frequency, and hence τ, is practically the same for all of the lines. But for the higher order lines (β, γ, δ ...) we have larger values of n and hence smaller values of $(1 - \beta)$. The case ($\tau \ll 1$, TE) is therefore approached for the higher order lines as illustrated in Fig. 13, using observations by Davies.[38] Higher order lines have larger error bars since they are of considerably weaker intensity. The values of T_e obtained using (81) for these lines are in agreement with values deduced from observations at optical wavelengths (Hjellming and Davies[41]).

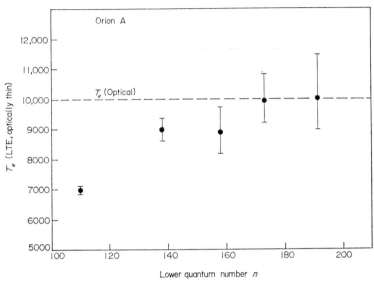

FIGURE 13

12.8. Densities and Emission measures for uniform density models

Given T_e, and assuming constant density, all line-to-continuum ratios depend on two parameters, N_e and emission measure $E = N_e^2 \sigma$. One may obtain values of these two parameters which give agreement between theory and observation for all of the observed lines (Hjellming and Churchwell,[42] Hjellming and Davies,[41] Hjellming and Gordon[43]). The results obtained are large ($N_e \approx 10^4 \text{ cm}^{-3}$, $E \approx 10^7 \text{ cm}^{-6} \text{ pc}$).

12.9. Line profiles for variable density models

For $N_e \approx 10^4 \text{ cm}^{-3}$, the calculated widths of the high lines are much larger than the observed widths. Brocklehurst and Seaton[37] have therefore considered models of variable density. Profiles obtained for a typical spherical model, based on Orion A, are shown in Fig. 14. In such models, the high lines are formed in the outer regions of low density, and the computed line widths and profiles compare well with those observed in Orion (Fig. 15). The theory also predicts lines with sharp cores and very broad wings. It is supposed that the observers have failed to detect the broad wings (the usual procedure is to subtract from the observed profile a frequency-dependent "baseline", assumed to be due to frequency variation of instrument sensitivity).

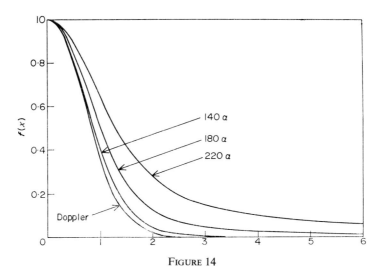

FIGURE 14

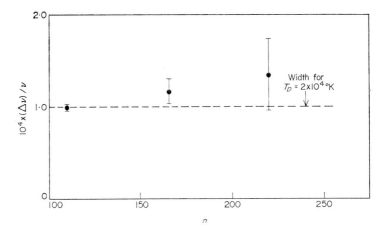

FIG. 15. Observed valves of $\Delta\nu/\nu$ for $n\alpha$ lines in Orion.[39]

The model also predicts that the 110α and 138β transitions should have comparable widths. This agrees with the observations of Davies[38] presented in Fig. 16. On a constant density model, the two lines have the same Doppler widths, but differ by a factor of two in collisional widths.

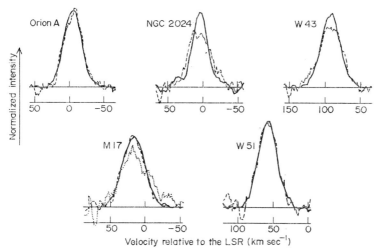

FIG. 16. Comparisons of lines profiles, for five nebulae.[38] Full lines, 110α; broken lines, 138β; dotted line, 158γ.

XIII. HELIUM AND CARBON RECOMBINATION LINES

Recombination lines of helium, and of a heavier atom, almost certainly carbon, have been detected. If it is assumed that the relative abundance of H and He is constant throughout the nebula, that the helium is all singly ionized, and that departures from LTE are identical for hydrogen and helium, then for any radio transition

$$\frac{I(\text{He})}{I(\text{H})} = \frac{N(\text{He}^+)}{N(\text{H}^+)} = \frac{N(\text{He})}{N(\text{H})}.$$

Average abundance ratios of 0.08 ± 0.01 are obtained (Palmer et al.,[44] Churchwell and Mezger,[45] Reifenstein et al.,[46] Davies[38]). This value compares reasonably with the mean optical ratio of 0.10 ± 0.01. There is a small discrepancy, probably due to the fact that the nebulae may contain regions in which H is ionized but He is neutral.

Palmer et al.[47] discovered a new emission line at a frequency slightly above those of the H and He 109α transitions. Because of uncertainties in the Doppler shift of the emitter, absolute identification was not possible. The suggestion that the "anomalous" emitter was a heavy element was not consistent with models of H^+ regions. Dupree and Goldberg[48] suggested, however, that the line was formed in a H^0 region surrounding the H^+ region. Large departures from TE would explain the significant line intensity. A further consequence of this theory is that recombination lines of ionized hydrogen in H^0 regions should be detectable.

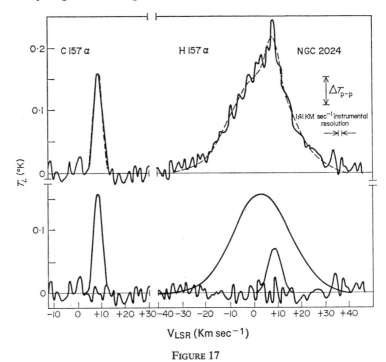

FIGURE 17

These suggestions were confirmed for Orion B, and the emitter identified as carbon, in a striking experiment by Ball et al.[49] Observations of the H 157α line and neighbouring frequencies, revealed a very narrow anomalous line and a broad, but asymmetrical spiked hydrogen line (Fig. 17). The hydrogen line was decomposed into two Gaussians (lower graph), indicating that the narrow hydrogen component and the anamalous line arose from regions with physical conditions very different from those of the H^+ region ($T_e \approx 10^2$–10^3 °K). Ball et al. showed that the narrow H line leads to a velocity in agreement with that of neutral hydrogen obtained from 21 cm. observations, indicating emission from an H^0 region along the line of sight of Orion B. Assuming no relative velocity between the neutral hydrogen and anomalous emitter, the observed frequency shift could only be caused by an element of atomic weight 11·7 (+0·9, −0·8). It is believed that this indicates a firm identification of carbon.

The experiment also afforded the first measurement of the degree of ionization in H^0 regions. Ball *et al.* deduce a value

$$N(H^+)/N(H) \approx 3 \times 10^{-4}$$

Other observations (Palmer *et al.*,[44] Zuckermann and Palmer[50,51]) indicate that the carbon line at lower frequencies becomes stronger relative to the corresponding H^+ line. Pedlar[52] finds the C 220α line to be 1/3 as strong as the H 220α line. This may be explained on the basis of the above discussion on line broadening. Lines from low temperature H^0 regions would not be broadened at low frequencies, whereas lines emitted from H^+ regions would have an appreciable fraction of their energy in the wings.

APPENDIX

A.1. Definition of intensity of radiation

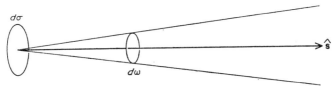

FIGURE A1

Radiant energy in direction $\hat{s}$, crossing area $d\sigma$ normal to $\hat{s}$, in solid angle $d\omega$, in frequency range dv, per unit time, is

$$I_v \, d\sigma \, d\omega \, dv. \tag{A1}$$

A.2. Equation of radiative transfer

In length ds, intensity absorbed is $\kappa_v I_v \, ds$, intensity emitted is $j_v \, ds$. The equation of radiative transfer is

$$\frac{dI_v}{ds} = -\kappa_v I_v + j_v. \tag{A2}$$

Note that $j_v \, dv$ is the emission per unit volume per unit solid angle.

Defining $\tau = \int_0^s \kappa_\nu \, ds$ and $S = j_\nu/\kappa_\nu$, and imposing the boundary condition $I_\nu(o) = 0$, the solution of (A2) is

$$I(\tau) = e^{-\tau} \int_0^\tau e^t S(t) \, dt.$$

It should be noted that the sign convention for s in the main text differs from that used here. (In the main text we consider the intensity in the direction towards the observer but measure s in the direction away from the observer—the signs in (37) therefore differ from those in (A2)).

A.3. Emission and adsorption in a spectrum line

For the emission line $n' \to n$,

$$j_\nu = h\nu_0 N_{n'} \frac{A_{n'n}}{4\pi} \phi_\nu \tag{A3}$$

where ν_0 is the frequency at the line centre, $N_{n'}$ the number of atoms per unit volume in upper level, $A_{n'n}$ the Einstein spontaneous transition probability, and ϕ_ν the normalised line profile factor ($\int \phi_\nu \, d\nu = 1$).

Using the Einstein relations between the coefficients for spontaneous emission, absorption and stimulated emission,

$$\kappa_\nu = \frac{c^2}{2\nu_0^2} \left[\frac{N_n}{\omega_n} - \frac{N_{n'}}{\omega_{n'}} \right] \omega_{n'} \frac{A_{n'n}}{4\pi} \phi_\nu \tag{A4}$$

where ω_n is the statistical weight of level n. Note that stimulated emission is counted as negative absorption and is in the direction of the stimulating radiation. Introducing the oscillator strength $f_{n'n}$,

$$\kappa_\nu = \frac{\pi e^2}{mc} f_{n'n} \omega_n \left[\frac{N_n}{\omega_n} - \frac{N_{n'}}{\omega_{n'}} \right] \phi_\nu. \tag{A5}$$

A.4. Laws of thermodynamic equilibrium (TE)

(i) *Maxwell law for electron energies*
Number of electrons per unit volume with energy $E = \frac{1}{2}mv^2$ in range dE is $dN_e = N_e f(E, T) \, dE$ where

$$f(E, T) = 4\pi m^2 (2\pi mkT)^{-3/2} v \exp(-E/kT) \tag{A6}$$

such that
$$\int f(E, T) \, dE = 1.$$

(ii) *Boltzmann law for level populations*

$$\frac{N_{n'}}{N_n} = \frac{\omega_{n'}}{\omega_n} \exp\{-E_{n'n}/kT\} \tag{A7}$$

where $E_{n'n} = (E_{n'} - E_n)$. This law must be satisfied if the level populations are determined entirely by electron collisions, and the electrons have a Maxwell distribution. It follows that the collisional rate coefficients, q_{12} and q_{21}, must be such that $(q_{12}/q_{21}) = (\omega_2/\omega_1) \exp(-E_{21}/kT)$.

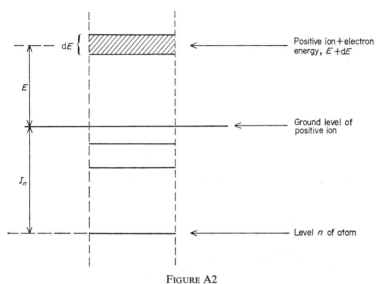

FIGURE A2

(iii) *The Saha law for ionization equilibrium*
Per unit volume we have: N_n atoms of a given chemical element, in level n; N_+ ions of the same element, in the ground state; N_e electrons. The Saha equation is

$$\frac{N_n}{N_e N_+} = \frac{\omega_n}{2\omega_+} \left(\frac{h^2}{2\pi mkT}\right)^{3/2} \exp(I_n/kT) \tag{A8}$$

where I_n is the minimum energy required to ionize an atom in level n.

(iv) *The Planck equation*
For *TE*, $I_\nu = B_\nu(T)$ where

$$B_\nu(T) = \frac{2h\nu^3}{c^2} (\exp(h\nu/kT) - 1)^{-1}. \tag{A9}$$

(v) *Kirchoff's law*

For TE, $dI_\nu/ds = 0$. From (A2) it follows that

$$j_\nu = B_\nu \kappa_\nu. \qquad (A10)$$

Note that the Einstein relations are obtained using (A3), (A7) (A9) and (A10).

A.5. Derivation of the Saha equation

By analogy with (A7)

$$\frac{N_+ \, dN_e}{N_n} = \frac{\omega_+ \, d\omega_e}{\omega_n} \exp\{-(E + I_n)/kT\} \qquad (A11)$$

where $d\omega_e$ is the statistical weight for electrons per unit volume with energy E, $E + dE$:

$$d\omega_e = 2h^{-3} \, d^3\mathbf{p} \qquad (A12)$$

(factor 2 for electron spin). Now $d^3\mathbf{p} = 4\pi p^2 \, dp = 4\pi m^2 v \, dE$ since $E = \tfrac{1}{2}mv^2$. Hence using $dN_e = N_e f(E, T) \, dE$ and (A6), (A11) gives (A8).

REFERENCES

1. Seaton, M. J., *Rep. Prog. Phys.* **23**, 313 (1960).
2. Seaton, M. J., *Adv. Atom. Mol. Phys.* **4**, 331 (1968).
3. Osterbrock, D. E., *Ann. Rev. Astr. and Ap.* **2**, 95 (1964).
4. Flower, D. R. and Seaton, M. J., *Mem. Soc. Roy. Sci. Liège* **17**, 251 (1969).
5. Wurm, K., "Die Planetarischen Nebel", Akademie-Verlag, Berlin (1951).
6. Dufay, J., "Nebuleuses Galactiques et Matière Interstellaire", Albin Michel, Paris (1954).
7. Aller, L. H., "Gaseous Nebulae", Chapman and Hall, London (1956).
8. Zanstra, H., *Publ. Dom. Astr. Obs. (Victoria)* **4**, 209 (1931).
9. Bowen, I. S., *Ap. J.* **67**, 1 (1928).
10. Danks, A, C., *Astr. and Ap.* **9**, 175 (1970).
11. Osterbrock, D. E. and Flather, E., *Ap. J.* **129**, 26 (1959).
12. Flower, D. R., *Astrophys. Lett.* **2**, 205 (1968).
13. Flower, D. R., *M.N.R.A.S.* **146**, 171 (1969).
14. Flower, D. R., *M.N.R.A.S.* **146**, 243 (1969).
15. Brocklehurst, M., *M.N.R.A.S.* **148**, 417 (1970).
16. Brocklehurst, M., *M.N.R.A.S.* **145**, 153, 471 (1971).
17. Percival, I. C. and Richards, D., *Astrophys. Lett.* **4**, 235 (1969).
18. Percival, I. C. and Richards, D., *J. Phys. B.* **3**, 316 (1970).
19. Berman, L., *M.N.R.A.S.* **96**, 890 (1936).
20. Aller, L. H., Bowen, I. S. and Minkowski, R., *Ap. J.* **122**, 62 (1955).
21. Aller, L. H. and Minkowski, R., *Ap. J.* **124**, 110 (1956).

22. Seaton, M. J., *M.N.A.R.S.* **120**, 326 (1960).
23. Aller, L. H., Bowen, I. S. and Wilson, O. C., *Ap. J.* **138**, 1013 (1963).
24. Miller, J. S., *Ap. J. (Letters)* **165**, L101 (1971).
25. Osterbrock, D. E., *Q.J.R.A.S.* **11**, 199 (1970).
26. Capriotti, E. R., *Ap. J.* **150**, 95 (1967).
27. Robbins, R. R., *Ap. J.* **151**, 511 (1968).
28. Robbins, R. R., *Ap. J.* **160**, 519 (1970).
29. Griem, H., *Ap. J. (Letters)* **156**, L103 (1969).
30. Drake, G. W. A., *Phys. Rev. A.* **3**, 908 (1971).
31. Oster, L., *Rev. Mod. Phys.* **33**, 525 (1961).
32. Dupree, A. K. and Goldberg, L., *Ann. Rev. Astr. and Ap.* **8**, 231 (1970).
33. Goldberg, L., *Ap. J.* **144**, 1225 (1966).
34. Menzel, D. H., *Nature* **218**, 756 (1968).
35. Griem, H., *Ap. J.* **148**, 547 (1967).
36. Brocklehurst, M. and Leeman, S., *Astrophys. Lett.* **9**, 35 (1971).
37. Brocklehurst, M. and Seaton, M. J., *Astrophys. Lett.* **9**, 139 (1971).
38. Davies, R. D., *Ap. J.* **163**, 479 (1971).
39. Pedlar, A. and Davies, R. D., *Nature (Phys. Science)* **231**, 49 (1971).
40. Churchwell, E. and Edrich, J., *Astr. and Ap.* **6**, 261 (1970).
41. Hjellming. R. M. and Davies, R. D., *Astr. and Ap.* **5**, 53 (1970).
42. Hjellming, R. M. and Churchwell, E., *Astrophys. Lett.* **4**, 165 (1969).
43. Hjellming, R. M. and Gordon, M. A., *Ap. J.* **164**, 47 (1971).
44. Palmer, P., Zuckerman, B., Penfield, H., Lilley, A. E. and Mezger, P. G., *Ap. J.* **156**, 887 (1969).
45. Churchwell, E. and Mezger, P. G., *Astrophys. Lett.* **5**, 227 (1970)
46. Reifenstein, E. C. III, Wilson, T. L., Burke, B. F., Mezger, P. G. and Altenhoff, W., *Astr. and Ap.* **4**, 357 (1970).
47. Palmer, P., Zuckerman, B., Penfield, H., Lilley, A. E. and Mezger, P. G., *Nature* **215**, 40 (1967).
48. Dupree, A. K. and Goldberg, L., *Ap. J. (Letters)* **158**, L49 (1969).
49. Ball, J. A., Cesarsky, D., Dupree, A. K., Goldberg, L. and Lilley, A. E. *Ap. J. (Letters)* **162**, L25 (1970).
50. Zuckerman, B. and Palmer, P., *Astr. and Ap.* **153**, L145 (1968).
51. Zuckerman, B. and Palmer, P., *Astr. and Ap.* **4**, 244 (1970).
52. Pedlar, A., Ph. D. Thesis, University of Manchester (1971).

Introduction to Molecular Spectra

H. M. FOLEY

Columbia University, New York, U.S.A.

The study of atomic states and spectra proceeds along well-known lines, relying heavily on the concept of an average central symmetric field dominated by the nuclear field, in which the individual electrons are described by angular momentum quantum numbers. The fact that the nucleus is a particle in motion can be ignored (or taken into account by the "reduced mass" of the electron), and the resulting description (Hartree–Fock) provides a very good basis for analysis of spectra and for further refinements (fine structure, hyperfine structure, electron correlation effects).

In the case of molecules the situation is very different, for two related reasons. The electrons now move in an average potential field which is not even approximately central. For diatomic molecules the field is axially symmetric, and the only "good" quantum number related to this symmetry is the component of angular momentum along the symmetry axis. For general polyatomic molecules even this symmetry is lacking, and only certain reflection and inversion symmetries remain. The nuclear motion is now important; indeed the analysis proceeds by establishing a coordinate system "fixed" to the (equilibrium) nuclear positions and describing the electron motion (wave function) within this frame of reference. This procedure is justified by the fact that the heavy particle motions are slow and the electron quantum states adjust adiabatically to the rotations and vibrations of the nuclei. Because of this the rotation and vibration of the molecule can be discussed separately from electronic motions. The approximations that have been thereby made, however, give rise to certain typical "relative motion" effects which have no analog in atomic spectra, and which are important, even in the radiofrequency spectra.

In our study of molecular spectra, the principal works of reference are provided by the three volumes of Herzberg[1–3], and the volume by Townes

I. THE MOLECULAR WAVE EQUATION

We discuss here, for simplicity, the case of a diatomic molecule with a single electron. The extension to many electrons is evident, and will be given below. The Hamiltonian of the system in fixed "laboratory" coordinates has the form

$$H = KE \text{ (nuclei)} + KE \text{ (electron)} + V_{el}(\mathbf{x}_0, \mathbf{X}_A, \mathbf{X}_B) + H_{mag}$$

V_{el} includes the electrostatic interaction between the nuclei and the electrons. In the many-electron case V_{el} includes the electron–electron interaction. H_{mag} includes the internal magnetic interactions, the most important of which is the electron spin–orbit interaction. This small term will be discussed later. For the time being we ignore spin entirely.

We now express the Hamiltonian and the Schrödinger equation in the "molecular" coordinate shown below.

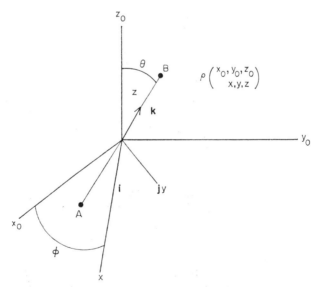

Nuclear coordinates $\mathbf{X}_A, \mathbf{X}_B; \mathbf{X}_B - \mathbf{X}_A = \boldsymbol{\rho}$
Electron coordinates x_0, y_0, z_0 in laboratory frame
  x, y, z in $\mathbf{i}, \mathbf{j}, \mathbf{k}$ frame
The vector $\mathbf{i}$ (x axis) lies along the 'line of nodes' established by AB.

The coordinate transformation is

$$x_0 = x \cos \phi - y \cos \theta \sin \phi - z \sin \theta \sin \phi$$
$$y_0 = x \sin \phi + y \cos \theta \cos \phi + z \sin \theta \cos \phi \qquad (1)$$
$$z_0 = \quad\quad\quad - y \sin \theta \quad\quad + z \cos \theta.$$

We note for future reference that this linear transformation between (x, y, z) and (x_0, y_0, z_0) can be written in terms of a rotation matrix $D_{ij}(\theta, \phi)$, or as a dyadic operator $\mathbf{x}_0 = \mathbf{D}(\theta, \phi) \cdot \mathbf{x}$. The required transformation on the Hamiltonian can be found by either (a) inserting the above coordinate transformation in the classical Lagrangian, then defining canonical momenta as $\mathbf{p} = \dfrac{\partial L}{\partial \dot{\mathbf{x}}}$ etc. or (b) writing out the Schrödinger operator in laboratory center of mass coordinates

$$-\frac{\hbar^2}{2\mu\rho^2} \left[\frac{\partial}{\partial \rho}\left(\rho^2 \frac{\partial}{\partial \rho}\right) + \frac{1}{\sin \theta}\frac{\partial}{\partial \theta}\sin \theta \frac{\partial}{\partial \theta} + \frac{1}{\sin^2 \theta}\frac{\partial^2}{\partial \phi^2} \right] - \frac{\hbar^2}{2m}\nabla_0^2 + V_{el} \qquad (2)$$

where

$$\mu = \frac{M_A M_B}{M_A + M_B}$$

and then transforming the partial derivatives by the above coordinate transformation. Either way we arrive at the following wave equation in terms of the internal electron coordinates

$$\left\{ -\frac{\hbar^2}{2m}\nabla^2 + V_{el}(\mathbf{x}, \rho) - B\left[\frac{\partial}{\partial \rho}\rho^2\frac{\partial}{\partial \rho} + \frac{1}{\sin \theta}\left(\frac{\partial}{\partial \theta} - iL_x\right)\sin \theta\left(\frac{\partial}{\partial \theta} - iL_x\right) \right.\right.$$
$$\left.\left. + \frac{1}{\sin^2 \theta}\left(\frac{\partial}{\partial \phi} - iL_y \sin \theta - iL_z \cos \theta\right)^2 \right]\right\} \Psi(\mathbf{x}, \rho, \theta, \phi) = E\Psi \qquad (3)$$

or

$$\{H_{ad} + BH'\}\Psi = E\Psi, \qquad B = \frac{\hbar^2}{2\mu\rho^2}.$$

Here $L_x = yp_z - zp_y$ etc. are the internal angular momenta of the electron. We note that the first two terms, H_{ad}, include all of the potential energy of the molecule and the electron's kinetic energy in the (moving) molecular coordinate system. The remaining terms, all with the factor B, include the kinetic energy of the heavy nuclei and the remaining kinetic energy terms of the electrons due to "relative motion" effects.

This equation is (nearly) exact. We now approximate by writing the wave function as a product of factors which are respectively functions of electron, vibration, and rotation coordinates as

$$\Psi \approx \Phi_{el}(\mathbf{x}, \rho) R(\rho) F(\theta\phi)$$

and in a first approximation we neglect all nuclear motion ($B \to 0$). Thus we set $H_{ad}(\mathbf{x}, \rho) \Phi_{nel} = E_{nad}(\rho) \Phi_{nel}$, the Schrödinger equation for fixed nuclei. In the adiabatic Hamiltonian H_{ad} the nuclear coordinate ρ enters only as a parameter. For many electrons it has the form

$$\left\{ -\frac{\hbar^2}{2m} \sum_i \nabla_i^2 + \sum_{i \neq j} \left(\frac{e^2}{r_{ij}} - \frac{Z_A e^2}{r_{iA}} - \frac{Z_B e^2}{r_{iB}} \right) + \frac{Z_A Z_B e^2}{\rho} \right\} \Phi_n(\mathbf{x}_i, \rho)$$
$$= E_n(\rho) \Phi_n(\bar{x}_j, \rho). \quad (4)$$

We consider this equation as solved. A typical solution is an eigenfunction of L_Z, i.e., $L_Z \Phi_{n\Lambda} = \hbar \Lambda \Phi_{n\Lambda}$, but not of L^2. We substitute for $H_{ad}\Phi_n$ in the complete wave equation (3) and average over x (i.e., $\langle \Phi | \rangle$). This gives a Schrödinger equation for the nuclear motion in an effective potential $E_{nad}(\rho)$. The resulting differential equation separates into rotation and vibration eigenvalue equations

$$\left\{ \frac{1}{\sin \theta} \frac{\partial}{\partial \theta} \sin \theta \frac{\partial}{\partial \theta} + \frac{1}{\sin^2 \theta} \left(\frac{\partial}{\partial \phi} - i\Lambda \cos \theta \right)^2 + \Gamma \right\} F(\theta, \phi) = 0$$

$$\left\{ B \frac{\partial}{\partial \rho} \rho^2 \frac{\partial}{\partial \rho} - E_{nad}(\rho) + B\Gamma - \bar{u}(\rho) + E \right\} R(\rho) = 0$$

(5)

where Γ is a separation parameter and

$$\bar{u}(\rho) = B \langle \Phi | L_x^2 + L_y^2 | \Phi \rangle.$$

We note that the average values of L_x and L_y are zero. The solutions of these rotation and vibration equations will be discussed later.

II. PROPERTIES OF ELECTRONIC STATES

2.1. H_2^+—single electron

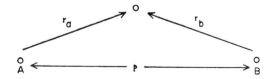

The adiabatic Hamiltonian is

$$H_{ad} = \frac{p^2}{2m} - \frac{e^2}{r_a} - \frac{e^2}{r_b} + \frac{e^2}{\rho}$$

(the last term is an additive constant in H_{ad}). The solution is of the form $\Phi = f(r_a, r_b)\, e^{i\Lambda\Psi}$. $\Lambda = 0, \pm 1, \pm 2 \ldots$ for $\sigma, \pi, \delta \ldots$ states. Although this Schrödinger equation can be solved exactly, we discuss approximate solutions here.

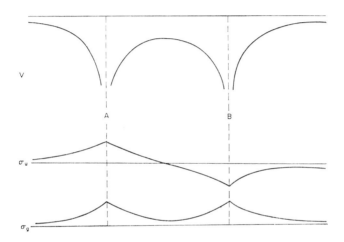

For large separation $\rho \gg a_0$ (Bohr radius) we have approximately $H + H^+$ i.e., $\Psi_{1s}(r_a)$ or $\Psi_{1s}(r_b)$ in the ground state. These are degenerate and form symmetric and antisymmetric combinations of the form

$$U_{\sigma u} = N^{-\frac{1}{2}}(e^{-\alpha r_a} + e^{-\alpha r_b})$$

$$U_{\sigma g} = N^{-\frac{1}{2}}(e^{-\alpha r_a} - e^{-\alpha r_b})$$

where N is the normalization constant in each case. With respect to the nuclei and the potential field V these functions have shapes and relative energies as indicated. Note the inversion symmetry character g or u with respect to the center. This is characteristic of all *homonuclear* electron states. The energy of σ_g is lower, and when e^2/ρ is added only σ_g is a bound state.

Clearly each asymptotic (H + H$^+$) state gives two molecular states, g and u. From atomic p states, $m_l = 0, \pm 1$, we get σ and $\pm \pi$ states. The $2p\sigma$ states combine with the $2s\sigma$ states (Stark effect) to give four states of various energies. The $\pm 2p\pi$ states remain degenerate.

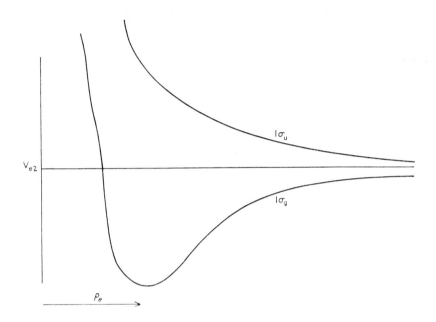

2.2. H_2—two electrons

In this case we must recall electron spin and the Pauli principle. A systematic procedure is to employ an antisymmetrized product (determinant) of individual electron orbitals (wave functions) with spins. A variation of the energy with respect to these orbitals leads to the Hartree–Fock equations for the molecular wave functions. The previous H_2^+ wave functions $U_{\sigma g}$ and $U_{\sigma u}$ are examples of molecular orbitals. Some possible H_2 wave functions, including spin, are:

$$U_1(^1\Sigma_g) = U_{\sigma g}(1)U_{\sigma g}(2)\frac{\alpha(1)\beta(2) - \alpha(2)\beta(1)}{\sqrt{2}} = U_{\sigma g}(1)U_{\sigma g}(2)\chi_0^{\;0}$$

$$U_2(^1\Sigma_g) = U_{\sigma u}(1)U_{\sigma u}(2)\chi_0^{\;0}$$

(these two $^1\Sigma_g$ functions may be combined in forming a state, as with Heitler–London).

Also:
$$U_1(^1\Sigma_u) = \frac{1}{N^{\frac{1}{2}}}[U_{\sigma g}(1)U_{\sigma u}(2) + U_{\sigma g}(2)U_{\sigma g}(1)]\chi_0^0$$

$$U_2(^1\Sigma_u) = \frac{1}{N^{\frac{1}{2}}}[U_{\sigma g}(1)U_{\sigma u}(2) - U_{\sigma g}(1)U_{\sigma u}(2)]\chi_1^{M_s}.$$

Note that the total wave function has g or u inversion symmetry. Also, for Π states we can have

$$U(^1\Pi_g) = \frac{1}{N^{\frac{1}{2}}}[U_{\sigma g}(1)U_{\pi g}(2) + U_{\sigma g}(2)U_{\pi g}(1)]\chi_0^0$$

which is two-fold degenerate. Most diatomic molecules have $^1\Sigma$ ground states. Some important exceptions are:

$$O_2(^3\Sigma_g), \quad NO(^2\Pi), \quad OH(^2\Pi).$$

2.3. Symmetry with respect to interchange of nuclei in homonuclear molecules.

The Hamiltonian is invariant to this operation, which results in $\pm\Phi_{el}$ if the state is nondegenerate. Referring to the diagram of coordinate axes, interchanging nuclei A and B has the effect

$$x \to -x$$
$$y \to +y$$
$$z \to -z$$

for each electron. Also $\phi \to \pi + \phi$ and $\theta \to \pi - \theta$, which does not affect Φ_{el}. Note that this operation is *not* the same as the g/u inversion, which is $x \to -x, y \to -y, z \to -z$. Indeed if we add the operation of reflection with respect to the xz plane ($y \to -y$), we find the relationship: Exchange of Nuclei × Reflection (y) ≡ Inversion. We denote the reflection property of the wave function by $\pm$. From this it follows that the combinations $(g+)$ or $(u-)$ are symmetrical (SN), and $(g-)$ or $(u+)$ are antisymmetrical (AN), with respect to exchange of nuclei.

2.4 Further remarks on symmetry of electronic states

Individual σ molecular orbitals all have $+$ reflection symmetry. Thus there is a direct relationship in this case $g \to SN$, $u \to AN$. Individual $\pm\pi$ states have the property on reflection that $\pm\pi \to \mp\pi$, since the reflection reverses the azimuthal angle Ψ, and $e^{i\Psi} \to e^{-i\Psi}$. These states are degenerate in the approximation employed to this point, however, and linear combinations of these states may be formed. We shall see later that higher order rotational interactions remove the π degeneracy, and the eigenstates are just the linear combinations with $\pm$ symmetry. Thus, consider the following interesting two-electron wave function which approximately describes the ground state of O_2

$$U(^3\Sigma_g^-) = \frac{1}{N^{\frac{1}{2}}}[U_{+\pi g}(1)U_{-\pi g}(2) - U_{+\pi g}(2)U_{-\pi g}(1)]\chi_1^{M_s}.$$

For isotopic substitutions into homonuclear molecules, e.g., *HD*, the electron wave-function keeps the same symmetry properties g/u, as are possessed by the exact homonuclear case. The consequences for the molecular state of the Bose or Fermi statistics of the nuclei, however, do not apply in the isotopic case. Note further, that the general heteronuclear diatomic molecule has no inversion symmetry g/u, but does possess the $\pm$ reflection symmetry. Thus the ground state of HCl is $^1\Sigma^+$ etc.

III. ROTATION AND VIBRATION

Referring to the eqns (5), the eigenfunctions of the first (rotation) equation have the form $F_{\Lambda M}^J(\theta, \phi) = T_{\Lambda M}^J(\theta)\, e^{iM\phi}$. Here J is the quantum number of total angular momentum and M is its z_0 (laboratory) component. The corresponding eigenvalue is $\Gamma_{J\Lambda} = J(J+1) - \Lambda^2$. For Σ states $F_{0M}^J = Y_J^M(\theta, \phi)$. Electron spin has been neglected in deriving these equations. In the case of strong spin–orbit coupling the quantum number Λ is replaced by Ω, the sum of orbital and spin z components. These matters will be discussed later.

The second (vibration) equation is the one-dimensional Schrödinger equation which determines the vibration wave functions and energies. The small term $\bar{u}(\rho) = B\langle \Phi | L_x^2 + L_y^2 | \Phi \rangle$ is usually considered absorbed into the "effective" potential $E_{ad}(\rho)$. The term $B(\rho)\Gamma_{J\Lambda}$ represents the centrifugal potential. If we give ρ its equilibrium value ρ_e in a bound state, this term becomes the rigid rotator energy

$$E_R = \frac{\hbar^2}{2\mu\rho_e^2}[J(J+1) - \Lambda^2].$$

The vibration equation then takes the form

$$-\frac{\hbar^2}{2\mu}\frac{d^2 P_v}{d\rho^2} + V_n(\rho)P_v(\rho) = [E - E_n - E_R]P_v(\rho) = E_v P_v(\rho) \qquad (6)$$

with $\rho R(\rho) = P(\rho)$. This equation gives the vibrational states in the effective potential $E_{n_{ad}}(\rho)$ for each electronic state (n). In the approximation that $V_n(\rho) \equiv \frac{1}{2}k(\rho - \rho_e)^2$, expanded about its equilibrium position, the vibrations are harmonic, and $E_v = \hbar\omega_e(v + \frac{1}{2})$; $\omega_e^2 = k/\mu$. Thus in a first approximation the vibration and rotation energy in any electronic state is given by

$$E_{vJ\Lambda} = \hbar\omega_e(v + \tfrac{1}{2}) + B[J(J + 1) - \Lambda^2]$$

$$v = 0, 1, 2, \ldots \quad J = \Lambda, \Lambda + 1, \Lambda + 2 \ldots .$$

The pattern of levels in two electronic states is shown in the diagram. The effect of the centrifugal term $B(\rho)\Gamma$ is the rotational distortion of the states, corresponding to the larger moment of inertia in the higher J states.

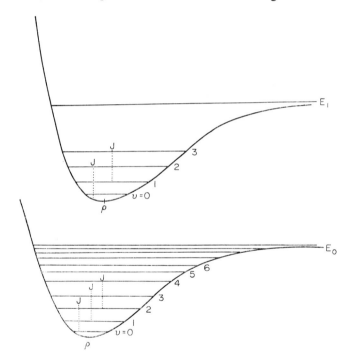

IV. SYMMETRY AND STATISTICS

The total wave function must be antisymmetric with respect to interchange of the coordinates of any pair of electrons, and symmetric or antisymmetric for interchange of identical nuclei, according as the nuclear spin is integer or half integer. For the electrons it suffices to construct the electron wave function $\Phi_{n\Lambda}(\mathbf{x}, \rho)$ with the proper antisymmetric character, e.g., determinantal wave functions, much as in atomic structure. The nuclear symmetry problem, however, is more complicated. The (moving) coordinate system for the electrons is defined in terms of nuclear locations; hence coordinate transformations of the nuclei also affect the electron wave function. We have already seen that the electron states $u-$, $g+$, are SN, and that $u+$, $g-$ are AN. (For $\Lambda \neq 0$ the Λ degeneracy is raised and $\pm$ states are realized.)

For the explicit nuclear coordinate part of the wave function the vibrational state is of no account, since its coordinate ρ is always positive and is unaffected by interchange. The symmetry of the rotational wave functions is as follows. For $\Lambda = 0$ $F_{\Lambda M}^J \to Y_J^M(\theta, \phi)$, which has the well-known symmetry $(-1)^J$ to the transformation $\theta \to \pi - \theta$, $\phi \to \pi + \phi$ which corresponds to $A \rightleftarrows B$. The symmetry character of the total molecular wave function to interchange of the nuclei in Σ states is accordingly found by combining the rotational factor $(-1)^J$ with the above AN or SN character of the electronic wave function. Taken together with the possible nuclear spin states and the Boson or Fermion character of the nuclei, these properties determine the statistical weights of the rotational levels. Thus for a $g+$ state with $I = 0$ only $J = 0, 2, 4$, are populated. (In the oxygen ground state $^3\Sigma_g^-$ accordingly, only *odd* values of the molecular rotational quantum number are found.) For $I = \frac{1}{2}$ the total nuclear spin values 0, 1 are respectively antisymmetric and symmetric with respect to interchange of spin coordinates. For Σ_g^+, then, the even J values are populated by total spin zero, and the odd J values by spin one, as shown in the accompanying diagram. The statistical weights of the states are given on the right side of the diagram as the product of the factor for nuclear spin and the $(2J + 1)$ rotation weight factor. For $I = 1$ the total spin states 0, 2 are symmetric and the spin 1 is antisymmetric. Thus spin states 0, 2, are both found in even J states and the spin state 1 in the odd J states. Examples are the ground states of D_2 and N_2.

J	$I = 0$	$I = \frac{1}{2}$	$I = 1$
s 4	——— 1×9	——— 1×9	——— 6×9
a 3	- - - - - -	——— 3×7	——— 3×7
s 2	——— 1×5	——— 1×5	——— 6×5
a 1	- - - - - -	——— 3×3	——— 3×3
s 0	——— 1×1	——— 1×1	——— 6×1

For electronic states $\Lambda \neq 0$, e.g. Π, Δ states, the 2-fold "Λ-doubling" degeneracy is raised, as we have pointed out, in higher approximations. Hence for each value of J there are two states, with $\pm$ reflection symmetry. The discussion above for $g+$, $u-$, etc. holds for each of these nearly degenerate states, and the allowed states and their statistical weights follow accordingly (see Herzberg[1]).

4.1. Parity

In this operation the effect on the total molecular wave function of inversion of *all* particles through the origin is examined. All nondegenerate states must have a definite (± 1) parity quantum number. This operation may be regarded as taking place in two steps; (1) the nuclei are inverted through the origin, electron coordinates fixed in space. For homonuclear molecules this is exactly the nuclear interchange operation discussed above, and it has a definite symmetry. (2) the electrons are inverted through the origin. For the homonuclear case this is the operation associated with g/u symmetry. Thus in this case the SN/AN symmetry, with the g/u symmetry, determines the parity. For the heteronuclear case the nuclear inversion produces in the rotational wave function a definite symmetry character [$(-1)^J$ for Σ states], but within the electronic wave function the transformation $x \to -x$, $x \to +y$, $z \to -z$ has no symmetry. The subsequent electron inversion $x \to -x$, $y \to -y$, $z \to -z$ then has the net effect of the $y \to -y(\pm)$ operator. Thus the common $^1\Sigma^+$ states all have parity $(-1)^J$.

In Π states the Λ-doubling pairs of states have opposite parity, which alternates in sign with increasing J (see Herzberg,[1] p. 261).

V. ELECTRIC DIPOLE MATRIX ELEMENT—SELECTION RULES

Radiation absorption and emission will be dealt with later on. Here we simply examine the kinds of electric dipole transitions which may take place and the typical spectra which result, making use of the molecular theory as developed thus far. The molecular wave function has the form

$$\Psi = \Phi_{n\Lambda}(\mathbf{x}_i, \rho) \frac{1}{\rho} P_{n\Lambda v}(\rho) F_M^J(\theta, \phi)$$

where the $\mathbf{x}_i$ are the "internal" electron coordinates and ρ is the internuclear separation. The electric dipole operator is

$$\mathbf{P} = -\sum_i e\, \mathbf{x}_{0_i} + \sum_N Z_n e\, \mathbf{R}_N = \mathbf{P}_e + \mathbf{P}_N \tag{7}$$

where the $\mathbf{x}_{0i}$ are electron *laboratory* coordinates. The general electric dipole matrix element has the form $\langle n\Lambda v JM|\mathbf{P}|n'\Lambda'v'J'M'\rangle$. Because P is a polar vector operator it has matrix elements only between states of opposite parity, and it has the further selection rules $J' = J$, $J \pm 1$ and $M' = M$, $M \pm 1$.

5.1. Electronic transitions

Here we consider transitions between *electronic* states, Δn, $\Delta\Lambda \neq 0$. In such transitions the nuclear part of the dipole operator $\mathbf{P}_N$ plays no role, as it does not depend on the electron coordinates. The electron operator, as we have seen, is of the form

$$\mathbf{P}_e = -\sum_i e\,\mathbf{r}_{0_i} = \mathbf{D}(\theta,\phi)\cdot\left\{-\sum_i e\,\mathbf{x}_i\right\}. \tag{8}$$

The dipole matrix element accordingly factors as

$$-e\langle J\Lambda M|\mathbf{D}|J'\Lambda'M'\rangle \times \langle n\Lambda v|\Sigma \mathbf{x}_i|n'\Lambda'v'\rangle.$$

The first factor gives the rotation selection rules, and the second those for vibration and electron quantum numbers. The second factor may be written out as an integral

$\langle n\Lambda v|\mathbf{P}_e(\text{int})|n'\Lambda'v'\rangle$
$$= -e\int d\rho\, P_{n'\Lambda'v'}(\rho)\, P_{n\Lambda v'}(\rho)\cdot\int \Phi_{n'\Lambda'}(\mathbf{x}_i,\rho)(\Sigma\mathbf{x}_i)\,\Phi_{n\Lambda}(\mathbf{x}_i,\rho)\,d\mathbf{x}_i.$$

In the approximation of the Franck–Condon Principle the second integral, over the electron coordinates, is considered independent of ρ. Then

$$\langle n\Lambda v|\mathbf{P}_e(\text{int})|n'\Lambda'v'\rangle = \langle v|v'\rangle\cdot\mathbf{R}_{n\Lambda n'\Lambda'}. \tag{9}$$

The factor $\langle v|v'\rangle$ is just the "overlap" integral between vibrational wave functions in *different* electronic states. From an initial state v there is in general a distribution in intensity to several final states v'. A simple physical interpretation of the Franck–Condon Principle can be given semi-classically. In the two electronic states concerned the molecule vibrates in somewhat different potentials. The electronic transition is conceived as taking place "instantaneously" on the time scale of vibrations. During the transition the heavy nuclei do not change their positions or their momenta. The new states of vibration are obtained, as shown in the diagram, by drawing vertical lines upward, ending at levels with unchanged kinetic energy. In particular, the most probable initial position for the oscillation is at the end points (velocity zero), and these transitions are the most probable. These ideas are contained formally in the adiabatic approximation, together with the factor $|\langle v|v'\rangle|^2$ for the vibrational intensity distribution. The overlap integral approximately conserves momentum between the vibrational states.

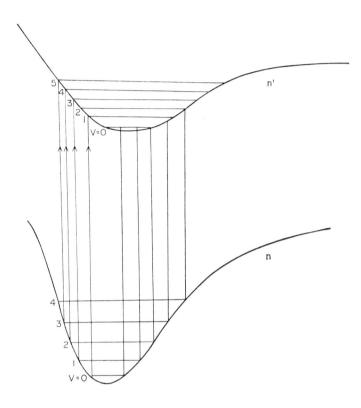

Franck Condon principle

The purely electronic part of the dipole matrix element $\mathbf{R}_{n\Lambda n'\Lambda'}$ has the following properties which give rise to selection rules. We have seen that for $^1\Sigma^\pm$ states the parity is $\pm(-1)^J$. Hence in a transition $n\Sigma^+ \to n'\Sigma^+$ the parity rule and the $\Delta J = 0 \pm 1$ rule together give $\Delta J = \pm 1$ *only*. Looking into the matrix element in more detail we see that only $\int \Phi_{n'\Sigma}^* z_i \Phi_{n\Sigma} dx_i \neq 0$, i.e. only the dipole moment *along* the molecular axis contributes to this transition. Also we see that $n\Sigma^+ \nleftrightarrow n'\Sigma^-$ at all, as all components vanish because of the axial reflection symmetry. In the homonuclear case, besides the above selection rules, we require $g \to u$ only, from the "internal parity" of the g/u symmetry. (This rule holds also for different isotopes, e.g., HD).

For identical nuclei we have the strict rule $SN \to SN$, $AN \to AN$. An allowed transition in this case would be $\Sigma_g^+ \to \Sigma_u^+$ with $\Delta J = \pm 1$.

For $\Lambda \neq 0$ we have seen that the nearly degenerate $\pm \Lambda$ doubling states have opposite parity. The vector property of $\mathbf{P}_e(\text{int})$ in the "internal" matrix element $\mathbf{R}_{n\Lambda n'\Lambda'}$ requires $\Delta \Lambda = 0 \pm 1$. As an example we show the transition $^1\Sigma^+ \to {}^1\Pi$ in a heteronuclear molecule. Here parity can be satisfied with $\Delta J = 0, \pm 1$. A complete rotational band is made up of an R branch $\Delta J = +1$, Q branch $\Delta J = 0$, P branch $\Delta J = -1$ as shown.

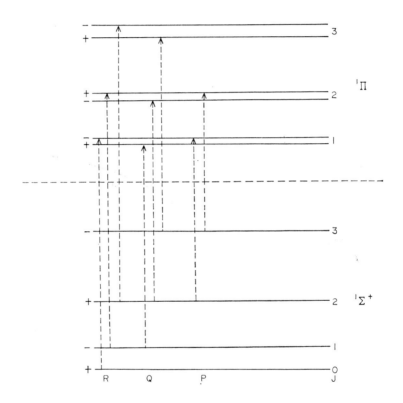

VI. VIBRATION–ROTATION AND ROTATION SPECTRA
($\Delta n = \Delta \Lambda = 0$)

Now the nuclei do contribute to the dipole matrix element as

$$\langle n\Lambda vJM|\mathbf{P}|n\Lambda v'J'M'\rangle = \langle n\Lambda v| - \Sigma e\mathbf{x}_i|n\Lambda v'\rangle\langle JM\Lambda|\mathbf{D}|J'M'\Lambda'\rangle$$
$$+ \langle n\Lambda v|Z_A\rho_A + Z_B\rho_B|n\Lambda v'\rangle\langle JM\Lambda|\mathbf{D}|J'M'\Lambda'\rangle \quad (10)$$

After averaging over the electronic coordinates we are left with

$$\int P_{n\Lambda v'}(\rho)\{\langle n\Lambda|\mathbf{P}_e(\text{int}) + \mathbf{P}_N(\text{int})|n\Lambda\rangle\} P_{n\Lambda v} \, d\rho \cdot \langle J\Lambda M|\mathbf{D}|J'\Lambda'M'\rangle.$$

The bracketed term in the first factor is the total dipole moment of the molecule $\mu(\rho)$ in the given electronic state $n\Lambda$, for any value of ρ. We expand this about the equilibrium separation ρ_e

$$\mu(\rho) = \mu(\rho_e) + \frac{\partial \mu}{\partial \rho}(\rho - \rho_e) + \frac{1}{2}\frac{\partial^2 \mu}{\partial \rho^2}(\rho - \rho_e)^2 + \ldots. \tag{11}$$

The first term is a constant, the permanent dipole moment of the molecule, the second term represents the "vibrating" dipole moment, the third term is usually neglected. (We note that for homonuclear molecules $\mu(\rho) = 0$, there is no permanent moment and no vibrating moment.)

Clearly the permanent moment has only matrix elements $\Delta v = 0$, and it gives rise to the *pure rotation spectrum*. From parity we have the selection rule $\Delta J = \pm 1$ in a Σ state, indeed the matrix element is simply of the form

$$\left\langle JM \left| \begin{matrix} \mu_e \cos\theta \\ \mu_e \sin\theta \cos\phi \\ \mu_e \sin\theta \sin\phi \end{matrix} \right| J'M' \right\rangle \quad \text{so that} \quad \Delta M = 0 \pm 1.$$

Since the rotation levels are given approximately by

$$E_J = BJ(J+1) - DJ^2(J+1)^2$$

(the added second term represents centrifugal distortion) the transition frequencies for $J+1 \to J$ are given by $\hbar\omega = 2B(J+1) - 4D(J+1)^3$. A typical pure rotation spectrum is shown in the diagram.

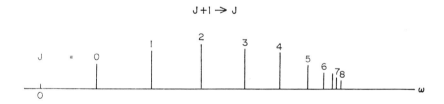

The frequency scale is determined by

$$B = \frac{\hbar^2}{2I}, \qquad I = \mu\rho_e^2.$$

For H_2 $B \cong 61 \text{ cm}^{-1}$, in the far infrared, but for CO $B \cong 2 \text{ cm}^{-1}$ in the microwave region. We note that isotopic substitution changes the moment of inertia by a known amount, and shifts the rotation spectrum accordingly.

For Π states with Λ-doubling, besides the pure rotation spectrum $\Delta J = \pm 1$, $+ \leftrightarrow -$, there is also the interesting low frequency transition $\Delta J = 0$ between the nearly degenerate components of opposite parity.

Simultaneous changes in v and J give the *vibration–rotation spectrum*. The matrix element has the typical form (z_0 lab component)

$$\int P_v^*(\rho) s \, P_{v'}(\rho) \, ds \langle J \Lambda M | \cos \theta | J' \Lambda' M' \rangle,$$

where $s = \rho - \rho_e$ the displacement from equilibrium. To a fairly good approximation the vibration about equilibrium is harmonic. Then the first integral gives the well-known result $v' = v \pm 1$ for the oscillator. Anharmonicity makes weak "overtone" transitions possible. For Σ states we have also $\Delta J = \pm 1$. For $\Delta v = 1$, $\Delta J = \pm 1$ the emission or absorption gives an infrared band for which the following approximate formula holds.

$$\omega = \omega_c + 2B(J + 1) \quad R \text{ branch } J \to J + 1$$
$$\omega = \omega_c - 2BJ \quad P \text{ branch } J \to J - 1$$

For Π states there is also allowed an additional Q branch with $\Delta J = 0$, transitions between vibration states and between the nearly degenerate Λ-doubled states of opposite parity. In first approximation the whole Q branch intensity lies at ω_e.

VII. SPIN–ORBIT INTERACTION AND HIGHER ORDER ROTATIONAL EFFECTS

7.1. Spin–orbit interaction

We have neglected thus far all effects of electron spin; indeed our results actually describe only singlet states. The electrons, to a good approximation, form a total spin S, a good quantum number. Just as with atoms, the principal magnetic spin interaction is with the average electric field through which it moves, $SO = gu_0[\mathscr{E} \times \mathbf{v}/c] \cdot \mathbf{S}$. Here the molecular electric field has only axial symmetry $\mathscr{E} = \mathscr{E}_\perp + \mathscr{E}_z$, so that after averaging over the electron distribution the spin–orbit interaction has the form

$$SO = F_\perp \mathbf{L}_z \cdot \mathbf{S} + F_z \mathbf{L}_\perp \cdot \mathbf{S}.$$

Only $L_z = \Lambda$ has a non-zero expectation value, so we expect

$$SO = F_\perp \mathbf{\Lambda} \cdot \mathbf{S} = F_\perp \Lambda \Sigma$$

as the eigenvalue, where Σ is the component of S on the z-axis and has the usual possible values between $-S$ and $+S$ through integral steps. This, in fact, represents a possible vector coupling (Hund case a), but we will see below that the competing rotational effects give rise to more complicated couplings.

7.2. Rotation-electron coupling

We discuss here certain non-adiabatic rotation effects on the electron states. We neglect the non-rigid character of the molecule and consider the nuclei as fixed in their equilibrium positions. We begin with the general polyatomic case. The molecule thus defines a rigid rotator with three principal axes and corresponding moments of inertia. We take these axes as the coordinate axes fixed in the molecule. Using operators instead of the previous differential equation formalism, the kinetic energy of rotation of this rigid rotator is

$$E_R = \frac{\hbar^2}{2I_A} R_x^2 + \frac{\hbar^2}{2I_B} R_y^2 + \frac{\hbar^2}{2I_C} R_z^2 \tag{12}$$

where $\mathbf{R}$ is the angular momentum of nuclear rotation. Now if $\mathbf{J}$ is the *total* angular momentum of the molecule (exclusive of nuclear spins), then $\mathbf{J} = \mathbf{R} + \mathbf{P}$, where P is the spin and orbital angular momentum of the electrons. Hence

$$E_R = \frac{\hbar^2}{2I_A}[J_x - P_x]^2 + \frac{\hbar^2}{2I_B}[J_y - P_y]^2 + \frac{\hbar^2}{2I_C}[J_z - P_z]^2. \tag{13}$$

The cross terms, $J_x P_x$, etc., produce several important molecular effects. The diatomic case is obtained from the above by setting $I_A = I_B$ and dropping the last term. Then

$$E_R = B(J_x - P_x)^2 - B(J_y - P_y)^2.$$

In general these problems can be solved, indeed reduced to well-known atomic structure problems, by the following 'trick' due to Van Vleck.[5]

The laboratory components of total angular momentum $J_{ox} J_{oy} J_{oz}$ have the usual commutation relations $J_{ox} J_{oy} - J_{oy} J_{ox} = iJ_{oz}$, etc. Likewise the internal electron angular momenta $\mathbf{P}_x = \mathbf{L}_x + \mathbf{S}_x$ obey such relations, as they were derived in coordinates with fixed nuclei. When the total angular momentum is projected onto the internal 'body' axes, however, as J_x, J_y, J_z, we find reversed commutators $J_x J_y - J_y J_x = -iJ_z$, etc. The essential reason for this is that in the relations of components between laboratory and body coordinate systems $J_x = \lambda_{11} J_{ox} + \lambda_{12} J_{oy} + \lambda_{13oz}$, etc., the direction cosines λ_{ik} depend on the orientation of the nuclei, hence are operators. Direct calculation verifies the above reversed commutator relations. (For diatomic molecules the xy axes are not strictly determined by the nuclear positions. This slightly troublesome point was discussed by Hougen.[6]) All the conventional angular momentum matrix elements and vector additions are based on the standard commutators. In order to treat the various internal angular momenta in the conventional way, we could substitute $J_x' = -J_x$, etc., restoring the commutation rules. In fact we will reverse $\mathbf{P}$, setting $\hat{P}_x = -P_x$, etc. The rotation energy then is $H_R = B[(J_x + \hat{P}_x)^2 + (J_y + \hat{P}_y)^2]$. Many molecular problems now become equivalent to well-known atomic problems, and their solutions can be written down immediately. Below we will work out examples of spin decoupling, leading to Hund's case a and case b, and also Λ-doubling.

7.3. Spin decoupling

We take as the effective hamiltonian for perturbation treatment

$$H = B[(J_x - S_x)^2 + (J_y - S_y)^2] + f\Lambda S_z \tag{14}$$

Note that we have neglected all orbital angular momentum components other than $L_z = \Lambda$. As above, we set $\hat{S}_x = -S_x$, $\hat{S}_y = -S_y$ and introduce $S(S+1) = \hat{S}_x^2 + \hat{S}_y^2 + \hat{S}_z^2$ and $J_z = \Lambda - \hat{S}_z$,

$$H \Rightarrow B[J(J+1) - \Lambda^2] + BS(S+1) - f\Lambda\hat{S}_z + 2B[J_x\hat{S}_x + J_y\hat{S}_y + J_z\hat{S}_z]$$

The first two terms are constant for fixed J, Λ, S, and can be disregarded. The remaining terms are equivalent to the Hamiltonian for an atomic S state in a magnetic field, and include a nuclear hyperfine interaction

$$H_{at} = 2\mu_0 H S_z + \gamma \mathbf{I} \cdot \mathbf{S} \tag{15}$$

The correspondences are as shown

$$-f\Lambda \to 2\mu_0 H$$
$$2B \to \gamma$$
$$\hat{S} \to S$$
$$J \to I$$

Solutions of the atomic problem exist from "weak field" to "strong field", depending on the ratio $\mu_0 H/\gamma$, in which the angular momentum vectors are either coupled to a resultant $\mathbf{S} + \mathbf{I} = \mathbf{F}$ (weak field) or precess independently about z with M_S and M_I as quantum numbers (strong field). The corresponding molecule vectors couple in exactly the same way

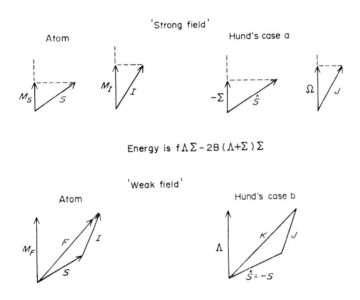

Energy is $f\Lambda\Sigma - 2B(\Lambda+\Sigma)\Sigma$

Note that $\mathbf{K} = \hat{\mathbf{S}} + \mathbf{J}$ or $\mathbf{J} = \mathbf{K} + \mathbf{S}$. $\mathbf{K}$ is the molecular angular momentum exclusive of spin, and is a good quantum number in case b. The energy in case b is

$$B[K(K+1) - J(J+1) - S(S+1)]$$
$$+ f\Lambda^2 \frac{[J(J+1) - K(K+1) - S(S+1)]}{2K(K+1)}$$

in exact analogy to the atomic case in weak field. As in the atomic case, all off-diagonal matrix elements are known, and the energy states can be worked out for intermediate cases between Hund's case a and case b, depending on the ratio of the spin–orbit coupling to the rotation constant B. An example of such an energy level diagram is shown for the lower rotation states of a $^2\Pi$ state. In Hund's case a we have $^2\Pi_{3/2}$ and $^2\Pi_{1/2}$ states, where $\Omega = \Lambda + \Sigma$ is 3/2, 1/2, and in Hund's case b we have rotational states $K = 1, 2, 3$ split into doublets $J = K \pm 1/2$.

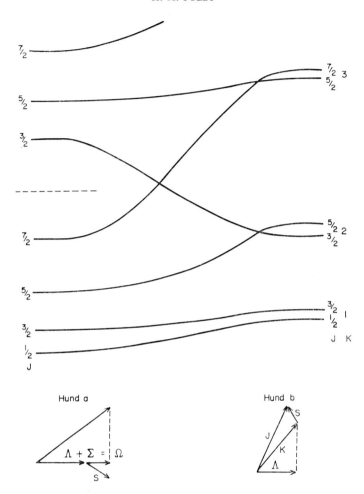

7.4. Λ-doubling

For simplicity we consider only singlet states, e.g., $^1\Pi$. Now

$$H_R = B[(J_x - L_x)^2 + (J_y - L_y)^2].$$

The operators L_x, L_y have matrix elements only between different electronic states, so in first approximation the rotational energy is given by $B[J_x^2 + J_y^2] = B[J(J+1) - \Lambda^2]$. We seek the terms which split the degenerate $\pm \Lambda$ states. Now the term $B(L_x^2 + L_y^2)$ gives only an added constant to all states; the interesting term is

$$H_{\text{pert}} = -2B[J_x L_x + J_y L_y] = 2B[J_x \hat{L}_x + J_y \hat{L}_y] \tag{16}$$

conveniently written as $H_{pert} = B[J_+\hat{L}_- + J_-\hat{L}_+]$. Here $L_z = \Lambda$ is a good quantum number, and $L_\pm$ has a strict selection rule $\Delta L_z = \pm 1$. Hence the expectation value of H_{pert} vanishes, and it has matrix elements from a Π state only to Σ or Δ states. Thus, in removing the 2-fold Λ degeneracy this perturbation gives effects only in second order, of the form

$$\frac{\langle \pm\Pi | H | \Sigma \rangle \langle \Sigma | H | \pm\Pi \rangle}{\Delta E_{\Sigma\Pi}}. \tag{17}$$

The diagonal values in this 2×2 matrix are the same, hence the effect of the off-diagonal elements $+\Pi \to -\Pi$, is to split the degenerate states into symmetric and antisymmetric combinations of the basis $\pm\Pi$ states. This produces the Λ-doubling phenomena referred to above repeatedly.

Molecular Λ-doubling corresponds exactly to the following rather special atomic problem. Consider an atom in a strong Stark field, so that the potential is cylindrical $V(\mathbf{r}_\perp, z)$. Then the effect of spin–orbit coupling $f\mathbf{L}\cdot\mathbf{S}$ in a state $M_J = 0$ in removing the 2-fold Stark degeneracy is the same as the Λ-doubling.

VIII. ROTATION OF POLYATOMIC MOLECULES

In the rotation of polyatomic molecules, as described in the previous section, we imagine x, y, z axes fixed in the molecule, and oriented along the principal axes, as determined in the equilibrium positions of the nuclei. Three angles are necessary to fix the orientations as shown in the figure:

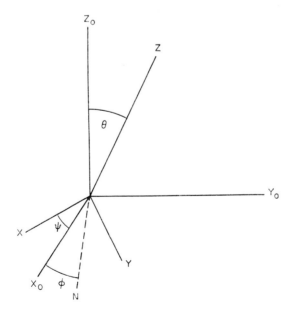

ϕ and ψ are respectively the angles made with the nodal line N by the x_0 and x axes. The energy of this rigid rotator is

$$H_R = \frac{J_x^2}{2I_x} + \frac{J_y^2}{2I_y} + \frac{J_z^2}{2I_z}. \qquad (18)$$

Any symmetry axis of the molecule is necessarily a principal axis. If any axis (z-axis) has 3-fold or higher symmetry, then two moments of inertia are equal $I_x = I_y = I_B$, and the Hamiltonian becomes

$$H_R = \frac{J^2}{2I_B} + J_z^2 \left(\frac{1}{2I_C} - \frac{1}{2I_B} \right) \qquad (19)$$

the symmetric rotator. Examples are CH_3CN and NH_3. (Note that H_2O has a 2-fold axis, and $I_x < I_y < I_z$, but $I_x + I_y = I_z$ because it is planar.) The energy eigenvalue is obviously

$$E_{JK} = \frac{J(J+1)\hbar^2}{2I_B} + \hbar^2 K^2 \left(\frac{1}{2I_C} - \frac{1}{2I_B} \right)$$

where $K = J_z$, and has values $-J \leqslant K \leqslant +J$. The wave function is similar to the diatomic case with $\Lambda \neq 0$

$$\Psi_{\text{Rot}} = F_{KM}^J(\theta) e^{iM\phi} e^{iK\psi}.$$

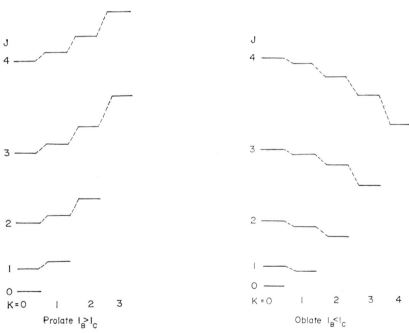

We note that $\pm K$ states are degenerate, corresponding to the two directions of rotation about the symmetry axis. Depending on the sign of $I_C - I_B$, we have the *prolate* or *oblate* cases, with corresponding energy levels, as shown. Levels of the same J are shown as connected. The linear molecule is a limit of the prolate case as $I_C \to 0$. Then all levels with $K \neq 0$ move upwards on the diagram and become inaccessible. The resulting rotation levels are then identical with those of a $^1\Sigma$ diatomic molecule.

Any nonplanar molecule has two non-equivalent configurations of the same energy which are achieved by inversion. They are effectively mirror images of one another. This situation is essentially different from the diatomic or linear molecule, in which the configuration obtained by inversion can also be obtained by an appropriate rotation. To distort the molecule adiabatically from one position to its inversion usually requires going over (or through) a potential maximum. Thus for a symmetric rotator there is an additional 2-fold degeneracy, so that there is a 4-fold degeneracy for given values of $J, |K|, M$.

We discuss here NH_3 as a typical symmetric rotator. Certain interesting features of the inversion symmetry will be pointed out. The x, y, z internal coordinates are chosen as follows. The three hydrogens lie in the xy plane and, initially, are regarded as distinguishable. The $+x$ axis is defined as passing through hydrogen 1, and the $+z$ axis is such that the hydrogens are counterclockwise viewed from $+z$. Note that the nitrogen atom does not automatically lie on the $+z$ axis; in fact, it has equilibrium positions on both $\pm z$. The rotational wave function $\Psi_R = F^J_{KM}(\theta)\, e^{iK\psi}\, e^{iM\phi}$ is thus independent of the location of N. On inversion of the molecule, $+z$ is unchanged in space, and the values of θ and ϕ are unchanged. However $\psi \to \psi + \pi$, introducing a factor $(-1)^K$ in ψ_R.

The vibrational motion of the N atom takes place in a double minimum potential. The atom may vibrate in either potential "hole", with a high frequency (950 cm^{-1}). Because of the finite height of the potential "hill", however, there is a certain low probability of "tunnelling" through. The resulting nearly degenerate states are odd and even functions $\phi_\pm(z)$. The two wave functions of the split lowest vibration state are shown in the diagram. In the case of the lowest state of NH_3 this splitting is 0·67 cm^{-1} or about 20,000 MHz. In higher vibrational states the splitting is larger, corresponding to the greater ease of tunnelling. The ϕ_- state always has the higher energy. An N atom initially in the ground vibration state, but with N definitely at $+z$ is therefore in a "packet" state formed from ϕ_+ and ϕ_-. The N atom then oscillates between the two positions at a frequency of about 20,000 MHz. (In more complex molecules the potential "hill" may be much higher and the "tunnelling lifetime" as long as 10^9 years, with interesting biological effects.)

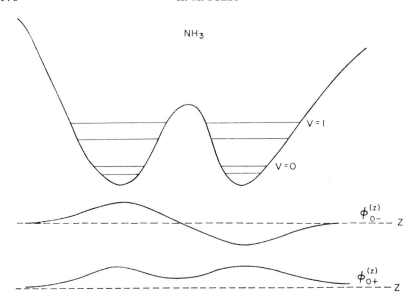

In a definite quantum state the wave function is a product of rotation and vibration functions. Thus in an inversion of the molecule the symmetry character is $\pm(-1)^K$. The lower rotational states of NH_3 are shown with $\pm$ parity signs

Each state with $K \neq 0$ is still 2-fold degenerate ($\pm K$). Actually the non-rigid character of the molecule (rotation–vibration interaction) acts to split these states into an additional fine structure.

The permanent dipole moment of NH_3 lies along the symmetry axis. Its components in the laboratory coordinates are independent of ψ; accordingly there is a selection rule $\Delta K = 0$, $\Delta J = 0 \pm 1$ in the rotation spectrum. The transitions $\Delta J = \pm 1$ lie in the far infrared at about 50 to 100 cm^{-1}, the transition $\Delta J = 0$, between the $\pm$ inversion levels, lies at about 2 cm^{-1} wavelength.

There are additional symmetry requirements due to the identity of the protons and the Exclusion Principle. We note that an interchange of the protons $2 \rightleftarrows 3$ reverses the z-axis. This transforms the rotation function Ψ_{JKM} to $(-1)^J \Psi_{JKM}$ and at the same time transforms the vibration function $\phi_\pm(z)$ to $\pm\phi_\pm$. Since no nuclear spin function can be antisymmetric with respect to all three pairs, for $K = 0$ this eliminates one of the inversion levels for each J (marked × in the above level diagram for NH_3). For $K \neq 0$ the $\pm K$ states must be combined with nuclear spin states to satisfy the Pauli Principle. (See Townes and Schawlow[4]).

8.1. Asymmetric rotators

Here $H_R = AJ_x^2 + BJ_y^2 + CJ_z^2$ with $A \geqslant B \geqslant C$. For $B = C$ we have a prolate rotator symmetric about x, for $A = B$ an oblate rotator symmetric about z. As we have seen, these two limiting cases of prolate and oblate rotators can be treated exactly. Now the matrix elements of the above asymmetric H_R can readily be found in either the prolate or oblate representation, and H_R can be diagonalized for any intermediate case. Thus in the oblate representation, with J^2 and $J_z = K$ diagonal, we have

$$\langle J\,K | J_z | J\,K \rangle = K$$
$$\langle J\,K | J_x + iJ_y | J\,K+1 \rangle = [J(J+1) - K(K+1)]^{\frac{1}{2}} \text{ etc.}$$

and accordingly for the Hamiltonian the matrix elements

$$\langle J\,K | H_R | J\,K \rangle = \frac{A+B}{2} J(J+1) + \left[C - \frac{A+B}{2}\right] K^2$$

$$\langle J\,K | H_R | J\,K+2 \rangle = \frac{A-B}{2} [J(J+1) - K(K+1)]^{\frac{1}{2}}$$
$$\times [J(J+1) - (K+1)(K+2)]^{\frac{1}{2}}.$$

For the nearly symmetric case $A = B + \delta$ we can obtain the corrected energies and wave functions by perturbation theory.

The general pattern of the lower rotational energy levels as B varies between fixed values of A and C is shown in the diagram. The left hand edge of the diagram shows the limiting prolate levels, the right side the oblate levels. A parameter

$$K = \frac{2B - A - C}{A - C}$$

may be considered to vary from -1 to $+1$ between prolate and oblate cases. The $\pm K$ degeneracy is lost in the intermediate case, indeed K is not a good quantum number, nor is any component of J, in the rotating frame, a constant of the motion. A convenient index of a level is the respective prolate and oblate values of K to which it goes in the limiting cases. Thus 2_{20}

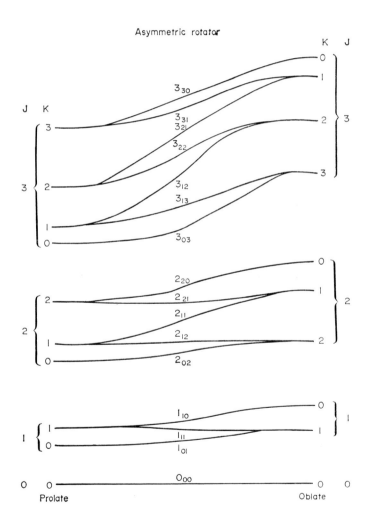

connects $J = 2$, $K = 2$ prolate with $J = 2$, $K = 0$ oblate. We note that there is no crossing of levels of a given J value. In the diagram states of different J values also do not overlap. This is the case for C not much less than A. In general, however, states of different J do overlap, as we already see, for example, in the symmetric prolate case for $A \gg C$. Thus, aside from the overall scale of energies the level pattern depends on the two parameters B/C and A/C. The diagram really shows the dependence of energy on B/C for a fixed value of A/C. Note that for planar molecules, e.g. H_2O, because of the relationship $I_x + I_y = I_z$, there is just one free parameter.

In the exact solution of the asymmetric rotator the only remaining symmetries are the reflection symmetry of the ellipsoid of inertia and the point group symmetries of the molecule. For non-planar molecules the two stereo isomeric states of opposite parity are virtually coincident in most cases, so that the $+ \leftrightarrow -$ parity selection rule plays no role. Selection rules do exist for particular molecules which depend on their point group symmetries. For example, the triangular molecule H_2O has a 2-fold axis of symmetry along which the permanent dipole must lie. With respect to 180° rotation about this axis all states have a $\pm$ symmetry character. Because the dipole lies along this axis only states with the same symmetry may combine optically. Asymmetric rotator spectra are complex, and show many accidental features. For H_2O the levels 6_{16} and 5_{23} lie only $1\cdot35\,\text{cm}^{-1}$ apart, and the dipole transition between them gives rise to a strong atmospheric absorption at this frequency, as well as an important interstellar line.

8.2. Normal vibrations

Vibration spectra of polyatomic molecules play little role in astrophysics, as the spectra lie in rather inaccessible regions of the infrared. They will be discussed only briefly here.

If we consider small displacements of the nuclei from their equilibrium positions, the energy of the molecule becomes, in first approximation, a quadratic function of these displacements, as

$$V = V_0 + \tfrac{1}{2} \sum_i^{3N-6} \sum_j \frac{\partial^2 V}{\partial q_i \, \partial q_j} q_i q_j. \tag{20}$$

The $3N - 6$ displacement coordinates must be chosen so as to conserve the linear and angular momenta of the molecule. The kinetic energy of the moving nuclei is also a quadratic form

$$KE = \tfrac{1}{2} \sum_i \sum_j a_{ij} \dot{q}_i \dot{q}_j. \tag{21}$$

The problem of diagonalizing the Hamiltonian is identical with the classical problem of finding the set of independent normal coordinates for such a system of coupled oscillators. A linear transformation $Q_n = \sum_i \lambda_{ni} q_i$ can be found, in which the vibration Hamiltonian takes the form

$$H_v = \sum_i \tfrac{1}{2}[P_i^2 + \omega_i^2 Q_i^2] \qquad (22)$$

a system of decoupled harmonic oscillators. The vibration wave function is the simple product

$$\Psi_v = \Psi_{v_1}(Q_1)\Psi_{v_2}(Q_2) \ldots \Psi_{3N-6}(Q_{3N-6})$$

and the energy is

$$E = \sum_i (v_i + \tfrac{1}{2})\hbar\omega_i.$$

These harmonic oscillator normal modes have the selection rule $\Delta v = \pm 1$. The presence of anharmonic terms in the potential, however, couples different normal modes and destroys their independence. The effect of this perturbation is to allow both overtone transitions $\Delta v = 2, 3$, and intercombinations in which two modes take part in a single transition.

As described previously the electric dipole moment of the molecule may be expanded as a sum of permanent moment components and terms proportional to the displacements from equilibrium. In terms of normal coordinates

$$\mu_x = \mu_x^0(R_{i0}) + \sum_j \frac{\partial \mu_x}{\partial Q_J} Q_J, \text{ etc.} \qquad (23)$$

The symmetries of the normal vibrations determine the nonvanishing dipole matrix elements. Thus for NH_3 only the component μ_z^0 of the permanent

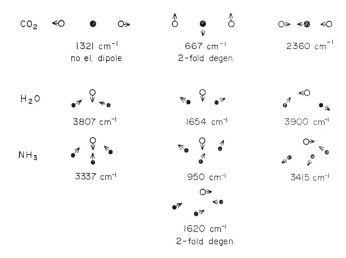

dipole exists, but some normal modes give perpendicular dipole components μ_x, μ_y. Diagram shows schematically the normal vibrations of CO_2, H_2O, NH_3, and their frequencies.

IX. MOLECULAR HYPERFINE STRUCTURE

Just as with atoms, the nuclear spin **I** may be coupled to the molecular angular momentum **J** to form a resultant angular momentum $\mathbf{F} = \mathbf{I} + \mathbf{J}$ of the system. The magnetic interaction arises from three terms:

(1) Orbital interaction $\quad H_0 = \dfrac{2\mu_0 \mu_N}{Ir^3} \mathbf{I} \cdot \mathbf{L} = \dfrac{2\mu_0 \mu_N}{Ir^3} \Lambda \mathbf{I} \cdot \hat{\mathbf{z}}$

(2) Tensor spin interaction $H_{ts} = \dfrac{2\mu_0 \mu_N}{I} \left[\dfrac{\mathbf{I} \cdot \mathbf{S}}{r^3} - 3 \dfrac{(\vec{I} \cdot \vec{r})(\vec{S} \cdot \vec{r})}{r^5} \right]$ (24)

(3) Scalar spin interaction $H_{ss} = \dfrac{16\pi}{3} \dfrac{\mu_0 \mu_N}{I} \Psi_e^2(0) \mathbf{I} \cdot \mathbf{S}$

where $\Psi_e^2(0)$ is the electron density at the nucleus.

The sum of these terms gives an effective Hamiltonian

$$H_{\text{mag}} = a \Lambda \mathbf{I} \cdot \hat{\mathbf{z}} + b \mathbf{I} \cdot \mathbf{S} + c(\mathbf{I} \cdot \hat{\mathbf{z}})(\mathbf{S} \cdot \hat{\mathbf{z}}).$$

Now for $^1\Sigma$ molecules the orbital and spin movements are paired off such that the net magnetic field at the nucleus is zero. These molecules show a weak magnetic hfs interaction due to molecular rotation and to nuclear spin–spin interaction, which is typically of order a few kHz. (The large hfs in $^1\Sigma$ molecules is usually the nuclear electric quadrupole interaction.)

For other states (paramagnetic) the magnetic hfs interaction is usually of order 100 MHz. There are clearly several possible couplings of the various vectors, but in practice the form of the magnetic interaction is simply $A(\mathbf{I} \cdot \mathbf{J})$ in most cases. (This of course gives an energy level pattern which follows the Landé Interval Rule.) Thus in Hund's case a, where S precesses about $\hat{\mathbf{z}}$ with component Σ we have $A = [a\Lambda + (b + c)\Sigma]\Omega/J(J + 1)$, which in NO, for example, is 74 MHz in the $^2\Pi_{3/2}$ state. The ground state of OH is a $^2\Pi$ fairly close to Hund's case a, and a similar formula holds. The $^3\Sigma$ state of O^{16}, O^{17}, however, is a good Hund's case b with $\mathbf{J} = \mathbf{K} + \mathbf{S}$. The vector coupling here gives

$$W_{J=K+1} = \frac{\mathbf{I}\cdot\mathbf{J}}{K+1}\left[b + \frac{c}{2K+3}\right]$$

$$W_{J=K} = \frac{\mathbf{I}\cdot\mathbf{J}}{K(K+1)}(b+c)$$

$$W_{J=K-1} = \frac{\mathbf{I}\cdot\mathbf{J}}{K}\left[-b + \frac{c}{2K-1}\right]$$

with $b = -102$ MHz and $c = 140$ MHz.

In Π states there is an additional effect from the *tensor* magnetic interaction, called "hyperfine doubling" by Townes and Schawlow, which is of some importance. The tensor interaction has matrix elements $\Delta\Lambda = \pm 2$,

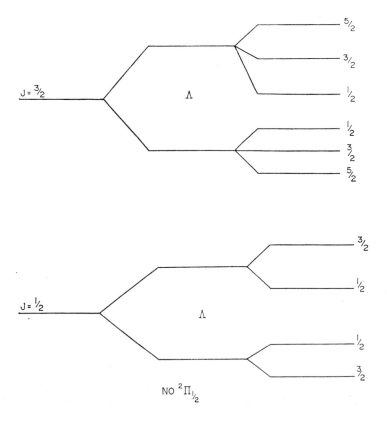

NO $^2\Pi_{1/2}$

hence it gives a direct first order coupling between the nearly degenerate Π states. Were it not already split by the rotational effects described in Section IV this interaction would remove the degeneracy. The tensor interaction slightly increases this splitting, but more important, its effect is opposite in the two Λ-doubled states. Indeed in NO it reverses the order of the hfs levels in the lower Λ component. In the accompanying diagram the lowest two rotational levels of $^2\Pi_{1/2}$ NO are shown as an example of a well developed magnetic hfs. Transitions take place according to all of the selection rules given previously, with the additional rule $\Delta F = 0, \pm 1$. The most intense lines are those in which the change in F is in the same direction as the change in J, e.g., for $J\ 3/2 \to 1/2$, $F\ 5/2 \to 3/2$ and $3/2 \to 1/2$ are the brightest lines.

In non-paramagnetic polyatomic, and in $^1\Sigma$ diatomic molecules, the largest hfs interaction, where it exists, is that of the nuclear electric quadrupole. This is proportional to the gradiant of the electric field components at the position of the nucleus. If this field has axial symmetry the interaction energy has the form

$$W_Q = \frac{eq\ Q}{r^3} \frac{3\cos^2\chi - 1}{2} \tag{25}$$

where χ is the angle between the nuclear spin and the symmetry axis, and $q = \partial^2 V_{el}/\partial z^2$, the second derivative of the electrostatic potential along the symmetry axis. For diatomic or linear molecules this can be written in terms of angular momentum operators as

$$W_Q = \frac{1}{2} \frac{eq_J Q}{I(2I-1)J(2J-1)} [3(\mathbf{I}\cdot\mathbf{J})^2 + \tfrac{3}{2}(\mathbf{I}\cdot\mathbf{J}) - I^2 J^2]$$

where q_J is now evaluated in the J state of the molecule, and includes an average of $\tfrac{1}{2}(3\cos^2\chi_J - 1)$, and χ_J is the angle between the axis and the direction of J, defined for $M_J = J$. Because of the quadratic dependence of the energy on $(\mathbf{I}\cdot\mathbf{J})$ the level spacings here do not follow the Landé rule, and the presence of quadrupole coupling is readily detected when a sufficient number of transitions have been observed.

In symmetric top molecules, e.g., NH_3, because the component K of angular momentum along the axis is fixed, the average of $\tfrac{1}{2}(3\cos^2\chi_J - 1)$ becomes

$$q\left[\frac{3K^2 - 1}{J(J+1)} - 1\right]\frac{J}{2J+3}$$

instead of q_J in the expression for W_Q. Thus each inversion level of NH_3 is split into a hyperfine structure by the N quadrupole interaction.

X. EMISSION AND ABSORPTION

We consider here the fundamental emission and absorption processes of electric dipole radiation and relate them, in an elementary way, to the astrophysical observations. For two levels i and j respectively upper and lower energy states, each with degeneracies $i\alpha$ and $j\beta$, the electric dipole matrix element has the form

$$\langle j\beta | \mu C_1^q | i\alpha \rangle$$

where the components of the dipole are written in spherical form

$$\mu C_1^0 = \mu_z, \quad \mu C_1^1 = \frac{1}{\sqrt{2}}(\mu_x + i\mu_y), \quad \mu C_1^{-1} = \frac{1}{\sqrt{2}}(\mu_x - i\mu_y).$$

In the usual case the degeneracy quantum numbers are the angular momentum components, as $\alpha = m_i$, etc. We define the dipole 'strength' of the transition

$$S_{ij} = \sum_\alpha \sum_\beta \sum_q |\langle j\beta | \mu C_1^q | i\alpha \rangle|^2$$

usually $m_i = q + m_J$, in which case we employ the reduced matrix element

$$S = \mu_{ij}^2 |\langle j \| C_1 \| i \rangle|^2. \tag{26}$$

10.1 Emission

The probability for spontaneous emission $i \to j$ is

$$A_{ij} = \frac{64\pi^4 \nu^3 S_{ij}}{3hc^3 g_i} \tag{27}$$

and is the same from all of the g_i sublevels. Thus $n_i A_{ij}$ is the number of photons of frequency ν emitted per second from 1 cm³ if n_i is the number density. This emission is isotropic. If from the initial state i there are several different possible final states, then the total decay rate of state i is

$$\sum_j A_{ij} = \frac{1}{T_i} \quad (T_i \text{ is the lifetime of state } i).$$

A simple example which gives the orders of magnitude of emission rates in the radio-frequency spectrum is the case of the rotation spectrum of a $^1\Sigma$ molecule. Then $S_{J,J-1} = \mu^2 J$, where μ is the permanent dipole moment, and

$$A_{J,J-1} = \frac{8\pi^3}{3\hbar} \frac{1}{\lambda^3} \frac{\mu^2 J}{2J+1}.$$

If we take $\mu = ea_0 \approx 10^{-18}$ in c.g.s. units and $\lambda = 1$ cm, then $A_{J,J-1} \cong 10^{-8}$ sec^{-1}. Note, however, the cubic dependence on the frequency. In rarified interstellar conditions the lifetimes of excited states may be longer or shorter than the collision times. In the latter case, collisions will not maintain thermal equilibrium with the kinetic temperature. A rather odd case is the strong microwave transition in H_2O at 0.74 cm^{-1} between two highly excited but close-lying states. The spontaneous decay rate for the microwave line is 2×10^{-9} sec^{-1}, however the initial state has a decay mode to a much lower state at a rate 0.65 sec^{-1}, so that the lifetime of the H_2O initial state is just 1.5 sec.

10.2 Absorption

If the radiation intensity $I(v)$ is incident on molecules in the lower state j the rate (per unit solid angle) of absorption transitions is $B_{ji}I(v)/c$ hence $n_j B_{ji} I(v)/c$ is the rate of photon absorption per cm^3. The well-known relation between the Einstein coefficients is derived by considering the special case of thermal equilibrium between the molecules and the radiation field. Introducing the stimulated emission rate $B_{ij}I(v)/c$ we have, in equilibrium

$$n_i \left[A_{ij} + I_p(v) B_{ij} \frac{4\pi}{c} \right] = n_j I_p(v) B_{ji} \frac{4\pi}{c}$$

where $I_p(v)$ is the Planck radiation intensity function and must refer to the same temperature as the kinetic motion of the molecules. From microscopic reversibility we have

$$g_j B_{ji} = g_i B_{ij}$$

$$A_{ij} = \frac{8\pi h v_{ij}^3}{c^3} B_{ij} \tag{28}$$

as the required relations between emission and absorption coefficients.

XI. ASTROPHYSICAL OBSERVATIONS

We do not, of course, observe transitions directly but only their effects on the radiation field at a detector (telescope). To understand these phenomena it is necessary to introduce the concept of line shape. While the spontaneous decay rate and collisions broaden spectral lines, in the interstellar spectra Doppler shifts are the principal mechanism which determines the line shape. For a molecular speed $v_\parallel$ along the line of sight the frequency emitted is $v = v_0[(1 + v_\parallel/c)]$ where v_0 is the frequency emitted at rest. Thus for a speed of 30 km/sec a 1 GHz line is shifted by 0·1 MHz. The velocity distribution is typically not thermal in nature, but seems to arise from large scale convection motions of the gas. We introduce the normalized distribution $P(v)$ to represent the statistical distribution of Doppler shifts. The reduction in incident intensity $I(v)$ due to absorption is given by

$$\frac{dI(v)}{dz} = -I(v)n_j B_{ji} P(v) \frac{hv}{c}.$$

This defines a photon mean free path $\tau(v)$ at each frequency, as

$$\frac{1}{\tau(v)} = n_j B_{ji} P(v) \frac{hv}{c}. \tag{29}$$

We imagine a gas cloud in which emission of these photons is taking place at a rate $\varepsilon(v) = n_i A_{ij} P(v)$ from each unit volume. This emission is observed by a telescope of detecting area a and solid angle Ω. We assume the cloud fills the

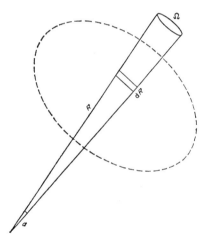

field of view. $\varepsilon a/4\pi R^2$ is the energy intercepted from unit volume in the field of view. The molecules at a distance R in the interval dR contribute a signal

$$\frac{\varepsilon}{4\pi} \frac{a}{R^2} R^2 \Omega \, dR$$

and the total number of photons received from the cloud per unit area, per unit solid angle and per unit time is

$$n_i A_{ij} P(v) \Delta R$$

Here we have taken the gas cloud to be optically *thin*, i.e., self-absorption is neglected. Integration across the line shape gives $n_i A_{ij} \Delta R$ as the observed quantity. If A_{ij} is considered known, then if the temperature can be determined, we have $n_i \Delta R$, the total number of molecules in the line of sight. If two lines from the same molecule are observed, e.g., $h \to i$, $i \to j$, then from $n_h/n_i = g_h/g_i \exp(-E_{hi}/KT)$ we can determine T and hence $n_i \Delta R$.

If conditions are such that $\Delta R > \tau(v)$ then self-absorption and stimulated emission become important, and the above linear relations cannot be employed. Indeed if $\Delta R \gg \tau(v)$ the emitted radiation intensity approaches the Planck curve, becomes insensitive to the density, and becomes a measure only of the temperature.

Similar effects are found in absorption spectra, which are usually observed against a hot star or other bright source. If the absorbing cloud is optically thin, then if we compare intensities off and on the spectra we obtain $\Delta I/I = -\Delta R n_j B_{ji} P(v) h\nu/c$. Integration across the line gives $n_j B_{ji} \Delta R$, hence the column density, as in emission.

In actual cases the situation is usually much more complicated, not only because of the non-linear self-absorptive effects referred to above, but the source and cloud often do not fill the field of view and have different solid angles. Possible maser action introduces new and sometimes bizarre effects.

XII. THE ASTROPHYSICAL MOLECULES

We give here an account of the molecular spectra that have been observed (so far) in the interstellar regions. With a few exceptions these have all been observed as radio-frequency spectra. With the notable exception of the 1420 MHz hfs line of H atoms, all the radiofrequency lines are molecular. This is to be expected, since the very low frequency transitions in atoms all take place within fine structure or hyperfine structures or as Zeeman splittings of atomic levels. The pairs of levels involved have the same parity, electric dipole radiation is forbidden, and the principal emissions or absorptions are of magnetic dipole character, hence very weak. With molecules, on the other hand, we have electric dipole transitions, as we have seen, in rotation spectra and in such special effects as Λ doubling and molecular inversion spectra, all of which are observed in the radiofrequency regions.

For simplicity, we discuss the spectra in the order of their complexity, which is neither the historic order of their discovery, nor that of their importance in astrophysics.

12.1 Diatomic Molecules

(i) Visible and Ultraviolet Spectra. Absorption spectra observed in transmitted starlight.

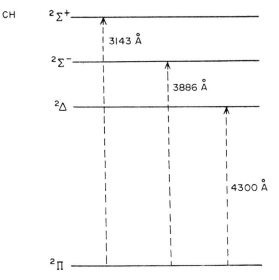

The only Franck–Condon transition seen is $0 \to 0$. The beginnings of *PQR* rotational branches are seen as three lines.

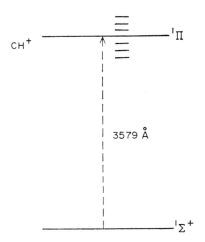

Here several *F–C* transitions seen originating in the lower state $v = 0$.

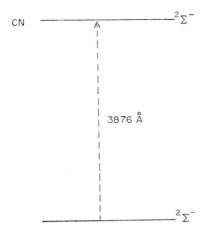

Only F–C $0 \to 0$ is seen

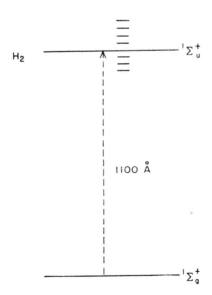

An historic discovery in 1970, as H_2 is by far the most common molecule. Some eight F–C transitions are seen in this transition which corresponds to $1s-2p$ in H.

(ii) Radiofrequency Lines. Some are seen in both emission and absorption.

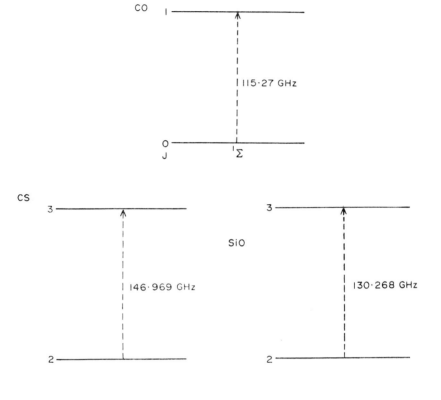

Three isotope lines are observed in $C^{12}O^{16}$, $C^{12}O^{18}$, $C^{13}O^{16}$. Rare isotope lines are "too intense" (saturation?).

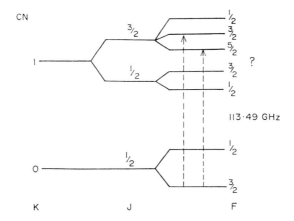

Two magnetic hfs components are observed in rotational transitions

$$(1\ 5/2) \rightarrow (0\ 3/2)$$
$$(1\ 3/2) \rightarrow (0\ 3/2).$$

The splitting is about 10 MHz. $^2\Sigma^+$ Hund's case b.

OH is intermediate between Hund's case a and case b. Some $O^{18}H$ lines have been seen. The brightness and anomalous intensities of the 1·72 GHz complex are such that maser action must be present.

12.2 Polyatomic Molecules

(i) Linear

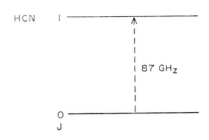

Two isotope lines are observed:

$HC^{12}N$ at 88·6 GHz.
$HC^{13}N$ at 86·3 GHz.

No hfs is seen.

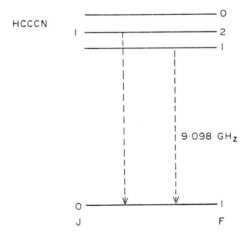

Two nuclear quadrupole hfs lines from N which are seen in emission.

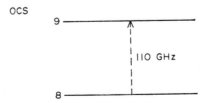

(ii) Symmetric Rotators

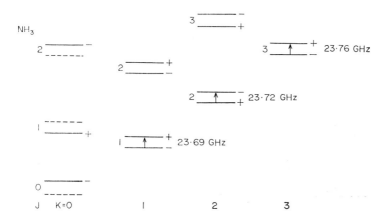

Note because $\Delta K = 0$ states $J = K$ are metastable with $\sim 10^{13}$ sec lifetimes and so are in thermal equilibrium. Thus no transitions are seen in states (2,1) (3,2) or (4,3) which have lifetimes 10–100 sec. No quadrupole hfs is seen, presumably due to too much Doppler broadening.

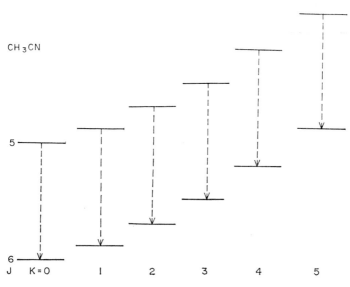

Six transitions with $\Delta K = 0$, all near 110·35 GHz, are seen.

(iii) Asymmetric Rotators

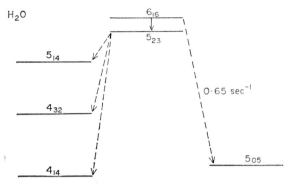

The $6_{16} \to 5_{23}$ line at $1 \cdot 35 \, \text{cm}^{-1}$ is $450 \, \text{cm}^{-1}$ above ground state and can radiate as shown. This line is extremely bright and localized, perhaps not truly "interstellar".

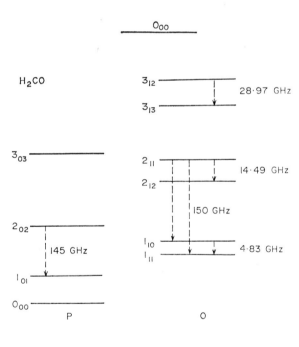

Formaldehyde is only a slightly asymmetric rotator. It occurs in para and ortho states, with no transitions between them.

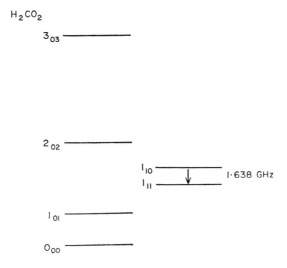

Formic Acid. Note the "K doubling" in ortho states. No hfs has been positively identified. The $1_{10} \rightarrow 1_{11}$ transition shows "super cooling".

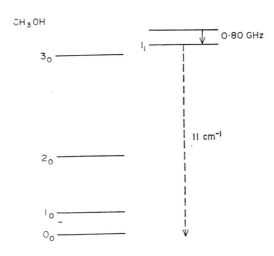

Methanol is slightly asymmetric. The rotational levels of methanol are more complex than is shown, because of the "triple minimum" of the potential of OH rotation about the axis.

XIII. LITERATURE

A list of instellar molecules and papers relating to them as given in the table below:

Molecule(s)	Reference(s)
CH, CH^+, CN	7–9
CO	9–10
H_2	11
CS	12
SiO	13
OH	14–16
HCN	17
HCCCN	18
NH_3	19–20
H_2O	21–23
H_2CO	24–26
H_2CO_2	27
CH_3OH	28
OCS	29
CH_3CN	30

REFERENCES

1. Herzberg, G., "Spectra of Diatomic Molecules" (Van Nostrand, 1950).
2. Herzberg, G., "Infrared and Raman Spectra of Polyatomic Molecules" (Van Nostrand, 1964).
3. Herzberg, G., "Electronic Spectra and Electronic Structure of Polyatomic Molecules" (Van Nostrand, 1967).
4. Townes, D. H. and Schawlow, A., "Microwave Spectroscopy" (McGraw-Hill, 1955).
5. Van Vleck, J. H., Rev. Mod. Phys. **23**, 213 (1951).
6. Hougen, J. T., J. Chem. Phys. **36**, 519 (1962).
7. Adams, W. S., Ap. J. **93**, 11 (1941).
8. Bates, D. R. and Spitzer, L. Jr., Ap. J. **113**, 441 (1951).
9. Jefferts, K. B., Penzias, A. A. and Wilson, R. J., Ap. J. (Letters) **161**, L87 (1970).
10. Penzias, A. A., Jefferts, K. B. and Wilson, R. W., Ap. J. **165**, 229 (1971).
11. Carruthers, G. R., Ap. J. (Letters) **161**, L81 (1970).
12. Penzias, A. A., Solomon, P. M., Wilson, R. W. and Jefferts, K. B., Ap. J. (Letters) **168**, L53 (1971).

13. Wilson, R. W., Penzias, A. A., Jefferts, K. B., Kutner, M. and Thaddeus, P., *Ap. J. (Letters)* **167,** L97 (1971).
14. Zuckerman, B. and Palmer, P., *Ap. J. (Letters)* **159,** L197 (1970).
15. Turner, B. E., Palmer, P. and Zuckerman, B., *Ap. J. (Letters)* **160,** L125 (1970).
16. Ball, J. A., Gottlieb, C. A., Meeks, M. L. and Radford, H. E., *Ap.J.qLetters?* **163,** L33 (1971).
17. Snyder, L. E. and Buhl, D., *Ap. J. (Letters)* **163,** L47 (1971).
18. Turner, B. E., *Ap. J. (Letters)* **163,** L35 (1971).
19. Cheung, A. C., Rank, D. M., Townes, C. H., Thornton, D. D. and Welch, W. J., *Phys. Rev. Letters* **21,** 1701 (1968).
20. Cheung, A. C., Rank, D. M., Townes, C. H. and Welch, W. J., *Nature* **221,** 917 (1969).
21. Cheung, A. C., Rank, D. M., Townes, C. H., Thornton, D. D. and Welch, W. J., *Nature* **221,** 626 (1969).
22. Meeks, M. L., Carter, J. C., Barrett, A. H., Schwartz, P. R., Waters, J. E. and Brown, *Science* **165,** 180 (1960).
23. Snyder, L. E., Buhl, D., Zuckerman, B. and Palmer, P., *Phys. Rev. Letters* **22,** 679 (1969).
24. Townes, C. H. and Cheung, A. C., *Ap. J. (Letters)* **157,** L103 (1970).
25. Zuckerman, B., Buhl, D., Palmer, P. and Snyder, L. E., *Ap. J. (Letters)* **160,** L485 (1970).
26. Thaddeus, P., Wilson, T. W., Kutner, M., Penzias, A. A. and Jefferts, K. B., *Ap. J. (Letters)* **168,** L59 (1971).
27. Zuckerman, B., Ball, J. A. and Gottlieb, C. A., *Ap. J. (Letters)* **163,** L41 (1971).
28. Ball, J. A., Gottlieb, C. A., Lilley, A. E. and Radford, H. E., *Ap. J (Letters)* **162,** L203 (1970).
29. Jefferts, K. B., Penzias, A. A., Wilson, R. W. and Solomon, P. M., in press.
30. Jefferts, K. B., Penzias, A. A., Wilson, R. W. and Solomon, P. M., in press.

Non-Equilibrium Processes in Interstellar Molecules

M. M. LITVAK

Smithsonian and Harvard Observatory, Cambridge, Mass., U.S.A.

I. INTRODUCTION

The observation[1] in 1963 of the hydroxyl radical OH by means of the absorption of microwave radiation in two ground state transitions represented a decisive step in astrophysics. Not only had the first interstellar molecule been observed in large clouds throughout the galaxy, but also new spectral line detection techniques[2] had proven successful. In 1965, maser emission[3] from OH confirmed the suspicions engendered from the galactic centre absorption data[4] that the populations of the molecular energy levels were not in thermal equilibrium. In 1968 and 1969, the discovery and analysis of microwave transitions of ammonia,[5] water vapour,[6] and formaldehyde[7] indicated that a wealth of new interstellar molecules awaited discovery and that thermal equilibrium was hardly likely.

The purpose of these lectures is to elucidate the likely causes of the non-equilibrium effects that appear so strikingly in the OH[8] and water masers[9] (anomalous emitters) and the OH[10, 11] and formaldehyde[12] antimasers (anomalous absorbers), and appear more subtley in the recently discovered cyanoacetylene[13] and methanol[14] molecules, for example. Table I lists those interstellar molecules known to us at this time by their microwave lines, their date of discovery, the quantum numbers for the levels involved in the microwave transitions, the line rest frequencies, the astronomical objects they were observed in, the estimated maximum projected densities and the Einstein coefficient for the spontaneous radiative transition. The rapid rate of discovery, especially of new millimeter wave transitions[15–20] has probably already made this list obsolete. Isotopes like ^{13}CO and ^{18}OH are not included.

Table I

Molecule	Year	Transition	Rest frequency	Objects	Max. proj. density	Einstein-A (transition)
Hydroxyl OH	1963	$J = 3/2, K=1$				
	1967	$F = 2 \rightarrow 2$	1667·358 MHz	W3, W49, W75, W51, NGC6334, Ori A,	10^{16} cm^{-2}	$7·71 \times 10^{-11}$ sec
		$1 \rightarrow 1$	1665·400			7·71 ...
		$2 \rightarrow 1$	1720·529	W28, W44, ON-3, etc.		0·94 ...
		$1 \rightarrow 2$	1612·231	NML Cyg, VY CMa, etc.	10^{18}	1·29 ...
	1971			NGC253, M82		
Ammonia NH$_3$	1968	$J_K = 1_1 \rightarrow 1_1$	23,694·48	Sgr A, B$_2$, Ori A, W3(OH), etc.	10^{16}	$1·7 \times 10^{-7}$
		$2_2 \rightarrow 2_2$	23,722·71	...		2·1 ...
		$3_3 \rightarrow 3_3$	23,870·11	...		2·5 ...
		$4_4 \rightarrow 4_4$	24,139·39	...		2·8
		$6_6 \rightarrow 6_6$	25,056·04	...		3·3
	1971	$2_1 \rightarrow 2_1$	23,098·79	Sgr B$_2$		
		$3_2 \rightarrow 3_2$	22,834·17			
Water Vapour H$_2$O	1969	$J_{K_-K_+} = 6_{16} \rightarrow 5_{15}$		W3, W49, Ori A, W75 Cyg 1, NGC 6334	10^{20}	2×10^{-9}
		$F = 7 \rightarrow 6$	22,235·041	IRC + 10406, NML Cyg,		
		$6 \rightarrow 5$	22,235·078	U Her, W Hyd		
		$5 \rightarrow 4$	22,235·121			
Formaldehyde H$_2$CO	1969	$J_{K_-K_+} = 1_{10} \rightarrow 1_{11}$	4829·660	W3, W49, M17, NGC2024, Sgr A & B$_2$, W33, NGC 6334, Cyg A, Cas A, Ori A, Cloud 2, etc.	10^{16}	3×10^{-8}
		$2_{11} \rightarrow 2_{12}$	14488·65	Sgr A & B$_2$, W51		
		$2_{02} \rightarrow 1_{01}$	145602·97	Ori A		3×10^{-7}
		$2_{12} \rightarrow 1_{11}$	140839·53	Ori A, Sgr A, W3(OH), W51		
		$2_{11} \rightarrow 1_{10}$	150498·36	Ori A		$5·3 \times 10^{-5}$

Molecule	Year	Transition	Frequency (MHz)	Sources	N (cm^{-2})	Abundance
Carbon Monoxide CO	1970	$J=1\to 0$	115271·2	Ori A, IRC+10216, W51	10^{19}	6×10^{-8}
Cyanogen CN	1970	$K=1\to 0$ $F=\tfrac{5}{2}\to\tfrac{3}{2}$ $\tfrac{3}{2}\to\tfrac{3}{2}$	113492· 113503·	Ori A, W51	10^{15}	$1\cdot 31\times 10^{-5}$
Cyanoacetylene HCCCN	1970	$J=1\to 0, F=1\to 1$	9097·09	Sgr B$_2$	10^{14}	$3\cdot 79\times 10^{-8}$
Hydrogen Cyanide HCN	1970	$J=1\to 0, F=1\to 1$ $2\to 1$ $0\to 1$	88,630·4157 88,631·8473 88,633·9360	W3(OH), Ori A, Sgr A (NH$_3$A), W49, W51, & DR21	10^{15}	2×10^{-7}
X-ogen (HCO$^+$?)	1970		89,190·	Ori A, W51, W3(OH), L134, Sgr A(NH$_3$A), NH$_3$-Cloud 3′ So. of Sgr A		
Methanol CH$_3$OH	1970	$J=1, K=1\to 1, \tau=1\to 1$	834·3	Sgr A & B$_2$	10^{16}	$2\cdot 72\times 10^{-12}$
	1971	$J=4, K=2\to 1, \tau=1\to 2$	24,933·7	Ori A	10^{16}	$8\cdot 40\times 10^{-8}$
		5, ...	24,959·08	...		8·74 ...
		6, ...	25,018·14	...		8·98 ...
		7, ...	25,124·88	...		9·21 ...
		8, ...	25,294·41	...		9·48 ...
		$5_1\to 4_0$	85,521·5			
		$4_1\to 3_0$	36,169·2			
Formic Acid HCOOH	1970	$1_{11}\to 1_{10}$	1638·805	Sgr B$_2$, A?	10^{14}	$4\cdot 96\times 10^{-11}$
Carbon Monosulphide CS	1971	$J=3\to 2$	146,969·16	**Ori A**, W51, DR21, IRC+10216	10^{14}	$6\cdot 5\times 10^{-5}$
Carbonyl Sulphide OCS	1971	$J=9\to 8$	109,463·	Sgr B$_2$		1×10^{-6}

Table I—*continued.*

Molecule	Year	Transition	Rest frequency MHz	Objects	Max. proj. density	Einstein-A (transition)
Silicon Monoxide SiO	1971	$J = 3 \to 2$	130,268·	Sgr B_2	4×10^{13}	$1 \cdot 1 \times 10^{-4}$ sec
Hydrogen Isocyanide HNC	1971	$J = 1 \to 0$	90,665·	Sgr B_2		
Formamide NH_2CHO	1971	$J = 1, K = 1,$ $F = 2 \to 2$ $1 \to 1$ $J = 2, K = 1$ $F = 3 \to 3$ $2 \to 2$ $1 \to 1$	1539·85 1538·13 4619·01 ± ·03 4617·14 4620·03	Sgr A & B_2	10^{14}	$2 \cdot 2 \times 10^{-10}$
Isocyanic Acid HNCO	1971	$4_{04} \to 3_{03}$ $1_{01} \to 0_{00}$	87,925· 21,982·	W51, DR21		1×10^{-5} 1×10^{-7}
Methyl Cyanide CH_3CN	1971	$J = 6 \to 5, K = 0, 1$ $2, 3$ $4, 5$	110,383; 110,381 110,375; 110,364 110,350; 110,331	Sgr B_2, A	10^{14}	$1 \cdot 4 \times 10^{-4}$
Methyl Acetylene CH_3CCH	1971	$J = 5 \to 4, K = 0$	85, 457·29	Sgr B_2		2×10^{-6}
Acetaldehyde CH_3CHO	1971	$1_{10} \to 1_{11}$ $2_{11} \to 2_{12}$	1065·075 3195·	Sgr A & B_2 ...	10^{15}	
Thioformaldehyde CH_2S	1971	$2_{11} \to 2_{12}$	3139·38	Sgr B_2	10^{15}	

The competition between radiative and collisional rates, much as for laboratory lasers, is the key to the non-equilibrium. Estimates of rates between microwave doublet levels, far-infrared rotational states, and near-infrared vibrational states are discussed, though accuracy is severly limited. The transport of line radiation at the various wavelengths, through fairly thick clouds, is a necessary complication. The observed large microwave optical depths usually imply still larger depths at the higher frequencies. Sources of radiation besides the now well-known 2·7°K cosmic background radiation often are associated with the molecules. Such is the case of infrared stars,[21-23] H II (ionized hydrogen) regions[24] and supernova remnants.[25, 26] Dynamic effects that are induced by large scale imbalances of pressure may be equally important. For example, shock waves in the above-mentioned objects or in self-gravitating clouds might provide not only the conditions of high gas temperature, followed by rapid cooling, that are important for chemical formation of organic molecules like formaldehyde, but also might convert considerable mechanical and gravitational energy into line radiation that is resonant with molecular transitions.[27, 28] Pumping action, like that shown in Fig. 1, might

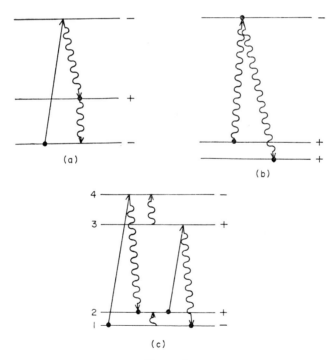

FIGURE 1

result in maser or anti-maser action in lower microwave transitions. All these special effects are to be distinguished from the relatively ubiquitous and static effects of the cosmic background radiation (predominantly at millimeter wavelengths[29]) and collisions in typical interstellar molecular clouds.

Astrochemistry, the science of chemical reactions among interstellar molecules and atoms, is discussed with respect to the formation rates of molecules like OH and formaldehyde in shockwaves and on surfaces of interstellar grains (those sub-micron sized particles of graphite, silicate, or whatever, that are closely associated with high molecular abundances).

II. OBSERVATIONS OF MASER MOLECULES

2.1. Hydroxyl and water vapour masers

(i) *Pumping models*

The first of the molecules detected by microwave emission, the hydroxyl radical, has spurred hundreds of observations and studies of the hyperfine-split components of its ground state microwave lines, of the lines of the excited rotational states[30-36] and, recently, of the vibrationally-excited states.[37] The unusual ratios of the intensities of the hyperfine-split components, the high degree of polarization (usually circular), the brightness temperatures measured in $10^{10}°K$ or more, and other properties attest to maser amplification. This is seen in or near such diverse astronomical objects as dark dust clouds, infrared stars H II regions, infrared nebulae and supernova remnants. A common ingredient is the unusual obscuration by interstellar grains, which is beneficial to the molecules but not to the optical astronomers.

Table IIa summarizes data on the fluxes observed from the dominant ground state lines. In anticipation of the later discussion on pump requirements, Table IIb lists estimates of the photon fluxes that are available for optical pumping at the various resonant wavelengths of OH. Under the most efficient conditions the number of microwave photons emitted per second should be balanced by the number of absorbed pump photons per second.[38, 39] The microwave photon flux for a single emission point (a single Doppler feature) in W3 or W49 is about 3×10^{45} microwave photons sec^{-1}.

A simple calculation of pumping by collisions which excite the first rotational levels shows that optical pumping is not the only mechanism that is hard-pressed to account for the microwave flux. If the four level scheme of Fig. 1(c) is consulted, one can see that the collisions of OH with

Table IIa

Galactic OH Sources

Source	Distance (kpc)	Flux ($\times 10^3$)	Frequency (MHz)	Luminosity ($\times 10^{43}$)
W3	3	20	1665	200
W49	14	30	1665	6,000
W49	14	15	1667	3,000
W51	5	4	1665	100
W51	5	1	1612	20
W51	5	4	1720	100
W28	4	10	1720	200
NGC 6334	1	20	1665	20
NGC 6334	1	15	1667	10
Orion	0·5	2	1665	0·6
NML Cyg	0·5?	150	1612	40
VY CMa	0·5?	150	1612	40

In this table, distances are given in kiloparsecs. The third column shows the approximate microwave flux at the Earth, in units of 1,000 photons per square meter per second. The fifth column is the estimated isotropic luminosity in units of 10^{43} photons per second. (The total flux from the sum over all wavelengths is about 10^{45} photons per second.)

Table IIb

A Comparison of Optical Pumps

Pump Object	Temperature (°K)	Size (a.u.)	Excitation	Maser Frequency (MHz)
05 star (UV emitter)	50,000	0·1	10^{45}	1665, 1667
Protostar (UV shock)	4,000	1,000	10^{46}	1665, 1667
Near-infrared star	2,000	10	10^{44}	1612
Protostar (IR shock)	1,000	100	10^{45}	1612
Far-infrared nebula	100	10,000	10^{46}	1720

Columns 2 and 3 give the characteristic temperature and size of the object; 4 is the available excitation flux, in photons per second; 5 is the expected maser-produced frequency, in megahertz.

H, for example, might carry the OH population from level 1 to 4 more rapidly than from 2 to 3.[40] Radiative decay connects level 4 to level 2 only (a reversal of parity being required by the electric dipole selection rules) and level 3 to level 1. Collisional transitions here would be represented by straight arrows and radiative ones by wavy arrows in Fig. 1. In order for there to be nonthermal equilibrium, the reverse rate of *de*-excitation by collisions must not exceed that of radiative decay (taking into account the trapping of the radiation that slows this decay). This usually limits the hydrogen density in the OH maser region to less than about 10^6cm^{-3}. Then the total rate of microwave emission from a uniform spherical cloud of radius R cannot be more than $L \leqslant n_{\text{OH}}(4\pi/3) R^3 \Delta W$, where n_{OH} is the OH density and ΔW is the difference between the two collisional excitation rates. Since the rate for exciting rotation by collision with H is probably less than $3 \times 10^{-10} \, n \exp(-120°\text{K}/T) \sec^{-1}$ and the fractional difference

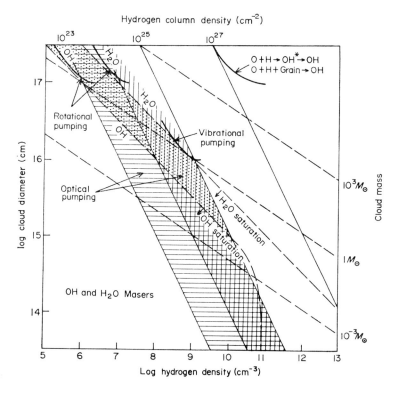

FIGURE 2

between the two cross-sections for pumping is probably less than ten per cent,[40] we have that

$$L \ll 30 \exp(-120°/T)(n_{OH}/n)R^3 4\pi/3.$$

Taking $L = 3 \times 10^{45}$ photons sec^{-1} and $T \leqslant 100°$K for the kinetic temperature, we have $R \gg 10^{15}(n_{OH}/n)^{-1/3}$ cm. Since n_{OH}/n is probably $\lesssim 10^{-6}$ (reference 25), we have $R \gg 10^{17}$ cm, which is even larger than a typical distance which separates different emission points in W3–OH (reference 41). The apparent size of an emission point is about 10^{15} cm or less.[41] A more detailed analysis of such collisional–radiative pumping is considered later. Figure 2 indicates the conditions of adequate maser action according to various models by plotting on logarithm scales the size $l \simeq 2R$ versus the density n (cm^{-3}) for probably reasonable values of kinetic temperature and relative concentrations of OH and H_2O to hydrogen.[43]

Table III lists the data from microwave emission from excited OH states and the various explanations or models offered. Various associations have tentatively been suggested between the spectral lines from the excited states and certain Doppler features in the ground state emission from the same source. For example, the $\Pi_{3/2}$ ($J = 5/2, F = 3 \to 3$) 6035 MHz lines in W75B seem somewhat correlated with $\Pi_{3/2}$ ($J = 3/2, F = 2 \to 2$) 1667 MHz lines with respect to Doppler velocity (if only the high frequency Zeeman components are present) and with respect to time-variations (but not always in phase). The $\Pi_{1/2}$ ($J = 1/2, F = 1 \to 0$) 4765 MHz lines appear correlated with 1720 MHz lines in W3(OH), W49, Sgr B_2, and NGC 6334N, with respect to Doppler velocity. Time variations of this excited state have yet to be monitored.

(ii) *Long-baseline interferometry*

Radio interferometry by means of two or more widely-spaced antennas has shown that the strongest emitting points at 1665 MHz in the well-studied source W3(OH) are arranged along two or more arcs that are part of a circumference having about a 10^{17} cm diameter.[41] The angular size attributed to a typical emitting point is about 10^{-2} arc sec (or 5×10^{14} cm). Measurements of apparent OH angular sizes in W49 and an infrared star[22] (at 1612 MHz) yield larger angular sizes ($\approx 10^{-1}$ arc sec) corresponding to apparent spot diameters of 10^{16} cm and 5×10^{14} cm, respectively, the IR star being only about 500 pc away.

The brightness temperature T_B for the equivalent isotropic blackbody is given by $kT_B/\lambda^2 = P/\Omega$, where P is the received flux (erg cm^{-2} sec^{-1} Hz^{-1}) in one sense of polarization, and Ω is the apparent solid angle subtended by the source. The actual physical size of the emitter-cloud is probably much larger than the radiation solid angle since the actual

Table III

OH excited state	Energy above ground state (cm^{-1})	Frequency (MHz)	Relative line-strength	Signal strength (f.u.)	Source	Models considered
$^2\Pi_{3/2}$ ($J=5/2$)	84					Zeeman splitting $B=10^{-2}G$. Population $\approx 0.1 \times$ ground state population.
$F=3\to 3$		$6035.085 \pm .005$	20	79, 20 2, 47 4, 3	W3(OH), W75B W49, NGC 6334N Sgr B_2, NML Cyg	Far IR + near UV pumping ($N_{OH}l \approx 10^{16}$ cm^{-2})
$2\to 2$		$6030.739 \pm .005$	14	26, 2	W3(OH), NGC 6334N	Collisional excitation of rotation ($T_K \approx 100°K$, emitting area $\sim l_{17}$ arc$_2$ sec; $N_{OH}l \approx 10^{17}$ cm^{-2}).
$2\to 3$		$6016.741 \pm .008$	1			Far IR coupling of ground state and excited state population inversions.
$3\to 2$		$6049.084 \pm .008$	1			
$^2\Pi_{1/2}$ ($J=1/2$)	126					
$F=1\to 1$		4750.656	2			Far IR + near UV pumping (correlation of 4765 and 1720 MHz emission).
$1\to 0$		4765.562	1	3, 1, 0.7, 0.3	W3(OH), W49, Sgr B_2, NGC 6334N^2	
$0\to 1$		4660.242	1	0.7	Sgr B_2	Far IR coupling

$^2\Pi_{1/2}$ ($J=3/2$)	188			Not detected yet	Collisional excitation of rotation, anti-inversion of doublet.
$F=2\to 2$		7820·125 ± 0·005	9	<0·2 W3(OH)	
$1\to 1$		7761·747	5		
$1\to 2$		7831·962	1		
$2\to 1$		7749·909	1		
$^2\Pi_{3/2}$ ($J=7/2$)	202				Collisional excitation of rotation ($T_K > 100°$K), far IR coupling to $^2\Pi_{3/2}$ ($J=5/2$, $F=3\to 3$) population inversion. Far IR pumping (153 cm^{-1}) via $^2\Pi_{3/2}$ ($J=9/2$), overlap of hyperfine split IR lines in upper doublet only.
$F=4\to 4$		13441·371	35	19 W3(OH)	
$3\to 3$		13434·608	27		
$3\to 4$		13441·963	1		
$4\to 3$		13434·015	1		
$^2\Pi_{1/2}$ ($J=5/2$)	289			Not detected yet	Collision excitation of rotation→anti-inverted doublet.
$F=3\to 3$		8189·586 ± 0·005	20	<0·3 W3(OH)	
$2\to 2$		8135·868	14		
$2\to 3$		8118·052	1		
$3\to 2$		8207·401	1		

Table III—*continued.*

OH excited state	Energy above ground state (cm^{-1})	Frequency (MHz)	Relative line-strength	Signal strength (f.u.)	Source	Models considered
$^2\Pi_{3/2}$ ($J=9/2$)	355					
$F=5\to 5$		23826·6	54	Not detected yet		
$4\to 4$		23817·6	44			
$5\to 4$		23805·4	1			
$4\to 5$		23838·8	1			
Vibrationally-excited ($v=1$) $^2\Pi_{3/2}$ ($J=3/2$)	3568					Near infrared pumping.
$F=2\to 2$		1538·80	9			
$1\to 1$		1537·06	5			
$2\to 1$		1586·76	1			
$1\to 2$		1489·10	1			

angular distribution of the emission from any point on the output surface is probably highly directional. However, the coverage of directions in the sky, owing to all the points of the surface, might be nearly 4π steradian. Another description of the radiation pattern is that the emission occurs in spherical wavefronts emanating from centres that occupy a region which subtends the angle measured by the interferometers. This central region might correspond to a "hot spot" in which appreciable unsaturated amplification of the brightness temperature occurs. Typical values of T_B are $10^{12}\,^\circ$K for OH and $10^{15}\,^\circ$K for H_2O. Yet, the larger region surrounding the "hot spot" might contribute the bulk of the energy, being induced to emit primarily along the ray directions that emerge only from the "hot spot(s)".

The interferometer maps of OH and also of $H_2O^{(42)}$ have shown that the different Doppler-shifted features of the spectrum usually are separated in the sky, giving rise to the notion of emission from separate cloudlets or

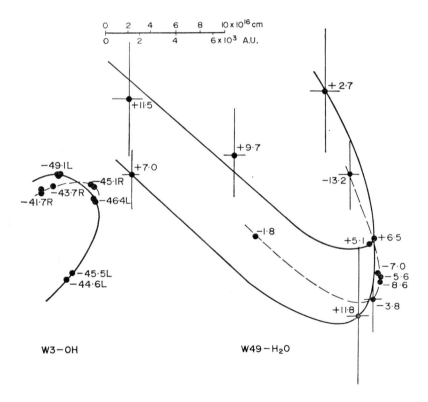

FIGURE 3

possibly protostars of a budding star cluster. In lieu of a map of W3–H_2O or of W49–OH, it is perhaps a bit speculative to deduce dynamic and magnetic properties of the OH–H_2O regions on the basis of separate W3–OH and W49–H_2O maps. Nevertheless, the picture of a large rotating disk of about $100 M_\odot$, containing cloudlets of about $1 M_\odot$ at its rim, rotating with a period of 33,000 years emerges from the observed monotonic increase of 10 km/sec in radial velocity along a line of length of $\approx 10^{17}$ cm, interpreted as the disk looked at edge-on.[43] Some outflow from one edge (caused by Coriolis forces) accounts for the scattered points away from the disk. The separation of positive and negative radial velocity features into two families differing consistently by about 16 km/sec in W49–H_2O suggests that two parallel disks, rotating with nearly the same period, are

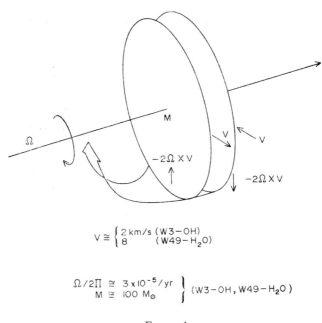

FIGURE 4

shearing with respect to each other (see Figs. 3 and 4). However, if one supposes that the separation of two families of emission points in W3–OH according to left- or right-handedness of the nearly-pure circular polarization corresponds to a change of magnetic field direction, then the differential velocity of 16 km/sec might be due to a rotational discontinuity[44] in which the velocity and magnetic vectors both rotate through some angle across the front. The change in velocity ΔV is accompanied by a change in magnetic field strength ΔB given by

$$\Delta V = \Delta B \, (4\pi\rho)^{-1/2}$$

where ρ is the mass-density, assuming the usual good coupling between the charged and neutral particles. Taking $\Delta B \approx 10^{-2} G$, corresponding to the field required for Zeeman splittings in OH like those seen, though highly distorted, in the spectra from W49, and taking $\Delta V \approx 16 \times 10^5$ cm/sec, we obtain $\rho \approx 5 \times 10^{-18}$ g/cm^3 or a particle density $n \approx 3 \times 10^6$/cm^3. This density is comparable to that expected in the OH regions. The magnetic field jump might actually be somewhat higher in the H$_2$O region, say 10^{-1} G, thereby bringing the density to 3×10^8/cm^3, comparable to that expected in the H$_2$O maser region itself. The Zeeman splitting factor for H$_2$O is about 10^{-3} times that for OH so that direct evidence of such high fields is lacking in the H$_2$O spectra. The rotational discontinuity, which is stable to a variety of perturbations, and has no discontinuity of density, moves with respect to the gas at the Alfvén speed determined by the continuous component B_n of the magnetic field perpendicular to the approximately planar wave front. This speed is approximately 2 km/sec with $B_n \approx 10^{-2} G$ and $n \approx 10^8$/cm^3.

Where do these density estimates come from among the OH and H$_2$O maser properties? Basically, 10^6 and 10^8/cm^3 are the densities that correspond to collisional de-excitation rates that quench trapped fluorescence from rotation levels in OH and H$_2$O respectively, that lie just above or just below the maser levels. Quenching of fluorescence in OH ought to be avoided in order to allow the rotational radiation to select hyperfine states by unequal radiation trapping among the hyperfine components of the far infrared radiation. In both OH and H$_2$O strong quenching of fluorescence will probably destroy the desired population inversion.

(iii) *Physical conditions.*

For a steady state the populations N_u and N_l of the states u (upper) and l (lower) must satisfy the relation

$$N_u W_{ul} = n_l W_{lu}$$

where W_{ul} and W_{lu} are the transition rates (or transition probabilities per unit time) between the states due to radiative and collisional processes.

We may write

$$W_{ul} = W_{ul}^P + W_c + W_m + A_{ul}$$
$$W_{lu} = W_{lu}^P + W_c e^{-h\nu/kT_k} + W_m$$

where W_c and $W_c e^{-h\nu/kT_k}$ represent the transition rates for direct collisional deexcitation and excitation, for effective kinetic temperature T_k, W_m and A_{ul} are the (microwave-) stimulated and spontaneous radiative transition rates, and the W^P are the (indirect) pump rates. We then define the *fractional population inversion* by the relation

$$\frac{N_u - N_l}{N_u + N_l} = \frac{W_{lu} - W_{ul}}{W_{lu} + W_{ul}} = \frac{W_{lu}^P - W_{ul}^P - W_c^* - A_{ul}}{W_p + 2W_c - W_c^* + 2W_m + A_{ul}}$$

where

$$W_p = W_{lu}^P + W_{ul}^P$$

and

$$W_c^* = W_c(1 - e^{-h\nu/kT_k}).$$

The microwave transition rate W_m is given by

$$W_m = kT_B A \Omega_m / 4\pi h\nu$$

for both senses of polarization, where T_B is the brightness temperature, A is the Einstein coefficient for spontaneous emission, and Ω_m is the solid angle over which this temperature acts on a molecule. For saturation, $2W_m$ must exceed the pump rate and the collision rate. The effective pump rate W_p, due either to optical effects or to special collisions, is comparable to or greater than the thermal collision rate, W_c, across the microwave transition, if there is to be population inversion. We take $W_c \simeq 10^{-9} n$ sec^{-1}.

As summarized in Table IV, the hydrogen density n is restricted to less than 10^{15} cm^{-3}, by keeping the pressure-induced full width of the H_2O transition below 200 kHz, assuming that there is line narrowing due to maser amplification by a factor of five. There is very little line broadening due to saturation for this case. The broadening parameter used here was $2 \times 10^{-10} n$ Hz cm^3 for the full width. The broadening parameter for OH was taken to be about three times larger. Keeping the full width below 5 kHz implies densities less than 10^{13} cm^{-3}. Similarly, the kinetic temperature must be less than 100°K, assuming that there is negligible line narrowing for a well-saturated, Doppler-broadened case. For some residual line-narrowing, by a factor x ($5 > x > 1$), the temperature must be less than $100 x^2$ degrees. The factor of five is the square root of the unsaturated microwave optical depth, estimated from an amplification factor of about 10^{12}.

Table IV gives the ratio W_m/W_c as an estimate of the degree of saturation for the two cases of density: the pressure-broadening limited one and the far-infrared fluorescence-limited one. The latter is the density for which the trapped fluorescence from the excited rotational state just above or just below the maser states is quenched by collisional de-excitation. This criterion is irrelevant to H_2O with respect to selection of hyperfine states because the infrared Doppler widths exceed the hyperfine splittings, but is relevant to obtaining non-equilibrium populations.

Table IV

Characteristics of a single emission point

	W49–H_2O	W49–OH	W3–OH
Observed			
T_B (°K)	10^{15}	5×10^{10}	5×10^{12}
l_0 (cm)	3×10^{13}	6×10^{15}	3×10^{14}
L (phot/sec)	10^{48}	10^{46}	10^{45}
Press.-broad. limit			
n (cm^{-3})	10^{15}	10^{13}	10^{13}
$\Delta l/l_0$	<1	0	<1
$M/M_\odot$	10^{-2}	10^2	1
W_m/W_c	1	10^{-2}	1
Fluor.-quench. limit			
n (cm^{-3})	10^8	10^6	10^6
$\Delta l/l_0$	10^3	20	300
$M/M_\odot$	1	1	1
W_m/W_c	40	100	50
V_{ff} (km/sec)	3	1	1
L (phot/sec)	10^{48}	10^{45}	10^{45}

Under saturated conditions the output of microwave photons per second, L, is given by

$$L \approx \tfrac{1}{2} \int \Delta n_m \, W_p \, dV + \Delta v \, \frac{kT_{B_0}}{h\nu} \, 2\pi \, A_0/\lambda^2$$

where W_p is the pump rate per molecule, dV is a differential element of volume of the *saturated* region, Δn_m is the unsaturated difference of the population density in the maser transition, T_{B_0} and A_0 are the brightness temperature and frontal area of the unsaturated core, respectively, and Δv is the emission bandwidth at frequency v and wavelength λ. The observed brightness temperatures T_B are the result of significant unsaturated amplification over a length l_0. The product $\Delta n_m l_0$ enters into the unsaturated microwave optical depth τ_m

$$\tau_m/(\Delta n_m l_0) \approx \begin{cases} 10^{-13} \text{ cm}^2 \text{ for OH} \\ 10^{-15} \text{ cm}^2 \text{ for H}_2\text{O} \end{cases}$$

From values of $T_B \approx 10^{12} - 10^{15}$°K we estimate that $\Delta n_m l_0 \approx 3 \times 10^{14}$ cm^{-2} for OH and 3×10^{16} cm^{-2} for H$_2$O. However, Δl, the radial distance for saturated amplification, might be much greater than l_0. For cases of no quenching of fluorescence Δl will be comparable to the typical separation of the points 3×10^{16} cm and l_0 will be comparable to the smallest H$_2$O interferometer sizes, 3×10^{13} cm. In Table IV we will take $l_0 + \Delta l$ to give one solar mass surrounding each point. Then

$$\Delta l/l_0 \approx 300 \quad \text{and} \quad \Omega_m/4\pi \approx l_0^2/(l_0 + \Delta l)^2 \approx 10^{-5}.$$

The luminosity L, mainly due to the saturated emission contribution, is approximately proportional to $\Delta n_m l_0$ times $n(\Delta l)^2$ when $\Delta l/l_0 \gg 1$ and when we use W_c as an estimate for W_p and use an inverse-square law for the radius dependence of n and Δn_m. The quantity l_0 in Table IV is the observed interferometer size as possibly caused by plasma scattering. However, W_m can still be calculated from the observed value of $T_B l_0^2$, the same as the 'unscattered' value. For W49–OH, the calculated luminosity seems to be a factor of ten too small, indicating that W_p, n or Δl ought each to be increased somewhat, or that the assumption of nearly isotropic emission has led to an overestimate of the 'observed' luminosity. The free-fall velocity V_{ff}, proportional to $n^{1/2} (l_0 + \Delta l)$, is also included in Table IV.

As already noted, the OH and H$_2$O masers may be associated with protostars (stars in the process of formation) since these H II regions are known to contain young stars. However, many cases of OH and H$_2$O emission are known to come from infrared stars which are highly-evolved (M-type).[21-23, 45, 46] The characteristic these emitters most often share with the H II region ones is the presence of considerable dust that obscures the regions. However, at this point, not all the H II region emitters appear to be significant infrared emitters, except perhaps at the very long infrared wavelengths. Not only does the dust protect the molecules from photo-dissociating ultraviolet radiation and perhaps provide surfaces on which to form, but the high dust concentration also indicates high gas densities which increase molecular formation rates, microwave optical depths, and probably excitation rates (either collisional or radiative) of low-lying energy levels. However, the infrared fluxes from the infrared stars themselves and their envelopes provide extra pump rates. Such a case may be the 1612-MHz OH maser emission.[47] Details of maser pumping mechanisms are discussed below.

One other class of molecular clouds is the dark dust clouds that show very little of the 21-cm hydrogen atom line, weak 18-cm OH emission[48-50] (anomalous in the satellite lines)[51] and anomalous 6-cm formaldehyde absorption.[12] Probably no case of formaldehyde microwave emission has been found anywhere,[52, 53] but in this case, since there is no apparent background signal to be absorbed except for the 2·7°K cosmic background, there must be anti-maser action to 'cool' the microwave population differences below the 2·7°K value. Details of these cases will be considered below.

The OH emitters with the highest brightness, usually near H II regions, are almost always polarized, and usually highly circularly. However, particular Doppler-shifted features in a few such regions show elliptical, linear, or no polarization. This is true of all four of the hyperfine-split transitions of the ground state. The H_2O maser emission is rarely polarized and then only linearly-polarized.

Maser amplification, under the proper conditions of saturation, can result in a high degree of polarized emission. Laboratory experiments on laser oscillators in weak or no magnetic field have verified to a fair degree the predictions of non-linear optics that for laser electric dipole transitions between states of the same angular momentum quantum numbers, the output has nearly pure circular polarization.[54, 55] Cavity anistropy will produce some ellipticity. Similarly, for laser transitions between states of angular moments differing by unity, excluding $1 \leftrightarrow 0$, the output has linear polarization. Similar polarization characteristics are believed to occur in traveling-wave amplifiers of monochromatic signals. However, the OH and H_2O cases do not directly fit into these cases because, first, the interstellar signals are broadband, the widths probably being considerably larger than the radiative or collisional linewidths[56, 57] of the molecular states, and second, the Zeeman splitting due to the interstellar magnetic field also is likely to be larger than these same linewidths. Thus, it seems the interstellar maser polarization is not easily explained by analogy with laboratory lasers.

2.2 Asymmetric molecules—weak masers

Some of the more recently discovered molecules like methyl alcohol[14] and formic acid[58] are probably masers with very small microwave optical depths. These molecules are seen in emission despite the extremely high background continuum temperature T_c of the galactic center (Sgr A: $T_c \approx 5000°K$

and Sgr B_2: $T_c \approx 250°K$, at 834 MHz). Furthermore, the antenna temperature (about 0·5°K) does not depend much on the spatial details of the continuum temperature, indicating that the excitation temperature, given by

$$T_x = -\left(\frac{h\nu}{k}\right)\bigg/\ln(N_u g_l/N_l g_u),$$

is negative. Here N_u and N_l are the population densities for the upper and lower maser levels, and g_u and g_l are their degeneracies. For a uniformly-excited cloud that subtends a fraction Ω_c of the effective antenna beam angle, the antenna temperature T_a is approximately given[59] by

$$\Delta T_a = T_a - T_c = \Omega_c (T_x - T_c)(1 - e^{-\tau}),$$

where $\Omega_c = \Omega_{\text{source}}/\Omega_{\text{beam}}$ and $\Omega_{\text{beam}} = \lambda^2/A_a$ (A_a is the effective antenna area) and where τ is the microwave optical depth when microwave saturation can be neglected. Saturation occurs when the signal is sufficiently strong to change the populations significantly. For masers, τ and T_x are negative, that is, $N_u/g_u > N_l/g_l$, the condition of population inversion.

For small τ

$$\Delta T_a \approx \Omega_c (|T_x| + T_c)|\tau|.$$

When $|T_x| \gtrsim T_c$, the antenna temperature will not depend very strongly on T_c. However, if T_x were positive, T_x would have to be greater than T_c in order for emission to obtain. Such high values are unphysical unless the populations of the microwave levels are almost inverted anyhow. As a function of increasing pump intensity the excitation temperature heads toward ∞ and then $-\infty$ and then -0, for complete emptying of the lower state.

Other asymmetric-top molecules may be expected to be weak masers. This was found from calculations taking a temperature T_R of about 2·7°K for the radiation, at all but the very lowest frequencies, and kinetic temperatures of about 10–30°K.

Excited rotational levels of OH have shown fairly strong emission, the brightness temperatures being unknown since radiation emission angles are not available yet from interferometry. With the exception of the $J = 1/2$ Λ-doublet, the higher rotational states in the $\Pi_{1/2}$ fine-structure ladder of rotational levels have not been detected. However, emission from the $J = 5/2$ and $7/2$ states have been detected in the $\Pi_{3/2}$ ladder. This might suggest that the $\Pi_{3/2}$ ladder might have Λ-doublets that are all inverted because of one and the same pump mechanism. However, reasons can be found for obtaining population inversion for each of the maser doublets because of special properties of the doublet. For example, there is a very small hyperfine splitting for the upper Λ-doublet state of $J = 7/2$ but not for the lower state. Trapping of the infrared radiation between the $J = 7/2$ and the next state, $J = 9/2$, transfers population from the $F = 4$ to the $F = 3$ state in the lower doublet state but no such transfer occurs in the upper doublet state. Thus, a $F = 4 \to 4$ maser would be expected, as observed.

Widespread microwave absorption by OH has been found throughout the galaxy. Nowhere does the absorption appear completely in equilibrium, as attested to by the unequal satellite line signals at 1612 and 1720 MHz though they have equal line strengths. Generally, however, the agreement of the absorption profiles with those of hydrogen 21 cm is satisactory. The anomalies in the satellite lines might be attributed to collections of dense cloudlets that subtend very small angles compared to the beam angle.

Widespread formaldehyde 6-cm absorption closely associated with the OH, has also been mapped. Unlike OH, however, no emission at this wavelength has yet been conclusively observed,[53] although several reasons for expecting emission have been offered. When the density is sufficiently high, about 10^6 cm^{-3} or more, the effects of millimeter wave pumping, if any, might be small compared to collisional effects which may cause emission and even maser action.[28, 29]

Complementary data to that at low frequencies is the millimeter wave emission observed at 2mm from formaldehyde, hydrogen cyanide, carbon monoxide and carbon monosulphide. The CO molecules come to equilibrium at the kinetic temperature because of the small radiative decay constant of the emitting rotational state compared to collisional de-excitation rates at even moderate densities, but such appears not to be the case for the others. Densities of almost 10^6 cm^{-3} have been deduced.

III. NON-EQUILIBRIUM CONDITIONS

3.1 Formaldehyde—anti-maser

The four levels of Fig. 1c have steady-state populations given[28] by the following ratios in terms of the rates W_{ij} for $i \to j$:

$$\frac{N_1}{N_2} = \left[W_{21} + \frac{(W_{23}W_{34} + W_{24}W_3)W_{41} + (W_{24}W_{43} + W_{23}W_4)W_{31}}{W_3 W_4 - W_{34}W_{43}} \right]$$

$$\times \left[W_{12} + \frac{(W_{13}W_{34} + W_{14}W_3)W_{42} + (W_{14}W_{43} + W_{13}W_4)W_{32}}{W_3 W_4 - W_{34}W_{43}} \right]^{-1}$$

and

$$\frac{N_3}{N_4} = \left[W_{43} + \frac{(W_{41}W_{12} + W_{42}W_1)W_{23} + (W_{42}W_{21} + W_{41}W_2)W_{13}}{W_1W_2 - W_{12}W_{21}} \right]$$

$$\times \left[W_{34} + \frac{(W_{31}W_{12} + W_{32}W_1)W_{24} + (W_{32}W_{21} + W_{31}W_2)W_{14}}{W_1W_2 - W_{12}W_{21}} \right]^{-1}$$

where $W_1 = W_{12} + W_{13} + W_{14}$, $W_2 = W_{21} + W_{23} + W_{24}$, etc.

Other ratios may be obtained by interchanging the appropriate indices. Often the rates W_{42}, W_{31}, W_{24} and W_{13} are very large compared to the rest. For example, these may be the millimeter-wave transitions between adjacent rotational levels. In formaldehyde these W's are $\approx 10^{-4}$ sec^{-1}, compared to collision rates of 10^{-6} to 10^{-4} sec^{-1} for densities from 10^3 to 10^5 cm^{-3}.

Then,

$$\frac{n_1}{n_2} \approx \frac{W_{21} + (W_{24}/W_{42})(W_{41} + W_{43}) + W_{23}}{W_{12} + (W_{13}/W_{31})(W_{32} + W_{34}) + W_{14}}$$

and

$$\frac{n_3}{n_4} \approx \frac{W_{43} + (W_{42}/W_{24})(W_{23} + W_{21}) + W_{41}}{W_{34} + (W_{31}/W_{13})(W_{14} + W_{12}) + W_{32}}$$

Usually, the energy splitting E_{43} is greater than E_{21} so that $W_{13}/W_{31} > W_{24}/W_{42}$ since these ratios are proportional to their appropriate Boltzmann factors (the degeneracy factors are the same if K-doublet states are involved). The temperature in the Boltzmann factor is usually the millimeter-wave radiation temperature ($\approx 2.7°$K), although analogous cases arise in the far infrared, where the temperature might be higher ($\approx 100°$K). When the radiation temperature is lower than the kinetic temperature, anomalous absorption in the $1 \to 2$ transition can result from collisional[40] or optical pumping if $W_{23} > W_{14}$. The terms with W_{24}/W_{42} and W_{13}/W_{31} are small. However, anomalous absorption $1 \to 2$ arises from an excess of millimeter radiation in the $2 \to 4$ transition compared to the $1 \to 3$ transition, if these latter terms are not too small.[29] It is important to note that the $3 \to 4$ transition will also be anomalously absorbing if $W_{23} > W_{14}$, because of the large terms containing W_{42}/W_{24} and $W_{31}/W_{12} \approx 10$ in the expression for n_3/n_4. However, the pumping by excess $2 \to 4$ radiation will easily cause less absorption and even population inversion for the $3 \to 4$ transition, a case not yet observed. Infrared pumping[28] for anomalous absorption involves the excitation of the lowest rotational levels of vibrational modes that have their oscillating dipole moment along the b-axis (in the plane of the molecule but perpendicular to the symmetry axis). This vibration is most likely the asymmetric C–H stretching mode (≈ 2874 cm^{-1}).[60] A lower frequency bending mode also has a b-axis dipole moment. However, another bending mode of similar frequency has a c-axis dipole moment which produces pumping of the opposite sign, thereby partially cancelling the other mode's effect. Figure 5 illustrates the pumping scheme whereby excitation to the vibration–rotation levels (indicated by asterisks) is followed by rapid cascading to the ground vibrational state. The $J = 1, 2$, and 3 ($K = 1$) doublets are all anti-inverted by this pumping. Pump rates of 10^{-7} to 10^{-5} sec^{-1} are provided by the resonance radiation from the shock front. The

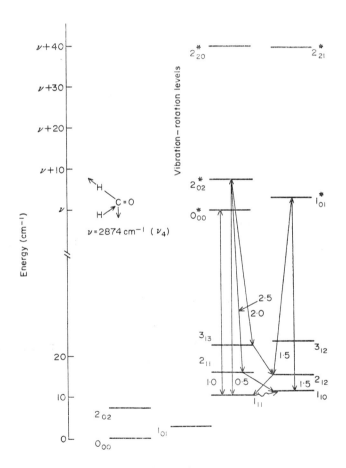

FIGURE 5

intensity of the ith infrared line is given by

$$I_i \approx \int \Omega \, dr \, hv_i \, n_i \, W_i/4\pi\delta v$$

where v_i is the infrared frequency, δv is the Doppler line width due to fluid velocities mainly, and n_i is the number density of H_2CO in the rotational state that is excited to emit the appropriate resonance line. The rate, W_i, per H_2CO molecule, of exciting this vibrational state by collisions in the front is proportional to $\exp(-hv_i/kT)$ times a molecular cross-section with hydrogen that is about 0·1 times the kinetic cross-section. This is estimated from data on C_2H_4–H_2 collisions at 288°K. The solid angle Ω is that subtended by the source at the molecules being pumped.

Upon multiplying I_i by the appropriate Einstein B-coefficient, one obtains the rate of infrared absorption to be

$$W_i' = B_i I_i \approx \int d\tau_i \, \Omega \, W_i/4\pi \, (\delta v/\delta v_T),$$

where $\tau_i = \int dr \, n_i \, B_i \, hv_i/\delta v_T$ is the infrared optical depth at line centre in the hot layer, and where δv_T is the Doppler width due to thermal molecular motion. The pump rate due to p infrared lines, which have pump cascade efficiencies of $\eta_i \simeq \pm 0·5$, may be compared with the rates W_C due to thermal collisions, which have an efficiency $-hv_J/2kT_0$, where v_J is the microwave frequency for the doublet transition in the level with angular momentum J, and T_0 is the kinetic temperature of the cold molecules ($\lesssim 10$°K) at some large radius r_0 from the center of the cloud. The amount of anomalous absorption depends heavily on the ration of these two rates, which is given by

$$C_{JJ}^- = \frac{\tau_S r_S^2 n(r_S) k T_0}{2(\delta v/\delta v_T) r_0^2 n(r_0) h v_J} \approx 2·5 \, \tau_S \, T_0 \, (\delta v/\delta v_T)^{-1} \quad \text{for} \quad J=1$$

where

$$\tau_S = \sum_{i=1}^{p} \eta_i \int d\tau_i \exp(-hv_i/kT) \, (\sigma_i v)_S/(\sigma v)_0$$

is an effective H_2CO optical depth. The quantities $(\sigma_i v)_S$ and $(\sigma v)_0$ are products of cross-section and relative thermal velocity for excitation of a vibrational state and of the microwave doublet, respectively. The optical depth $\tau_S \approx 10^{-15} N_{H_2CO}$ (the column density of H_2CO in the shock-heated layer) for $T \approx 1000$–3000°K and $T_0 \approx 10$°K. The H_2CO-to-hydrogen ratio is $\approx 10^{-8}$, according to recent surveys. Cases of collapsing spherical shock waves with radiation cooling give $\approx 10^{22}$ cm^{-2} for the projected density of hydrogen in the hot layer. Thus τ_S is probably close to unity if the ratio of H_2CO to hydrogen is $\gtrsim 10^{-7}$ in the shock heated-layer and if some molecular destruction follows during the cooling process. Because $\delta v/\delta v_T \gtrsim 5$ for the

Table V

Typical dust cloud properties

No. stars within	2
Total extinction	> 8 magnitudes
Distance	200 pc
Radius	1 pc
Density of dust	$> 10^{-10}$ cm^{-3}
H_2 density	> 500 cm^{-3}
Total mass	$> 100\, M_\odot$
(Gravitational energy)/ (Twice thermal energy)	> 1
Kinetic temperature	5–10 °K
H_2CO 6-cm temperature	≈ 2 °K

shocks considered here in typical dark clouds that are gravitationally-unstable, we have $C_{11}^- \gtrsim 5$, which leads to very low values for T_1 (the excitation temperature for the $J = 1$ doublet). Table V lists average properties of these clouds. An expression for $(n_3 + n_4)/(n_3 - n_4)$ is obtained by adding one to and then subtracting one from the expression for n_3/n_4 and then taking the ratio of the results. Similarly, an expression for $(n_1 + n_2)/(n_1 - n_2)$ is obtained from the n_1/n_2-expression. The excitation temperature T_2 for the $J = 2$ doublet is defined by $\exp(h\nu_2/kT_2) = n_3/n_4$ (the degeneracies of the two doublet states are the same). Then

$$T_2 \approx \frac{h\nu_2}{2k} \coth\left(\frac{h\nu_2}{2kT_2}\right) = \frac{h\nu_2}{2k} \frac{n_3 - n_4}{n_3 - n_4} \approx T_0 \frac{1 + C_{21}^+ + C_{22}^+}{1 + C_{21}^- + C_{22}^-}$$

where

$$C_{21}^{\mp} \approx \left[\frac{W_{42}}{W_{24}}(W_{23} + W_{21}) \mp \frac{W_{31}}{W_{13}}(W_{14} + W_{12})\right](W_{43C} \mp W_{34C})^{-1}$$

and

$$C_{22}^{\mp} \approx (W_{41} \mp W_{32} + W_{43P} \pm W_{34P})(W_{43C} \mp W_{34C})^{-1}.$$

The infrared pumping terms are mainly W_{14}, W_{41}, W_{23}, and W_{32}. The next higher doublet ($J = 3$) contributes to W_{34} and W_{43} (denoted by subscript P) by millimeter-wave fluorescence that is excited following infrared pumping. The microwave-induced rates are also included in these terms. The important collisions term are mainly contained in W_{12}, W_{21}, W_{34}, and W_{43} (denoted by subscript $_C$). However, the collisional contributions to

W_{14}, W_{41}, W_{23}, and W_{32} may not be negligible, even though these involve "forbidden transitions".

The excitation temperature T_1 for $J = 1$ is obtained in a similar manner,

$$T_1 \approx T_0 \frac{1 + C_{11}^{+} + C_{12}^{+}}{1 + C_{11}^{-} + C_{12}^{-}}$$

where

$$C_{11}^{\mp} = \left[\frac{W_{24}}{W_{42}}(W_{41} + W_{43}) \mp \frac{W_{13}}{W_{31}}(W_{32} + W_{34})\right](W_{21C} \mp W_{12C})^{-1}$$

and

$$C_{12}^{\mp} = (W_{23} \mp W_{14} + W_{21P} \mp W_{12P})(W_{21C} \mp W_{12C})^{-1}.$$

Here W_{12P} and W_{21P} consist of only the microwave-induced rates. We use the approximations that $W_{42}/W_{24} \gtrsim W_{31}/W_{13} \approx \frac{3}{5} \exp[E_{31}/k(2\cdot 7°K)] \approx 10$ for clouds that are optically-thin to millimeter waves. Then we find that

$$C_{21}^{+} \approx 10 \gg C_{22}^{+}$$
$$C_{21}^{-} \approx (C_{11}^{-} + 1)10 \gg C_{22}^{-}$$
$$C_{12}^{-} \ll C_{11}^{-} \approx 0\cdot 5\, T_0 \text{ (cf. p.224)}$$
$$C_{12}^{+} \ll C_{11}^{+} \approx 0\cdot 25.$$

Then, in many cases,

$$T_1 \approx T_0 (1 + C_{11}^{+})/(1 + C_{11}^{-}) \gtrsim 2\cdot 1°K \quad \text{(for } T_0 \gtrsim 10°K\text{)}.$$

When infrared and collisional pumping is unimportant then one should not neglect C_{12}^{-}, which is responsible for the millimeter-wave pumping effects of Solomon and Thaddeus. For sufficiently thick clouds, the radiation trapping of millimeter waves that are internally-generated by collisional excitation of rotation greatly increases the ratios W_{24}/W_{42} and W_{13}/W_{31} over their values at $2\cdot 7°K$. Since the second ratio stays larger than the first ratio, as cases of greater trapped intensity are considered, the tendency is for "heating" of the doublets rather than "cooling" (anomalous absorption) regardless of the characteristics of the $2\cdot 7°K$ background radiation.[28,61]

Anomalies in the OH so-called normal emission in the same dark clouds suggest that infrared pumping might be occurring for this molecule, too. Previous calculations[47] concerning optically-thin OH clouds showed that 1720 MHz emission might be amplified while the 1612 MHz would be anomalously-absorbed, as observed.[51] Infrared OH lines from the shock-heated layer would pump the cold OH left behind the shock. The microwave characteristics of the cloud for both H_2CO and OH are dominated by the outer regions of the cloud, because of the large angle subtended and because of the large microwave optical depths for the cold regions. Direct evidence for the shock-heated layer might be in detection of broad thermal microwave emission from optically-thin OH and H_2CO or in infrared molecular lines, especially of H_2 at 12μ.[62]

3.2. Collisions across K-doublets and for exciting rotation

Electrons and ions interact with the electric dipole of the molecules that are observed by their microwave lines. With electrons, the cross-sections are nearly proportional to the line strengths between the states coupled by collisions. Ions, however, penetrate into the molecule's electron cloud and cause significant cross-sections for dipole-forbidden collision-induced transitions. Collisions with neutrals are somewhere between these two cases, with perhaps some importance given to the dipole selection rules for heavy neutrals. However, H, H_2, and He might have significant dipole-forbidden cross-sections.[63-67]

Let us first consider the electric dipole (dipole moment operator μ) interaction of the molecule with a point charge.[68] The potential energy is $V(\mathbf{r}) = e\mathbf{\mu} \cdot \mathbf{r}/r^3$, where $\mathbf{r}$ is the distance between the dipole and the charge. Between the states a and b we have the dipole matrix element $\mu_{ba} \exp(-i\omega_{ba}t)$ in the interaction representation. If we imagine that the long-range collisions hardly deflect the particles from straight line-paths, then $\mathbf{r} = \mathbf{b} + \mathbf{v}t$, where $\mathbf{b}$ is the vector distance of closest approach. We have, of course, $r^2 = b^2 + v^2 t^2$ ($-\infty < t < \infty$). According to the usual quantum mechanical procedures, the matrix element for inelastic scattering is

$$M_{ba} = \left\langle b \left| T \exp\left[-i \int_{-\infty}^{\infty} dt\, \hbar^{-1} \exp(-iH_0^x t)\, V(\mathbf{r}(t))\right] \right| a \right\rangle$$

where T denotes time ordering and the superscript x denotes the commutator. This use of a straight-line trajectory depends upon the interaction potential being smaller than kT. This obtains for impact parameters less than a critical impact parameter $r_D = (\mu e/kT)^{1/2}$. This quantity is usually larger than the deBroglie wavelength/2π of the projectile, below which impact parameter quantum-mechanical effects occur. Fourier-transforming the interaction in time demonstrates the importance of small values of the parameter $\omega_{ba} b/v$, corresponding to the diabatic limit. The adiabatic approximation corresponds to large values of $\omega_{ba}b/v$, for which transition rates are very slow. Another criterion worth discussing is whether the interaction is weak or strong, depending upon whether the argument I of the trigonometric function in M_{ba} is small or large, where

$$I = \frac{e}{\hbar} \int_{-\infty}^{\infty} \frac{\mu_{ba} \cdot (\mathbf{b} + \mathbf{v}t) e^{-i\omega_{ba}t} dt}{(b^2 + v^2 t^2)^{3/2}}.$$

For a given v, sufficiently small b is seen to give large I. If $I = \pi/4$ for $b = b_c(v)$, then the contribution to the total cross-section is approximately $(\frac{1}{2})\pi b_c^2$. To this is added the weak collision contribution when b is greater than b_c, given by the Born approximation. When ions are involved, the low velocities require small b for $\omega_{ba} b/v$ to be small, whether or not I is large. The cross-section is approximately $\frac{1}{2}\pi(v/\omega_{ba})^2$ provided I is large and $b_c \approx v/\omega_{ba}$. The exponential decay with respect to the quantity $\omega_{ba}b/v$ makes itself felt when integrating over all impact parameters and velocities. This general scheme of calculation can be used for other multipole–multipole interaction potentials. The hard sphere type collisions are handled differently, however, through the use of partial waves.

3.3. Degree of ionization

The degree of ionization in clouds containing molecules in significant numbers is probably very small. If only the wavelengths shorter than the hydrogen cut-off at 916 Å were absent, then the species with ionization potentials less than 13·6 eV would still be ionized. But further shielding from the ultraviolet radiation is possible, especially for H_2, since the fastest photodissociation occurs because radiation, such as $Ly\,\alpha$ continuum, excites the H_2 molecule to a state which fluoresces leaving the H_2 in an unstable state that dissociates. However, the "uv" radiation can be selectively absorbed in the outer layers of a dense enough cloud. Other sources of ionization are cosmic rays, X-rays, and thermal sources like shockwaves. The rays are also sources of heating. If the temperatures are found to be significantly higher than 3°K, suorces of heating are believed to be necessary. One possibility is the conversion of gravitational and kinetic energy into thermal energy by shockwaves. This is of course only temporary, taking about 10^5 years, the travel time over a parsac distance for a 10 km/sec front.

Table VI

Degree of ionization

Density (cm^{-3})	10^2	10^3	10^4
Metals only—ultraviolet	$\leqslant 4 \times 10^{-4}$	$\leqslant 4 \times 10^{-5}$	$\leqslant 4 \times 10^{-6}$
Hydrogen-cosmic rays and X-rays	$\leqslant 6 \times 10^{-4}$	$\leqslant 10^{-4}$	$\leqslant 10^{-5}$

Table VI indicates that for densities greater than 10^3 cm^{-3}, the degree of ionization is less than 10^{-4}, and very likely less than 10^{-5}.[69] These low values suggest that collisions due to electrons or ions are unimportant in the dark dust clouds.

3.4. Collisions with hydrogen

Collisions with neutrals have been very rarely calculated or measured. Kinetic cross-sections give fairly good estimates for excitation of rotation or K-doublets. However, non-equilibrium populations can result from details of this type of excitation because of preferences of exciting one over another K-doublet state. In classical terms, as is often the case, a molecule is induced to rotate more often about its axis of maximum moment of inertia rather than the others. For H_2O or H_2CO it is because of the light hydrogen atoms stuck out in the plane of each molecule.[70] However, quantum mechanical calculations indicate that the classical calculations might be deceptive when applied to low-lying energy levels at low temperature. The cooling effect in K-doublets expected from classical collisions is found to occur at kinetic temperatures only as high as 40°K in formaldehyde. Below this temperature a strong tendency for maser action is calculated, at least for hard spheres. The relavent cross-sections were calculated by Thaddeus[71] on a quantum basis for hard spheres in the impulse approximation and also in the perturbation approximation of small radii. The results were similar

Table VII

H_2CO Collision de-excitation rates (sec^{-1})

($T_k \approx 10°K$)

	$2_{12} \to 1_{11}$	$2_{11} \to 1_{10}$	$2_{11} \to 2_{12}$	$1_{10} \to 1_{11}$
Dipole-allowed types				
Dipole-charge (e^-)	$1 \times 10^{-6} n_e$	$1 \times 10^{-6} n_e$	$2 \times 10^{-5} n_e$	$2 \times 10^{-5} n_e$
Dipole-quadrupole (H_2)	$4 \times 10^{-10} n_{H_2}$	$4 \times 10^{-10} n_{H_2}$	$4 \times 10^{-10} n_{H_2}$	$4 \times 10^{-10} n_{H_2}$
Hard-sphere (including Van der Waals)	$2 \times 10^{-11} n_{H_2}$	$2 \times 10^{-11} n_{H_2}$	$3 \times 10^{-11} n_{H_2}$	$3 \times 10^{-11} n_{H_2}$
Dipole-forbidden types	$2_{12} \to 1_{10}$	$2_{11} \to 1_{11}$		
Dipole-induced dipole (H or H_2)	$4 \times 10^{-10} n_{H_2}$	$4 \times 10^{-10} n_{H_2}$		
Hard-sphere (including Van der Waals)	$7 \times 10^{-12} n_{H_2}$	$1 \times 10^{-11} n_{H_2}$		

in both approximations. Namely, strong tendencies for maser (not anti-maser) action at low temperatures are implied by this model. Recent calculations by Halket and Litvak indicate that the inclusion of the attractive van der Waals forces does not change this result much. However, it was found that the "forbidden" collisions (in W_{23} and W_{14}) may be dominated by the dipole-induced dipole collisions (of electrostatic origin) which show no preference for a K-doublet level, thereby greatly diluting the maser effect of hard sphere type collisions. Also we find that the "dipole-allowed" collisions (in W_{13} and W_{24}) are stronger than the "forbidden" ones, for both electrostatic and hard sphere type collisions. In a two temperature non-equilibrium (e.g., $T_R = 2.7°K$, $T_K = 10°K$) with collisions showing no preference for a K-doublet level, there is a tendency for maser action or at least heating of the $J = 1$ doublet and the opposite in the $J = 2$ doublet. Thus, the lack of evidence for H_2CO emission in the $J = 1$ transition would argue for optical pumping to off-set the effects of collisions. The classical and quantum collision calculating are outlined below. The results of our calculations on the electrostatic and hard-sphere type collisions are given in Table VII.

Selection of Hyperfine States

Because the duration of most collisions is quite short, between 10^{-13} and 10^{-12} sec, except for the formation of metastable complexes, the hyperfine states are usually left unperturbed.[27] That is, the forces necessary to move the nuclear spin $\mathbf{I}$ to a preferred direction, so that a particular $\mathbf{F} = \mathbf{J} + \mathbf{I}$ state is formed, are very weak. For example, if the $I = \frac{1}{2}$ spin states are equally populated either parallel or anti-parallel to $\mathbf{J}$ before the collision, they will remain so after the collision. Thus, collisions are not by themselves sources of pumping of the satellite lines of OH, for example. Although unequal trapping of hyperfine components of fluorescence induced by collisions might account for the selection of hyperfine states.

(i) *Classical collisional model for formaldehyde*[70]

Conservation of angular momentum: $\mathbf{I} \cdot \mathbf{\Omega} + m\mathbf{r} \times \mathbf{v} = \mathbf{I} \cdot \mathbf{\Omega}' + m\mathbf{r} \times \mathbf{v}'$

Conservation of energy: $\frac{1}{2}m(|\mathbf{v}|^2 - |\mathbf{v}'|^2) = \frac{1}{2}(\mathbf{\Omega}' \cdot \mathbf{I} \cdot \mathbf{\Omega}' - \mathbf{\Omega} \cdot \mathbf{I} \cdot \mathbf{\Omega})$

(neglecting center of mass motion)

$\mathbf{I} \equiv$ moment of inertia tensor

$\mathbf{\Omega} \equiv$ angular velocity before collision (primes indicate quantities after collision)

$\mathbf{v} \equiv$ velocity of projectile (of mass m)

$\Delta \mathbf{v} = \mathbf{v} - \mathbf{v}' = \gamma |\Delta \mathbf{v}|$ (γ is a unit vector in the direction of the momentum transfer)

$\Delta \mathbf{\Omega} = \mathbf{\Omega}' - \mathbf{\Omega}$.

Then $\mathbf{I} \cdot \Delta \mathbf{\Omega} = m\mathbf{r} \times \Delta \mathbf{v} = m\mathbf{r} \times \gamma |\Delta \mathbf{v}|$ and

$$m(2\mathbf{v} \cdot \gamma |\Delta \mathbf{v}| - |\Delta \mathbf{v}|^2) = 2\Delta\mathbf{\Omega} \cdot \mathbf{I} \cdot \mathbf{\Omega} + \Delta\mathbf{\Omega} \cdot \mathbf{I} \cdot \Delta\mathbf{\Omega}$$
$$= 2m\mathbf{r} \times \gamma \cdot \mathbf{\Omega} |\Delta \mathbf{v}| + m^2(\mathbf{I}^{-1} \cdot \mathbf{r} \times \gamma) \cdot (\mathbf{r} \times \gamma)|\Delta \mathbf{v}|^2$$

therefore, $|\Delta \mathbf{v}| = (2\mathbf{v} \cdot \gamma - 2\mathbf{r} \times \gamma \cdot \mathbf{\Omega})(1 + m\mathbf{r} \times \gamma \cdot \mathbf{I}^{-1} \cdot \mathbf{r} \times \gamma)^{-1}$

and $\mathbf{I} \cdot \Delta\mathbf{\Omega} = [2m(\mathbf{r} \times \gamma)(\mathbf{v} + \mathbf{r} \times \mathbf{\Omega}) \cdot \gamma](1 + m\mathbf{r} \times \gamma \cdot \mathbf{I}^{-1} \cdot \mathbf{r} \times \gamma)^{-1}$

The mean square transfer of angular momentum is

$$\overline{|\mathbf{I} \cdot \Delta\mathbf{\Omega}|_j^2} = \int\int \sum_i \left[\frac{2m(\mathbf{v} + \mathbf{r}_i \times \mathbf{\Omega}) \cdot \gamma}{1 + m(\mathbf{r}_i \times \gamma) \cdot \mathbf{I}^{-1} \cdot (\mathbf{r}_i \times \gamma)} \right]^2 |\mathbf{r}_i \times \gamma|_j^2 \, P_i p(\gamma) \, d^2\gamma p'(\mathbf{v}) \, d^3v$$

where P_i is the relative collision cross-section for the ith atom of the molecule $p(\gamma)$ is the probability distribution for the momentum transfer, and $p'(\mathbf{v})$ is the probability distribution for the projectile velocity.

(Subscript j denotes the jth component of the vector which is being squared).

According to the correspondence principle, the mean square angular momentum change about either the b- or c-axis of the molecular system ought to be proportional to the corresponding collision cross-section for a given projectile speed. Exciting the $1_{11} \rightarrow 2_{11}$ transition (W_{14}) adds rotation about the b-axis and exciting the $1_{10} \rightarrow 2_{12}$ transition (W_{23}) adds rotation about the c-axis (the second subscript, K_+, increased). For formaldehyde, the ratio is about 1:2 for the b:c cross-sections. A ratio of about 2:3 would just suffice for anti-maser pumping, provided that the density is between 10^3 and 10^5 cm^{-3}.

(ii) *Quantum collisional model for formaldehyde*[70, 71]

Assume a collision with each atom in the molecule produces a known scattering amplitude which adds to the others with the appropriate phase-factor $\exp(i\mathbf{K}(\theta) \cdot \mathbf{r}_i)$, where $\mathbf{K}(\theta)$ is the momentum transfer in the center-of-mass frame and $\mathbf{r}_i$ is the coordinate vector of the ith atom of the molecule. The differential cross-section for scattering, associated with 2^l-pole type selection rules for $J' \rightarrow J$ and $K_-' = K_- = 1$ is given by

$$\frac{d\sigma^{(l)}}{d\Omega} = \frac{k}{k'} \tfrac{1}{2}(1 + \varepsilon_l) \frac{\pi^2(2J+1)}{4^l \Gamma((2l+3)/2)^2} \binom{J' \, lJ}{-1 \, 0 \, 1}^2 \left| \sum_{i,j} e^{i\delta_{ij}} \sin \delta_{ij} \right|$$

$$P_j(\cos\theta)(2j+1)(k'r_i)^l Y_i^{(l)}|^2 \left(\frac{K(\theta)}{k'} \right)^{2l}$$

where $\begin{pmatrix} J' & l & J \\ -1 & 0 & 1 \end{pmatrix}$ is the Wigner 3j-symbol, δ_{ij} is the scattering phase-shift for the ith atom and the jth partial wave. The quantity

$$Y_i^{(l)} = Y_{l0}(\omega_i') + (-1)^{S'} \begin{pmatrix} J' & l & J \\ 1 & -2 & 1 \end{pmatrix} \tfrac{1}{2}[Y_{l,2}(\omega_i') + Y_{l,-2}(\omega_i')] \Big/ \begin{pmatrix} J' & l & J \\ -1 & 0 & 1 \end{pmatrix}$$

where $\varepsilon_l = (-1)^{S+S'+J'+l+J}$, and S or S' equal zero for the upper state of a K-doublet (positive parity for $J = 1$) and equal to unity for the lower state (negative parity for $J = 1$). (If spin is included, then N and N' replace J and J', respectively.) The primes on the quantum numbers and wave numbers denote the initial state. The arguments ω_i' of the normalized Legendre polynomials $Y_{l,m}$ are the angular coordinates of the ith atom with respect to the center-of-mass, $K(\theta)$ is the magnitude of $\mathbf{K}(\theta)$ and equals $[k^2 + k'^2 - 2kk' \cos\theta]^{1/2}$, where k' is the magnitude of the incident momentum (in units of $\hbar$), k is the magnitude of the final momentum and θ is the angle of scattering. For s-wave scattering only, the total cross-section is

$$\sigma^{(l)} \approx \frac{k}{k'} \tfrac{1}{2}(1 + \varepsilon_l) \frac{4\pi(2J+1)\begin{pmatrix} J' & l & J \\ -1 & 0 & 1 \end{pmatrix}^2}{[(2l+1)!!]^2} \left| \sum_i e^{i\delta_{i0}} \sin\delta_{i0} (k'r_i)^l Y^{(l)} \right|^2$$

$$\cdot 2\pi \int_0^\pi d\theta \sin\theta \, (K(\theta)/k')^{2l}$$

where the integral factor equals $\dfrac{(k+k')^{2(l+1)} - (k-k')^{2(l+1)}}{kk'(k')^{2l}(l+1)} \pi$. For H_2CO and $l = 2$, the H's do *not* project out enough to make Y_{20} negative to favor $S' = 1$, $J = 2$ *anti*-maser effects.

3.5. Radiative transition rates

(i) Optical pumping

Molecules exhibit fluorescence from rotational states, vibrational-rotational states in the ground and electronically-excited states. Continuum radiation in the far infrared is available from large clouds of hot grains, for example the Kleinmann–Low Nebula in Orion A. Internal sources of excitation like hot stars might be present. Near infrared radiation from the surface of cool stars can excite the vibrational states. Ultraviolet radiation from hot stars can excite the electronic states. However, line radiation from hot molecules can come from shock-heated layers and excite similar molecules that are relatively cold either upstream or downstream from the shock front. For moderate density ahead of the shock and moderate Mach number, the lines are in the infrared wavelengths. For higher densities and Mach numbers ultraviolet lines may also be strong, along with increased ionization.

Table II lists the photons/sec output that is available for the different optical pumping sources, assuming that 100% efficiency were possible.[38] In comparison, we list the photons/sec emitted by various OH maser sources. For saturated maser emission the pump photons must account for the microwave photons, allowing for the efficiency of the pump process. For unsaturated emission, many pump photons are wasted.

(*a*) *Ultraviolet pumping.* Hydroxyl and formaldehyde are ideally suited for pumping by long wavelength ultraviolet radiation ($\approx$ 3100–3600 Å). Both have sharp well-known resonance lines which usually do not lead to dissociation of the molecule. Certain vibration–rotation levels (higher than those involved here) of the ultraviolet state ($A^2\Sigma^+$ in OH and $^1A''$ (B_2) in H$_2$CO) are predissociated. Predissociation is the result of the crossing of the bound state potential energy curve with that of an unbound state. Through certain interactions, often somewhat subtle, the molecular wavefunction is a superposition of the bound and unbound wavefunctions near the crossing point. The molecule may be thought of as switching from the bound potential to the unbound one as the molecule executes vibration and rotation near the crossing point. The nature of this predissociation in OH is still somewhat controversial despite recent work. As shown in Fig. 6, the lowest unbound potential curve is likely to be a $^4\Sigma^-$ state. The interaction with the bound, ground state $X^2\Pi$ is probably weak, involving spin–orbit forces. The interaction with the $A^2\Sigma^+$ state is very weak, being second-order in the spin–orbit force. The radiative transitions from the ground state to this unbound state that would lead to photodissociation are forbidden because of the spin change for a transition from a doublet to a quartet state.[72] The next unbound state is the $^2\Sigma^-$ state, which allows photodissociation, but the oscillator strength is probably not very large since the corresponding transitions that are involved both in the separated O and H atoms and in the united atom, F, are forbidden transitions. Another higher state $B^2\Sigma^+$ has similar photodissociation properties, but at much higher photon energy ($\approx$ 13 ev instead of 8 eV). One might expect some shielding of these molecules against such 8–13 eV photons by various absorbing species (like the metals) in the outer portions of a cloud.

The reverse process, preassociation, by which O and H combine along the unstable curve ($^4\Sigma^-$) to switch to $A^2\Sigma^+$, which fluoresces in the ultraviolet lines to form $X^2\Pi$, has been a proposed pumping mechanism for the OH masers.[73] Aside from the difficulties of a slow rate of reaction, of the requirement for kinetic temperatures of > 1000°K and very high densities and large masses for the clouds, a calculation of the pumping process, including all the infrared cascading in the $^2\Pi$ state, seems to give anti-inversion rather than inversion.[74]

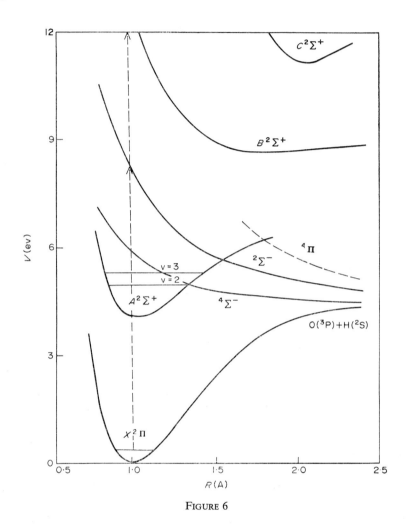

FIGURE 6

The ultraviolet pumping[11] of OH occurs mainly by six ultraviolet transitions shown in Fig. 7. These are all actually hyperfine-split. But the basic inversion of the ground state Λ-doublet can be calculated from the selective absorption of these six lines and the ensuing cascading from the $^2\Sigma^+$ levels. If the incident intensities of all the lines are equal, a thin cloud will be anti-inverted because of the $Q_1(1)$ transition despite the tendency of the $P_1(1)$ transition to produce inversion. Clouds of large optical depth will absorb out the intensity corresponding to $Q_1(1)$, leaving $R_{21}(1)$ and $R_1(1)$ to invert the Λ-doublet. Clouds having projected OH-densities of about 10^{16} cm^{-2} are good 1665 MHz-emitters, if the hyperfine-splitting of the lowest $^2\Sigma^+$ states are about 1 GHz or more (which seems possible[75]). Somewhat shorter clouds are fairly good 1667 MHz emitters. When ultraviolet resonance lines are emitted by a shock-heated layer of OH, then the $P_1(1)$ line might be sufficiently enhanced compared to the other lines to produce inversion even at the front edge of the clouds. Also, there are interesting cases for which moderate (5 km/sec) Doppler shifts of the lines between the pump source and the maser molecules cause pumping only in the transitions that produce inversion.

Shock fronts at $\approx 4000°K$ emit also infrared OH-resonance lines which

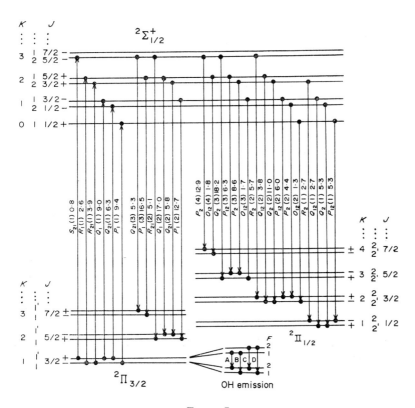

FIGURE 7

usually pump extra population in the $F = 1$ states in the ground state. The combined effect of the ultraviolet and infrared radiation is to favor 1665 MHz. A similar combination of one mechanism that inverts the Λ-doublet and another that transfers population from one hyperfine state to the other will have a similar effect. Far infrared radiation at 53 and 35µ, if somewhat stronger than the radiation at 100–300µ now observed in H II regions,[76] could also be the pump for the hyperfine preference leading to 1665 MHz.

Formaldehyde is pumped via a b-type ultraviolet transition[77] in a way similar in most details to the infrared pumping of this molecule that was discussed above. If the optical depths stay fairly low then anti-inversion in the microwave K-doublets is obtained. Perhaps unusual H_2CO microwave absorption near OH and H II regions is expected because of ultraviolet pumping.

(b) *Photodissociation of polyatomic molecules.* Recent estimates of the photodissociative lifetime of hydrogen molecules in unobscured regions is 100 years, and about the same (or a little less) for most polyatomic molecules such as water, ammonia, methane and formaldehyde.[78] The carbon monoxide lifetime is probably ten times longer.[78] In dust clouds the lifetimes increase by factors consistent with the absorption of the external ultraviolet by the dust—at least a factor of ten for each increment of the visual extinction by a factor of e. Carbon monoxide's lifetime increases more steeply, so that if the visual radiation is decreased by the dust by about ten (and at the important 1000 Å by 10^4), the CO lifetime against photodissociation is probably 10^6 years. In the darkest clouds methane and ammonia would have lifetimes hundreds of times longer than formaldehyde, but for X-rays and cosmic rays.

Since near-ultraviolet pumping has been discussed, the extinction of this radiation would be serious except that the extinction by dust inside the clouds might be greatly reduced by stellar winds, by radiation pressure from embedded stars, and by shock waves. A shock wave causes sputtering of the grains, in glancing collisions of gas particles with the surfaces of the grains that are not yet moving with the gas velocity behind the shock front,[79] thereby greatly reducing the cross-section of the grain. Likewise, radiation from a strong shock that enters dense regions can heat the grains to a few hundred degrees, enough to evaporate icy-mantles that cause most of the absorption. Silicates might be fairly transparent in the near-infrared and ultraviolet ranges.

(c) *Infrared pumping.* The very strong water vapor maser emission requires conditions of hydrogen densities somewhat below 10^9 cm^{-3} and projected densities of H_2O of perhaps 10^{20} cm^{-2} or more. The upper bound on the density is obtained from the condition that fluorescence and not collisions de-excite the rotational levels just above the upper maser level. The projected density is obtained from (1) an estimate of $\approx 10^{16}$ cm^{-2} for the projected density of the population difference in the maser transition obtained from the brightness temperature, and (2) an estimate of about 100°K for the effective rotational temperature that gives about 10^4 times as much total H_2O population as population difference in the maser levels. The water vapor emission is believed to be at least partially saturated, so that the Doppler widths provide an approximate upper bound on the kinetic temperature of a few hundred degrees. The important infrared pump routes[27] are shown in Fig. 8 that might be used if the H_2O molecules were close to an infrared source such as one of these shock fronts or an infrared star. Population inversion obtains only if the trapping of the far infrared radiation from the 6_{16} (upper maser) level to lower rotational levels is fairly important so that the lifetime of this level is comparable to that of the 5_{23} (lower maser) level.[27]

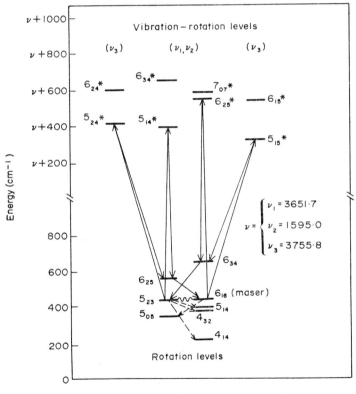

Infrared pumping of 22·2 GHz H_2O maser

FIGURE 8

Table VIII

collisional de-excitation rate (sec⁻¹)	$\int N_{OH} dr \simeq 10^{18}$ cm⁻²		frequency (MHz)	optical depth	Isotropic photons sec⁻¹	frequency (MHz)	optical depth
	collisions only (electric-dipole type)		1612	2	6×10^{41}	1612	2
						23818	6
						23827	4
10^{-3}	collisions and	$S_{NIR}/S_{Rot} =$ 3×10^{-4}	1612	3	30×10^{41}	1612	3
						23818	8
						23827	6
	near-infrared pumping	3×10^{-3}	1612	11	120×10^{41}	1612	11
			6017	2		1667	10
			4766	2		23818	8
						23827	6
	collisons only (electric-dipole type)		1612	3	1×10^{41}	1612	12
			4766	3		23818	15
						23827	13
10^{-4}	collisions and near-infrared	$S_{NIR}/S_{Rot} =$ 3×10^{-4}	1612	9	30×10^{41}	1612	11
			4766	> 2*		1667	?*
						13435	> 2*
						13441	> 2*
						23818	> 2*
						23827	> 2*
	pumping	3×10^{-3}	1612	37	120×10^{41}	1612	28
			6017	7		23818	33
			4766	3		23827	27

OH maser optical depths (greater than or equal to two) and the corresponding microwave frequencies for two collisional de-excitation rates, with and without near infrared pumping. The collision rates here are proportional to the dipole line strengths (neglecting the hyperfine interaction) but the cross-sections are proportional to the degeneracies of the final hyperfine states. Numerical values correspond to the de-excitation: $\Pi_{3/2}$ ($J = 5/2$, $F = 3$) → $\Pi_{3/2}$ ($J = 3/2$, $F = 1$). The collision rates across the Λ-doublet are about 1 to 4 times larger, depending on the hyperfine states. Infrared pumping is given for two choices of the vibrational line strength relative to the pure-rotational one, differing by a factor of ten. The low strength gives a total Einstein A-value for the $v = 1$ state of $11 \cdot 4$ sec⁻¹. The kinetic temperature is 200°K, the near-IR temperature is 2000°K (with a dilution factor of 10^{-2}). The projected density for OH is approximately 10^{18} cm⁻². The second column gives results obtained with overlap of the hyperfine splittings of the far-IR transitions that are less than approximately 18 MHz. Some 1667-MHz emission is evident here. The first column gives results obtained with no such overlap (except involving the $\Pi_{3/2}$ ($J = 7/2$)-state). Emission at 1612 MHz is evident here, with the estimated isotropic photon flux given. Asterisks indicate the possible influence of weak 118 cm⁻¹ maser emission, and the resulting computational uncertainties.

Observable fluorescence of about 10^{48} photons/sec in the ≈ 2000 and 4000 cm^{-1} v_2, v_1 and v_3 bands is expected. Other masers might also occur, $6_{25} \rightarrow 5_{32}$ (44 cm^{-1}) at higher rotational temperatures, $3_{13} \rightarrow 2_{20}$ (183 GHz) and $5_{15} \rightarrow 4_{22}$ (324 GHz) at lower rotational temperatures.

Infrared pumping in OH associated with Mira-type infrared stars gives maser properties shown in Table VIII, obtained by Dickinson and Litvak.

(ii) *Radiative transport.*

Non-equilibrium depends upon the leakage of the fluorescence radiation. Complete trapping of internally-generated radiation brings the populations to equilibrium at the exciting agency's temperature. Trapping is easily handled via trapping factors that are derived for various geometries, such as slabs or spheres. The useful and usually justified approximations are that the radiation is redistributed in frequency, at each fluorescence, over the whole Doppler-broadened linewidth, that there are two opposing streams of radiation and that the excitation conditions vary somewhat slowly over lengths corresponding to unit optical depth for the major lines.[47-80] With hyperfine components, the problem of interlocking lines often arises, and is important for the pumping of hyperfine state populations. A common excited state is coupled to two or more lower states by absorption and fluorescence. Because of different linestrengths, different amounts of trapping occurs for the different interlocking lines. With incident external radiation from one side of the cloud, the intensities belonging to the stronger absorptions do not penetrate into the cloud as well. Thus, in the back half of the cloud there is a tendency then to fill these absorbing states by fluorescence induced by the intensities that have penetrated. However, in the front half of the cloud, the backscattered intensity is high for the radiation that is most absorbed. This pumps population into the states that absorb least. That is, there is the opposite tendency of population transfer in the two halves of the cloud. The observer usually faces the back part, however. The spontaneous emission that is amplified in this part will probably dominate. Internal radiation, generated by excitation by collisions or by cascading from higher states that are being pumped, transfers population to the states that absorb least. This occurs throughout the cloud so that the position of the observer is not crucial. The near infrared pumping of OH is such a case.[47]

(iii) *Competition between collisions and radiation*

The steady state populations in the presence of trapped radiation and collisions are easily obtained from rate equations that are made self-consistent by iteration. The optical depths that appear in the trapping factors lead to populations that correspond to those used to calculate the optical depths. The incident radiation may be conveniently specified by a radiation temperature. If quite distinct types of radiation are present, then separate radiation temperatures may be used with dilution factors to take into account the solid angle subtended by the different radiation sources. The collision rates belong to a kinetic temperature that usually differs from this radiation temperature. If all the temperatures are the same and the radiation is isotropic then the population are in equilibrium despite trapping and selective absorption. However, rarely is the kinetic temperature as low as 2·7°K, despite fast mechanisms for cooling, such as collisions with the cold grains, radiation from fine structure lines of carbon and rotational lines of carbon monoxide. The rate of transferring population per second from the state i to state j is denoted by W_{ij}. Then

$$\frac{dn_i}{dt} = -n_i \sum_j{}' W_{ij} + \sum_j{}' n_j W_{ji}.$$

The n_i or n_j are the population densities. For steady-state the time-derivative is zero. The prime on the summation indicates the omission of the irrelevant $j = i$ term. The rate W_{ij} consists of the sum of a radiative and a collision term $[A_{ij}/(1 - \exp(-x_{ij}))] + w_{ij}$ where $x_{ij} = E_{ij}/(kT_R)$. Here E_{ij} is the energy difference between the states, taken to be positive. When E_{ij} is negative, indicating that level i is below level j, the sum of the two terms is:

$$[A_{ij}/(\exp(-x_{ij}) - 1) + w_{ij} \exp(E_{ij}/kT)] (g_i/g_j).$$

The g_i or g_j are the degeneracies of the states i and j respectively. When radiation trapping factors[47] are included, the A_{ij} of the $E_{ij} > 0$ term is multiplied by

$$(1 - \exp(-x_{ij})) \mathscr{L}(\tau_{ij}) + \exp(-x_{ij}) L(\tau_{ij})/2$$

but only if $\tau_{ij} \geqslant 0$, while the A_{ij} of the $E_{ij} < 0$ term is multiplied by $L(\tau_{ji})/2$ but only if $\tau_{ji} > 0$. Here τ_{ij} (or τ_{ji}) is the optical depth for the absorption $j \to i$ (or $i \to j$).

$$\tau_{ij} = c^2 \frac{A_{ij} g_i}{8\pi v_{ij}^2} \int_0^x \left(\frac{n_j}{g_j} - \frac{n_i}{g_i}\right) g(v - v_{ij}) \, dl$$

where v_{ij} is the transition frequency at line center, $\int g(v - v_{ij}) \, dv = 1$, and dl is an element of distance along the appropriate direction, such as perpendicu-

Table IXa

Formamide[81] and Formic Acid.

Relative Linestrength × Fractional Population Inversion ($T_R = 2 \cdot 7\,°K$)

	(no selection rules on collisions) case 1			(dipole selection rules on collisions) case 2		
	$T_k = 5\,°K$	$10\,°K$	$20\,°K$	$5\,°K$	$10\,°K$	$20\,°K$
$1_{10} \to 1_{11}$	$W/A^* =$					
	0·01–1	0·01–1	0·01–1	1	0·1–10	0·1–10
NH$_2$CHO	−4(−3)	−7(−3)	−8(−3)	−7(−4)	−2(−3)	−2(−3)
CHOOH	−2(−3)	−4(−3)	−4(−3)	−6(−4)	−2(−3)	−2(−3)
$2_{11} \to 2_{12}$	0·1	0·01–1	0·01–1	—	1–10	1–10
NH$_2$CHO	−3(−4)	−1(−3)	−2(−3)		−3(−4)	−6(−4)
CHOOH	—	—	−4(−5)†		−3(−4)	−4(−4)
$3_{03} \to 2_{12}$	—	10	1–10	1–10	1–100	1–100
NH$_2$CHO		−7(−6)	−3(−5)	−7(−5)	−2(−4)	−2(−4)
CHOOH		−1(−5)	−3(−5)	−3(−5)	−6(−5)	−7(−5)
$1_{01} \to 0_{00}$	—	0·1–1	0·1–1	—	—	1–100
NH$_2$CHO		−9(−3)	−2(−2)			−8(−5)
CHOOH		−1(−2)	−2(−2)			—
$3_{12} \to 3_{13}$	—	—	0·1	—	—	10
NH$_2$CHO			−2(−4)			−4(−5)
CHOOH			—			−7(−5)
$2_{02} \to 1_{01}$	—	—	1	—	—	—
NH$_2$CHO			−8(−5)			
CHOOH			−9(−4)			

* W/A is the ratio of collisional de-excitation rate (the same for all levels in case 1) to the Einstein-A coefficient for the $2_{11} \to 1_{10}$ transition. $A = 3 \cdot 4 \, 10^{-6}$ sec^{-1} (NH$_2$CHO) and $5 \cdot 4 \times 10^{-7}$ sec^{-1} (CHOOH). Case 2, $W_{ij} = W S_{ij}/0 \cdot 3 g_i$ for $i \to j$.

† For $W/A = 1$ only.

Number in parenthesis is the power of ten.

Table IXb

Methanol-$A^{(14)}$

Relative Linestrength × Fractional Population Inversion ($T_R = 2.7\,°K$)

	(no selection rules on collisions)			(dipole selection rules on collisions)		
	5°K	10°K	20°K	5°K	10°K	20°K
	$W/A^* =$			$W/A^\dagger$		
$1_{10} \to 1_{11}$	0·01–10	0·001–100	0·001–100	0·1–100	0·01–100	0·1–1000
	−3(−4)	−2(−4)	−6(−4)	−3(−4)	−6(−4)	−8(−4)
$2_{11} \to 2_{12}$	0·1–100	0·01–100	0·01–100	0·1–100	0·1–100	1–1000
	−2(−4)	−5(−4)	−9(−4)	−2(−4)	−6(−4)	−6(−4)
$3_{12} \to 3_{13}$	0·1–100	0·01–100	0·01–100	1–100	1–100	1–10000
	−3(−5)	−2(−4)	−6(−4)	−3(−5)	−3(−4)	−5(−4)

* W/A is the ratio of collisional de-excitation rate (the same for all levels) to the Einstein-A coefficients for the $2_{12} \to 1_{11}$ transition.

† The A is for the $2_{12} \to 1_{11}$ transition. The collisional de-excitation rate is $WS_{ij}/0.3g_i$ for $i \to j$. $A = 2.5 \times 10^{-6}\,\text{sec}^{-1}$.

Table IXc

Methanol-$E_1^{(82)}$

Relative Linestrength × Fractional Population Inversion

	(no selection rules on collisions)			
	$T_R = 2.7\,°K$		$T_R = 20\,°K$	
	10°	30°	30°	50°
$2_{21} \to 2_{11}$	$W/A^* =$			
	0·1–10	0·1–100	10–100	0·1–100
	−9(−3)	−1(−2)	−3(−3)	−5(−3)
$3_{22} \to 3_{12}$	0·1–10	0·01–100	10–100	0·1–100
	−6(−3)	−2(−2)	−3(−3)	−7(−3)

* W/A is the ratio of collisional de-excitation rate (the same for all levels) to the Einstein-A coefficient for the $2_{11} \to 1_{10}$ transition. $A = 2.6 \times 10^{-6}\,\text{sec}^{-1}$.

Methanol-E_1 (continued)

	$T_R = 2\cdot 7°$		$T_R = 20°$	
	10°	30°	30°	50°
$4_{23} \to 4_{13}$	0·01–10 $-3(-3)$	0·01–100 $-1(-2)$	10–100 $-2(-3)$	10–100 $-6(-3)$
$5_{24} \to 5_{14}$	0·01–10 $-1(-3)$	0·01–100 $-1(-2)$	10–100 $-2(-3)$	10–100 $-6(-3)$
$6_{25} \to 6_{15}$	0·01–10 $-3(-4)$	0·01–100 $-7(-3)$	10–100 $-8(-4)$	10–100 $-3(-3)$
$7_{26} \to 7_{16}$	0·01–10 $-5(-5)$	0·01–100 $-4(-3)$	10–100 $-4(-4)$	10–100 $-3(-3)$
$8_{27} \to 8_{17}$	0·01–10 $-8(-6)$	0·01–100 $-2(-3)$	10–100 $-2(-4)$	10–100 $-2(-3)$
$9_{28} \to 9_{18}$	0·01–10 $-9(-7)$	0·01–100 $-6(-4)$	10 $-2(-5)$	10–100 $-5(-4)$
$1_{01} \to 0_{00}$	1 $-2(-3)$	0·1–10 $-1(-2)$	—	0·1–10 $-1(-3)$
$2_{21} \to 1_{10}$	1 $-5(-4)$	0·1–10 $-2(-2)$	—	0·1–100 $-7(-3)$
$4_{04} \to 3_{12}$	0·01–10 $-1(-2)$	0·01–100 $-5(-2)$	10 $-3(-3)$	0·1–100 $-9(-3)$
$5_{05} \to 4_{13}$	0·01–10 $-2(-3)$	0·01–10 $-3(-2)$	10 $-5(-4)$	0·1–10 $-5(-3)$
$6_{06} \to 5_{14}$	0·01–1 $-3(-4)$	0·01–10 $-1(-2)$	—	0·1–10 $-4(-3)$
$7_{07} \to 6_{15}$	0·01–1 $-3(-4)$	0·01–10 $-5(-3)$	—	10 $-2(-4)$
$8_{08} \to 7_{16}$	0·01–1 $-1(-7)$	0·01–10 $-2(-3)$	—	10 $-2(-4)$

lar to the face of a slab geometry. The thinnest dimensions control the leakage of radiation. The symmetric transmission function[47] is

$$\mathscr{L}(\tau_{ij}) = \tfrac{1}{2}[L(\tau_{ij}) + L(\tau_{ij}')]$$

where

$$L(\tau) = \int_{-\infty}^{+\infty} \pi^{-1/2} \exp[-v^2 - \tau \exp(-v^2)]\,dv$$

and τ_{ij}', the complementary optical depth, applies from the point x to x_0, the end of the cloud.

Table IX lists the results of calculations for the two-temperature nonequilibrium populations of a few organic molecules without including radiation trapping. Partial optical depths $S_{ij}(n_j/g_j - n_i/g_i)$, where S_{ij} is the relative line strength for the transitions, are given only when amplification occurs. The optical depth being low, however, would prevent spectacular maser effects. However, with projected densities $\approx 10^{16}$ cm^{-2} and partial optical depths $\approx 10^{-3}$, the optical depths τ_{ij} may be almost unity, where

$$\tau_{ij} = \frac{(2\pi)^2 \mu_a^2}{3\hbar c} S_{ij}(n_j/g_j - n_i/g_i) Nl_0(v/\delta v)$$

μ_a is the a-component of the permanent electric–dipole moment ($\approx 10^{-18}$ e.s.u.), Nl_0 is the total molecular projected density, and $\delta v/v$ is the (full) fractional Doppler width ($\approx 10^{-5}$). The partial optical depths given in Table IX are the maximum values (normalized as if $\mu_a = 1$ and $\Sigma n_i = 1$). The range of density (given as the ratio of collision rate to a reference fluorescence rate) is that over which population inversion occurs. Typically, the partial optical depth is down by a factor of ten from its maximum value at either end of this density range. Some attempts are presently being made to use these calculations to account for the recent astronomical data.[14,58,81,82]

IV. NONLINEAR PROPERTIES OF MASERS

4.1. Line width

The observed interstellar maser signals seem to have linewidths comparable to the Doppler widths expected for cold clouds. The statistical properties seem to be Gaussian, i.e., similar to noise, despite the maser amplification. The effect of feedback due to backscatter or plasma reflection is probably very small so that simple travelling wave amplification is occurring. Laboratory optical lasers usually involve mirrors. The output is nearly monochromatic because amplitude fluctuations are very small. The residual linewidth, usually a few Hz, is due to frequency modulation via microvibrations of the optical cavity properties. Goldreich[83] and others have suggested that the interstellar masers are frequency-modulated signals from several tiny emission regions. The excursion of the frequency during modulation is comparable to the observed line width. Scattering from nearby plasma randomly superimposes these signals so that the output appears random. However, such a signal would have many coherent properties that do not occur for a Gaussian-like signal of the same linewidth. In particular, if the autocorrelation time

$$\tau = \int_0^\infty \langle E(t) E(0) \rangle \, dt / \langle E(0)^2 \rangle$$

of the signal from one such tiny spot is long compared to all characteristic times of the system, then the maser molecules will be responding to an almost monochromatic signal. The transition becomes saturated when

$$|\mu_{ab} E|^2 (\Gamma_{aa}^{-1} + \Gamma_{bb}^{-1})/2\hbar^2 \Gamma_{ab} > 1$$

where Γ_{aa} and Γ_{bb} are the collisional and/or radiative relaxation rates for the upper and lower states of the transition and Γ_{ab} is their average. μ_{ab} is the dipole moment matrix element and E is the quasi-monochromatic electric field amplitude. For a particular $\Delta M = 0, \pm 1$ transition, the appropriate spatial component of the electric field must be taken.

The more likely description of the interstellar signals is that they are Gaussian-type signals obtained by amplifying thermal continuum radiation or spontaneous emission. The randomness of phase remains despite the amplification and saturation. A proof of this has been offered by Icsevgi and Lamb.[84] Assuming Gaussian statistics several interesting cases of bandwidths have been calculated as a function of microwave intensity.[85] The method of calculation consists of writing the equation for the Fourier component of the off-diagonal density matrix element. This is then multiplied by an electric field Fourier component. This is then averaged with respect to random initial phases of the fields. The average of a pair of electric fields can be factored out of each term of the expression for the power source

$$-8\pi\omega Im \langle P(\omega) E(\omega')^+ \rangle / c.$$

By the Maxwell equations this source is approximately equal to $d\langle E(\omega) E(\omega')^+ \rangle / ds$, where ds is an element of distance along the path of a ray. Amplitude fluctuations produce additional affects over the usual saturation of the population difference. The latter occurs even for pure monochromatic signals. However, if one has even a quasi-monochromatic signal, whose bandwidth is small compared to the homogeneous widths (due to collisions and radiative decay), one obtains Raman-type interactions that frequency-broaden the response of the system. These are called Raman-type interactions because of the dependence upon the frequency difference of two interacting electric fields in a manner similar to that for standard Raman scattering. for example, A particularly important case arises when the signal is broadband with respect to the homogeneous width. This is expected where the inhomogeneous width due to Doppler broadening is also much larger than the homogeneous width. The convolution of these two distributions determines the width of the gain coefficient. Amplification occurs over this Doppler width. Since more amplification occurs at line center than on the sides of the line, the signal becomes somewhat narrower the more the resulting amplification. There is a simple derivation of the fact that the bandwidth decreases by a factor $\tau_m^{-1/2}$ where τ_m is the unsaturated microwave optical depth at line center. Suppose we parameterize the signal intensity by a Gaussian-shaped function

$$I(\omega, x) = I(x) \exp(-\omega^2/\Delta\omega(x)^2)$$

with line center at $\omega = 0$. $I(x)$ is the intensity at line center at a distance x along the amplifier ($x = 0$ at the output) and $\Delta\omega(x)$ is the bandwidth at x. Now

$$\frac{dI(\omega, x)}{dx} = \alpha_0 \exp(-\omega^2/\delta\omega_D^2) I(\omega, x)$$

therefore

$$\frac{dI(x)}{\alpha_0 \, dx} = I(x)$$

and

$$\frac{d\Delta\omega(x)^{-2}}{\alpha_0 \, dx} = \delta\omega_D^{-2}$$

upon equating the constant terms and the ω^2-terms on both sides of the equation. The exponential function on the left-hand side is expanded in powers of ω^2. Then

$$I(x) = I(0) \exp \tau_m \quad \text{and} \quad \Delta\omega(x)^{-2} - \Delta\omega(0)^{-2} = \delta\omega_D^{-2} \tau_m,$$

where

$$\int_0^x \alpha_0 \, dx = \tau_m.$$

For $\Delta\omega(0) = \infty$, corresponding to an input continuum signal,

$$\Delta\omega(x)/\delta\omega_D = \tau_m^{-1/2}.$$

When $\Delta\omega(0) = \delta\omega_D$, corresponding to spontaneous line emission at the input, then

$$\Delta\omega(x)/\delta\omega_D = (1 + \tau_m)^{-1/2}.$$

Saturation has been discussed above. The simple broadband case tells us that we multiply α_0 by $(1 + I(\omega, x)/I_s)^{-1}$, where I_s is the characteristic intensity above which saturation is very important.

$$I_s^{-1} = 4\pi^2|\mu_{ab}|^2 (\Gamma_{aa}^{-1} + \Gamma_{bb}^{-1})/\hbar^2 c.$$

With saturation

$$\frac{1}{I(x,\omega)} dI(x,\omega)(1 + I(x,\omega)/I_s) = \alpha_0 \, dx \exp(-\omega^2/\delta\omega_D^2)$$

or

$$\ln\{[I(x)/I_0]\exp(-\omega^2/\Delta\omega^2)\} + [I(x)/I_s]\exp(-\omega^2/\Delta\omega^2) - I_0/I_s$$
$$\approx \tau_m \exp(-\omega^2/\delta\omega_D^2).$$

$$\frac{I(x)}{I_s}\exp(-\omega^2/\Delta\omega^2) + \ln\frac{I(x)}{I_s} \approx \tau_m \exp(-\omega^2/\delta\omega_D^2) + \frac{\omega^2}{\Delta\omega^2} - \tau_s + 1$$

where $\tau_s = 1 + \ln(I_s/I_0)$ is the optical depth at which nearly exponential growth would bring the intensity up to the value I_s. The last three terms arise from the approximate expression for $I(x, \omega)$ in the logarithm term. Upon equating the constant terms, we obtain:

$$I(x)/I_s = \tau_m - \tau_s + 1 - \ln(1 + \tau_m - \tau_s) \quad \text{for} \quad \tau_m \geq \tau_s.$$

Upon equating the ω^2 terms we obtain

$$\frac{1 + I(x)/I_s}{\Delta\omega(x)^2} \approx \frac{\tau_m}{\delta\omega_D}$$

whence we have

$$\Delta\omega/\delta\omega_D \approx [\tau_m - \tau_s - \ln(1 + \tau_m - \tau_s) + 2]^{1/2}/\tau_m^{1/2}.$$

We note that $\Delta\omega/\delta\omega_D$ approaches unity as $\tau_m \to \infty$. For example, if $\tau_s \approx 24$, so that amplification by a factor of 10^{10} occurs at $\tau_m = \tau_s$, then the bandwidth is $\delta\omega_D/\sqrt{12}$, but at $\tau_m = 2\tau_s$, the bandwidth is a factor of about 2·4 times this. The bandwidth $\Delta\omega$ is about 0·7 $\delta\omega_D$. This agrees very well with the numerical solution (of the differential equation) shown in Fig. 9. If line narrowing prior to saturation brings $\delta\omega < \Gamma_{ab}$, then continued line narrowing occurs despite saturation.

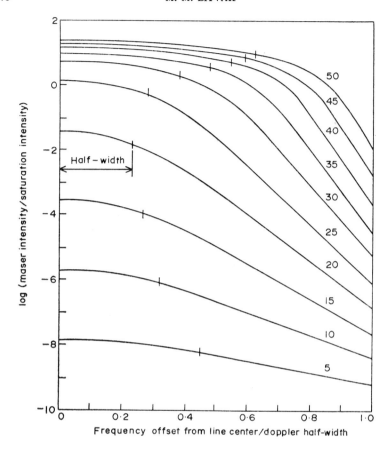

FIGURE 9

The other cases that may arise have to do with the relative sizes of various quantities having the dimensions of reciprocal seconds. These quantities are

(a) $W(\omega) = 2\pi|\mu_{ab}|^2 F(\omega)/4\hbar^2$, the induced transition rate, where

$$\langle E(\omega) E(\omega')^+ \rangle = F(\omega) \delta(\omega - \omega'), \quad I(\omega) = F(\omega) c/8\pi$$

(b) $W^{1/2}$, where $W = \int_0^\infty d\omega \, W(\omega)$

(c) Γ_{ab}, Γ_{aa}, and Γ_{bb}, the damping constants for the two levels

Table X

	Homogeneous Broadening	Inhomogeneous Broadening
	$(\delta\omega_D \ll \Gamma_{ab})$	$(\delta\omega_D \gg \Gamma_{ab})$
Quasi-Monochromatic $\delta\omega \ll \Gamma_{ab}$	$\dfrac{\Gamma_{ab}/\pi}{(\omega-\omega_{ba})^2+\Gamma_{ab}^2+2\Gamma_{ab}W(\Gamma_{aa}^{-1}+\Gamma_{bb}^{-1})/\pi}$	$\dfrac{g_D(\omega-\omega_{ba})\Gamma_{ab}}{\Gamma_{ab}^2+2\Gamma_{ab}W(\Gamma_{aa}^{-1}+\Gamma_{bb}^{-1})/\pi}$
Broadband $\delta\omega \gg \Gamma_{ab}$		
Moderate Saturation: $\Gamma_{ab}+W(\omega) \ll \delta\omega$	$\dfrac{[\Gamma_{ab}+W(\omega)]/\pi}{(\omega-\tilde{\omega}_{ba})^2+[\Gamma_{ab}+W(\omega)]^2}$	$\dfrac{g_D(\omega-\omega_{ba})}{1+(\Gamma_{aa}^{-1}+\Gamma_{bb}^{-1})W(\omega)}$
Extreme Saturation: $[\Gamma_{ab}^2+2W(\omega)\Gamma_{ab}]^{1/2} \gg \delta\omega$	$\dfrac{\Gamma_{ab}/\pi}{(\omega-\omega_{ba})^2+\Gamma_{ab}^2+2\Gamma_{ab}W(\omega)}$	$(\delta\omega_D \gg [\Gamma_{ab}^2+2W(\omega)\Gamma_{ab}]^{1/2})$ $\dfrac{g_D(\omega-\omega_{ba})\Gamma_{ab}}{[\Gamma_{ab}^2+2\Gamma_{ab}W(\omega_{ba})+(\Gamma_{aa}^{-1}+\Gamma_{bb}^{-1})W\Gamma_{ab}/\pi]^{1/2}}$

Note: $\tilde{\omega}_{ba} = \bar{\omega}_{ba} - P\int d\omega'\, W(\omega')/(\omega'-\omega)\pi$.

(d) $\delta\omega$, the signal bandwidth, the width of $W(\omega)$

(e) $\delta\omega_D$, the Doppler width $= \omega_{ba}(2kT_K/M_c^2)^{1/2}$ (where T_K is the kinetic temperature and M is the molecule's mass.

Table X lists the line shape functions for the different cases. $g_D(\omega - \omega_{ba})$ is the usual normalized Doppler (gaussian) function.

4.2. Polarization

One of the most difficult problems to solve is that of the polarization of the OH masers. The earliest suggestion is that it is caused by non-linear competition between oppositely-polarized modes.[54] The often-observed circular polarization is consistent with work in the laboratory with lasers involving $\Delta J = 0$ transitions and also the special case of $J = 0 \to 1$. With hyperfine splitting, we expect $\Delta F = 0$ and $F = 0 \to 1$ to act similarly. Perturbation calculations point to the instability of circular polarization for these lasers, provided that the Zeeman splittings are small compared to the homogeneous linewidth. The reason for these restrictions is clear from the following perturbation analysis:

$$\frac{dI_R}{dx} \approx I_R[a - bI_R - (c + c')I_L] - d|\langle E_R E_L^* \rangle|^2$$

$$\frac{dI_L}{dx} \approx I_L[a - bI_L - (c + c')I_R] - d|\langle E_R E_L^* \rangle|^2$$

where I_R and I_L are the intensities for right-handed and left-handed circular polarized signals at the same frequency. The coefficient is a the usual unsaturated gain coefficient, b is the self-saturation coefficient, c is the population *cross*-saturation coefficient, and c' is the Raman-type cross-saturation coefficient.[57,86] The coefficients b, c, c' and d are maximum at the Zeeman-splitting frequency associated with the sense of polarization of the saturating signal. Like self-saturation, cross-saturation depends upon populations being transferred by induced transitions. However, the Raman-type cross-saturation depends upon the statistical properties of the oppositely-polarized electric fields. The coefficient depends on

$$c' \propto \int_0^\infty d\omega_L\, K(\omega_L - \omega_R + \omega_B)\langle |E_R|^2 |E_L|^2 \rangle$$

where ω_L and ω_R are the frequencies belonging to I_L and I_R, respectively, ω_B is the Zeeman-splitting frequency and

$$K(\omega) = (\Gamma_{aa} + i\omega)^{-1} + (\Gamma_{bb} + i\omega)^{-1} \cdot \langle |E_R|^2 |E_L|^2 \rangle \approx I_R I_L (8\pi/c)^2$$

uncorrelated E_R and E_L. The integral is of order $\Gamma/\delta\omega$ smaller than for monochromatic signals. The term proportional to $|\langle E_R E_L^* \rangle|^2$ also contributes to the competition between modes. The equation dealing with the correlation function $\langle E_L E_R^* \rangle$ is written as

$$\frac{d\langle E_L E_R^* \rangle}{dx} \approx \langle E_L E_R^* \rangle [2a - b(I_R + I_L) - (c + c')(I_L + I_R) - 2d\langle E_R E_L^* \rangle].$$

If $\langle E_L E_R^* \rangle$ is zero at $x = 0$, as would be the case of incident thermal continuum, then $\langle E_L E_R^* \rangle$ stays zero for all $x > 0$. An analysis of the stability properties of this system is somewhat complicated by non-zero $\langle E_L E_R^* \rangle$. We will, for the moment, neglect it. In the phase space of I_R vs. I_L, two straight lines are the locus of points for which dI_R/dx and dI_L/dx vanish (see Fig. 10). The point of intersection of the two lines is an unstable point (saddle point) for the situation shown, when $c + c' > b$. This means cross-saturation is greater than self-saturation. The stronger mode then quenches the weaker one. If $b > c + c'$, however, then the point of intersection is stable and the modes become equal in strength. If some small coupling of the phase of

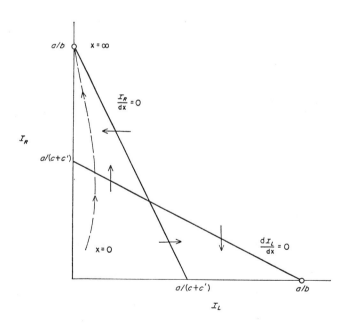

FIGURE 10

the left- and right-handed modes is possible, the signal becomes linearly-polarized. The coefficients are functions of the type of molecular transition. For $F = 1 \to 1$, for the pure monochromatic case, and for collisions that relax molecular orientation and alignment equally well, one finds that $b = 2c = 2c' = c + c'$. This is neutral stability, for which the two lines in the phase diagram coincide. However, generally, collisions among the various Zeeman sublevels relax molecular orientation faster than alignment, so that $b < c + c'$, and the system is unstable to circular polarization, as discussed above. For $F = 2 \to 2$, we have $b > c + c'$ without collisions among sublevels.

The Gaussian statistics of the signal in the amplifier is very likely a real problem.[56,57] The Raman-type cross saturation coefficient c' is very small so that $b > c + c'$. Another idea is that collision rates and far infrared rates are quite strong in equalizing the populations among the Zeeman sublevels. Then, $b \gtrsim c$, that is, the self-saturation coefficient is just a little bit larger than the population cross-section coefficient. The extra competition between left- and right-handed circular polarization to make the system unstable is possibly parametric down-conversion of the higher frequency Zeeman mode over the lower frequency mode. By Zeeman mode, we refer to those roughly oppositely-polarized signals that are resonant with a given molecule for $\Delta M = +1$ and -1. For $\Delta F = 0$ transitions, this simple description is adequate. For $\Delta F = \pm 1$, however, several Zeeman frequency splittings occur because of the different g_F-factor for upper and lower states. These are the satellite transitions that perhaps do not show as consistent circular polarization as the main lines ($\Delta F = 0$).

Parametric down-conversion consists of the elementary process, easily understood in quantum mechanics, which in one photon is converted into two other photons.[87] The sum of the frequencies of the generated photons equals the frequency of the initial photon. Similarly, photon momentum is conserved: the initial photon wave vector equals the sum of the final two wave vectors.

One of the final photons might be an electron cyclotron wave of frequency approximately $\omega_H \cos \theta$, that is close to the Zeeman frequency splitting of the microwave modes, where ω_H is the cyclotron frequency and θ is the angle of population of the cyclotron wave with respect to the local magnetic field.[88] The group velocity of a wave is perpendicular to its electric vector. The cyclotron wave has a large electric field along the wave vector. The group velocity is then nearly perpendicular to the wave vector. The conservation of momentum requires that the difference vector (momentum transfer) between the two microwave wave vectors be nearly perpendicular to each of them. The difference vector equals the wave vector of the cyclotron wave. Thus, the group velocities of the three waves are nearly parallel, a condition necessary for appreciable interaction over large distances. Non-zero temperature effects that modify the plasma wave properties are not discussed here.

The parametric gain coefficient for exponential growth is approximately given by[57]

$$K \approx (4\pi)^2 (2\pi)^{-8} \int d^4p\, d^4q\, \delta^{(4)}(p+q-k) \int d^4k' d^4p'\, \delta^{(4)}(k'-p'-q)|\mathbf{k}|$$

$$\delta\chi(k,p,q) : \delta\chi(q,k',-p') \langle \mathbf{E}(p)\, \mathbf{E}(p')^* \rangle 2\pi\delta(\omega(\mathbf{q}) - \omega_H \cos\theta_q)$$

$$\omega(\mathbf{q})/(2\cos^2\alpha_q \partial\omega(\mathbf{q})^2 \varepsilon/\partial\omega(\mathbf{q})^2$$

where

$$\langle \mathbf{E}(p)\, \mathbf{E}(p')^* \rangle = 8\pi k T_B \sqrt{\pi}\, \delta\omega g(\omega(\mathbf{p}) - \omega)(2\pi)^5\, \delta^{(4)}(p-p')\, \delta(p_0 - \omega(\mathbf{p})/c)$$

for travelling waves with random phases and a brightness temperature T_B, used in the Rayleigh–Jeans approximation, $\hbar\omega/kT_B \ll 1$. The factor $(n_q \cos\alpha_q)^{-2}$ comes from the propagator $\mathbf{R}(q)$ for the cyclotron wave. The Maxwell wave equation for the cyclotron wave is

$$\mathbf{R}(q)\cdot \mathbf{E}(q) = -4\pi(2\pi)^{-4} \int d^4k' d^4p'\, \delta^{(4)}(k'-p'-q)\, \delta\chi(-q,k',-p'):$$

$$\mathbf{E}(k')\, \mathbf{E}(-p')$$

where $\mathbf{R}(q) = \boldsymbol{\varepsilon} + q_0^{-2}\, \mathbf{q} \times (\mathbf{q} \times \mathbf{I})$ and where the right-hand side is the non-linear polarization induced by the two microwaves. If we introduce the unit polarization vector $\mathbf{e}(\mathbf{q})$ for the cyclotron wave, we have

$$\mathbf{e}(\mathbf{q})^* \cdot \mathbf{R}(q) \cdot \mathbf{e}(\mathbf{q}) = \mathbf{e}(\mathbf{q})^* \cdot \boldsymbol{\varepsilon} \cdot \mathbf{e}(\mathbf{q}) - q_0^{-2}|\mathbf{q} \times \mathbf{e}(\mathbf{q})|^2$$

$$= \cos^2\alpha_q(\varepsilon q_0^2 - |\mathbf{q}|^2)/q_0^2$$

That $|\mathbf{q}\times\mathbf{e}(\mathbf{q})|$ is $|\mathbf{q}|\cos\alpha_q$ follows from Fig. 11, where $\mathbf{v}_g$, the group velocity is along $\mathbf{E}\times\mathbf{B}$. $\mathbf{B}$ is perpendicular to $\mathbf{q}$ and $\mathbf{E}$, as usual.

A simplified reduction of the integrand in the expression for K yields

$$K \approx \frac{16\pi\, kT_B \sqrt{\pi}\, \delta\omega}{\omega(\mathbf{k})|\mathbf{k}|} \left(\frac{e}{mc}\right)^2 \int d^3p\, \delta(\omega(\mathbf{k}) - \omega(\mathbf{p}) - \omega_H \cos\theta_q)$$

$$\times \frac{\omega_H \cos\theta_q\, A(k,p,k-p) g(\omega(\mathbf{p}) - \omega)}{2\omega(\mathbf{p})^2\, D_q}$$

where

$$A(k,p,k-p) \approx |\mathbf{e}(\mathbf{k})^* \cdot \mathbf{e}(\mathbf{p})|^2\, |\mathbf{q}|^2/(4\pi)^2$$

and

$$D_q = \cos^2 d_q \, \partial\omega(\mathbf{q})^2 \, \varepsilon/\partial\omega(\mathbf{q})^2 \approx \omega_0^2/\omega H^2 \sin^2 \theta_q.$$

where ω_0 is the plasma frequency $(4\pi n_e e^2/m)^{1/2}$ and ω_H is the electron cyclotron frequency eB_0/mc.

The latter expression is obtained from the dispersion law for the cyclotron wave in a cold plasma

$$\omega(\mathbf{q}) \approx \omega_H \left[\cos^2 \theta_q + \frac{m}{M}\left(1 + \frac{1}{|\mathbf{q}|^2 r_e^2}\right) \right]^{1/2} \left(1 - \frac{1}{1 + |\mathbf{q}|^2 r_e^2}\right)$$

and then neglecting the m/M terms, where m/M is the ratio between electron and average ion masses, and $r_e = c/\omega_0$, the electron plasma length. It is assumed that $|\mathbf{q}| < \omega_H \cos\theta_q/c_T$ and λ_{De}^{-1}, where c_T is the ion–gas sound velocity and λ_{De} is the electron Debye length $(kT_e/4\pi n_e e^2)^{1/2}$. It is also assumed that the Zeeman splitting is sufficiently large that the modes of opposite polarization are fairly-well separated in frequency.

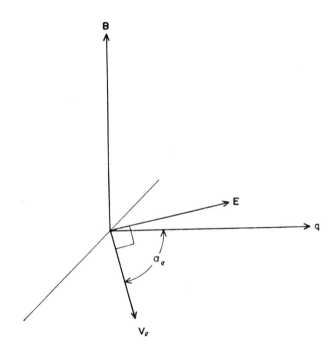

FIGURE 11

Upon using $d^3p = \sin\theta_{kp}\, d\theta_{kp}\, d\phi_q \omega(\mathbf{p})^2\, d\omega(\mathbf{p})/c^3$, and assuming that $|\mathbf{p}| \approx |\mathbf{k}| > r_e^{-1}$, $\omega(\mathbf{p}) \approx \omega(\mathbf{k})$ and $|\mathbf{q}| \approx 2|\mathbf{k}|\sin(\theta_k/2)$, we obtain

$$K \approx kT_B\left(\frac{e}{mc}\right)^2 \theta_m^4 |\mathbf{k}| |\mathbf{e}(\mathbf{k})^* \cdot \mathbf{e}(\mathbf{p})|^2 \frac{\omega_H^3 \sin^2\theta_q \cos\theta_q}{\omega_0^2 c^3}$$

where θ_{kp} is the angle between $\mathbf{k}$ and $\mathbf{p}$ and $\theta_m(\gtrsim 1)$ is the maximum value for θ_{kp} for which there is appreciable T_B. This formula embodies the corrections to Ref. 57 supplied by Goldreich and Kwan. For $\theta_m = 1$ rad., $\omega_H \approx \omega_0 \approx 3 \times 10^4$ sec^{-1}, $|\mathbf{e}(\mathbf{k})^* \cdot \mathbf{e}(\mathbf{p})|^2 \simeq 10^{-1}$, and $\theta_q \simeq \pi/4$, the parametric gain coefficient $X \simeq 10^{-17}$ cm^{-1} for OH with $T_B \approx 10^{14}$°K. However, Goldreich believes damping limits the real-values of $|\mathbf{q}|/|\mathbf{k}| \approx \theta_m$ to less than 10^{-1} rad. and furthermore, observations give $T_B \approx 10^{12}$°K. First, the limitation on θ_m may be avoided with $T_e \gg T$ (which is possible with a low degree of ionization and *dissociative* recombination) and electron drift, so that various dampings (Landau, cyclotron, and collision) are minimized to the point of *plasma instability* for $\mathbf{q}$ nearly perpendicular to $\mathbf{B}_0$. Second, higher T_B-values than observed are possible if there is OH absorption and plasma-scatter near the outer radii. An effective pump rate that approximately depends inversely on radius starting in the region just beyond the "hot spot", can keep $W(\omega) \leq \delta\omega$ despite T_B increasing with radius-squared. Though K depends inversely on radius-squared, so does probably the OH gain coefficient. Then, a maximum value of $K \approx 10^{-17}$ cm^{-1} is adequate, since cross-saturation differs from self-saturation by less than 10^{-1} percent owing to the far infrared radiation trapped between the ground and excited rotational levels.[57] Stronger dependences of pump on inverse-radius lead to analogous conclusions. Finally, the product of polarization unit-vectors is evaluated for non-orthogonal elliptically-polarized modes. The observed high degree of circular polarization could result from the *adiabatic* change to circularity as the surviving mode propagates through the very saturated region where the gyrotopy of the magnetoplasma dominates.[57]

Since the Zeeman–splitting factor is about 10^{-3} times smaller for H_2O than OH, the modes are not well-separated unless $B_0 > 10$ G. Furthermore, the H_2O transitions ($\Delta F = -1$) tend toward linear polarization, given cause (like molecular alignment) for a preferred direction. Thus, circular polarization in H_2O is not expected.

The most important hyperfine components of H_2O are separated in frequency by approximately the Doppler widths. Some H_2O line asymmetry might be interpreted as the $F = 6 \to 5$ emission accompanied by the $F = 7 \to 6$ with about one-tenth the intensity. The $F = 6 \to 5$ transition has a somewhat lower line strength. Hence, the hyperfine populations would have to cause a preference of this transition over the other. Also, the $F = 5 \to 4$ is not seen. The duration of collisions is too short and the widths of optical pump lines are too large to cause preference of populations among the hyperfine states. It may be that the $5 \to 4$ and $6 \to 5$ emission are being down-converted to the $7 \to 6$ emission (of lowest frequency).

4.3. "Hot spots" in spherical amplifiers

The simplest transport equation for the brightness temperature of the maser emission is given by

$$\frac{dT(s,b)}{ds} = \frac{T(s,b)}{1+F(r)} + T_x(r)$$

where ds is an element of optical depth along a ray (a chord of the spherical region) whose perpendicular distance from the center is b. That is, $s = \pm (r^2 - b^2)^{1/2}$, where r is the radial distance to a point on the ray (see Fig. 12), $-T_x$ is the excitation temperature that represents the spontaneous emission contribution. $F(r)$ represents the non-linear effect of saturation, where

$$F(r) = \frac{1}{2rT_p} \int_0^r ds\,(T_+ + T_-)$$

is the sum of the rates of emission (in either direction along the chord) divided by the saturation parameter written as $T_p(r)$, which is proportional to the pump rate. T_+ is the brightness temperature from one end of the cloud to a point the midpoint, while T_- is that from the end to the same point (over the shorter distance). The integral over s corresponds to integrating over the local angle of emission in order to obtain the total saturation rate.

Therefore,

$$T_+ = \int_{-(R^2-b^2)^{1/2}}^{s} ds'\, T_x(r') \exp\left[\int_{s'}^{s} \frac{ds''}{1+F(r'')}\right]$$

$$+ T_c \exp \int_{-(R^2-b^2)^{1/2}}^{s} \frac{ds''}{1+F(r'')}$$

$$T_- = \int_{s}^{(R^2-b^2)^{1/2}} ds'\, T_x(r') \exp\left[\int_{s}^{s'} \frac{ds''}{1+F(r'')}\right]$$

$$+ T_c \exp \int_{s}^{(R^2-b^2)^{1/2}} \frac{ds''}{1+F(r'')}$$

where T_c is the given boundary temperature. The quantity $r'' = + (s''^2 + b^2)^{1/2} = (s''^2 - s^2 + r^2)^{1/2}$. Combining the definition of $F(r)$ given above and these two expressions for T_+ and T_- results in an integral equation for $F(r)$ which has been solved numerically (see Fig. 12). Some analytic results have been obtained by replacing $F(r'')$ in the exp-factors by a quadratic function of r'', with unknown coefficients. $F(r) \approx F(0) + \frac{1}{2}F''(0)r^2$. A necessary and sufficient condition for a "hot spot" is that $F''(0) > 0$, so that saturation increases with radius. If the pump rate

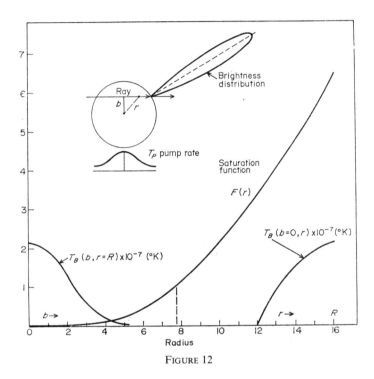

FIGURE 12

decreases from the center of symmetry with a characteristic distance h, for example

$$\frac{h^2}{2} = \left[\frac{-T_p''(0)}{T_p(0)}\right]^{-1},$$

then the condition that $F''(0) > 0$ implies generally that

$$h^{-2} + \tfrac{1}{3}[1 + F(0)]^{-2} > \tfrac{2}{3} R^{-1} [1 + F(R)]^{-1}$$

where all distances are radial optical depths, including the radius R of the spherical cloud. The approximate analysis mentioned above indicates that the diameter d of the "hot spot", over which there is appreciable increase of brightness temperature along a ray, is given by

$$d \approx 2[2F(0)(\tfrac{1}{3} + h^{-2})]^{-1/2} \lesssim \sqrt{6}/F(0)^{1/2}$$

where $h = \infty$ for the uniformly-pumped sphere. However, the apparent angle for the brightness distribution at one typical point on the surface is approximately $\sqrt{2}d^{1/2}/R$. This follows from the simple geometry that the

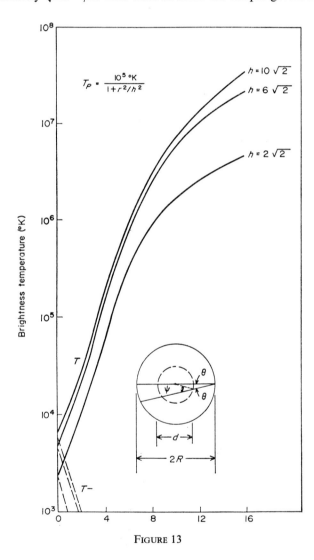

FIGURE 13

brightness is highest along the diameter and decreases for shorter chords that pass through the same output point, at an angle θ with respect to the to the diameter. From Fig. 13 we see that $T_B \propto \exp[d\cos(\psi)] \approx \exp(d) \exp(-d\psi^2/2)$, if we ignore the increase of T_B in the saturated region. However, $\psi \approx 2R\theta/d$. Then the full angle for e^{-1} of the maximum brightness is $\sqrt{2}\,d^{-1/2}\,d/R$. The apparent size (cm) of the emitter is approximately given by $\sqrt{d/\alpha}$, where α is the unsaturated amplification coefficient in the "hot spot".

V. FORMATION OF MOLECULES

5.1. Grains as catalysts

The low temperature of grains, except in the vicinity of high luminosity stars, is conducive to the adsorption of atomic species like H, C, O and N as well as molecular species like H_2, CO, and HCN which are evidently quite abundant. The main problem to be discussed here is the formation of the less abundant molecules like H_2CO, H_3COH, HCOOH, HC_3N, etc. These might be formed when various groups of molecules are also adsorbed, migrate on the grain surface, chemically combine, and evaporate by means of some translation energy derived from the heat of formation. Some of this heat of formation is lost to the grain lattice and some to rotational and vibrational energy of the product molecule. The crucial points deal with the supposedley-fast migration of the particles across the surface and the fast evaporation from the surface for the products of exothermal reactions[89-90]. The evaporation rate for one of the reactants, given by

$$W_{\text{evap}} = v_0 \exp(-D/kT)$$

must be less than the mean rate W_S of obtaining the other reactant particle (having a gas density n_R) anywhere on the surface

$$W_S = Sn_R(\overline{\sigma v})$$

assuming the surface has almost all sites available. S is the sticking efficiency. Here D may be a typical binding energy for the ideal surface or may be that for an enhanced site. Because the binding energy decreases when at least a monolayer forms, the grain temperature might be too high to keep most of the atoms from evaporating from the monolayer, except at sites of enhanced binding. It is convenient to define various characteristic temperatures

$$T_c = \frac{D/k}{\ln(v_0/W_S)} \approx 13°K \text{ or } 40°K \text{ for enhanced sites}$$

$$T_{tr} = \frac{D/k}{\ln\{[1 + v_0 t_D (N_S/N)^{1/2}] N/N'\}} \approx 25\text{--}50°K$$

$$T_{mono} = \frac{D_2/k}{\ln(N v_0/W_{S2})} \approx 11°K$$

The product $v_0 t_D \approx 10$, where v_0 is the characteristic lattice vibration frequency of the solid and t_D is the time for H to diffuse to neighbouring sites (mainly by quantum tunnelling). N is the number of regular sites for adsorption, while N' is the number of enhanced sites, and N_S is the total number of imperfection sites, W_{S2} is the rate of sticking for hydrogen molecules, while W_S is that for hydrogen atoms. When $T < T_{tr}$, the time for an adsorbed atom to find an enhanced site is less than the time for evaporation. When $T < T_c$ for enhanced sites there is a high recombination efficiency. For $T < T_{mono}$, a monolayer covers the surface and D drops to D_{mono}, unless the D corresponds to an enhanced site that is unaffected by the monolayer (a somewhat improbable case). Then T_c drops correspondingly, to $\approx 7°K$. For T between 7 and 11°K rapid evaporation of an atom might occur before the molecule is formed. On the other hand, even a small fraction of the binding energy might *not* be available for evaporation of the molecule that forms at an enhanced site since the probability is very high for the other adsorbed particles on the surface to take up this energy in translational motion along the surface. These atoms act like a nearly free two-dimensional gas, almost a liquid. Momentum transport is probably by phonon-like modes, accounting for the high "mobility" described in laboratory measurements of heat capacity of He adsorbed on argon. The motion perpendicular to the surface is like that of an oscillator. The packing of neighbouring adsorbed hydrogen molecules is very dense, with a high thermal conductivity and heat capacity. It seems unlikely that hydrogen *atoms* would be adsorbed without appreciable covalent bonding to the grain, i.e., chemisorption. These conditions would drain the translational motion associated with the excess energy of binding of the newly-formed hydrogen molecule. Another difficulty that arises is the possibility of a high activation energy for one atom to form a molecule with another atom bound to an enhanced site, if some chemisorption at that site is involved (which seems likely).

The rate of formation of H_2 has been estimated[69] to be

$$\frac{dn_{H_2}}{dt} \approx 10^{-17} n_H (n_H + 2n_{H_2})$$

where the total number density of hydrogen nuclei, $n_H + 2n_{H_2}$, is used to provide the number of grains per unit volume. More precisely, the cross-section times the density of grains has been estimated as approximately 3×10^{-22} cm^2 times $(n_H + 2n_{H_2})$.

Catalysis by grains at higher temperatures, such as found in the envelopes of stars, probably infrared stars, is also possible. The much higher densities make the high evaporation rate of little consequence, since the chemisorption "sticking" rate is also high. The pressures are of the order of 10^{-6} atmospheres (3×10^{13} particles cm^{-3}). Temperatures are approximately 1000°K. The grains probably contain oxides of transition elements (Fe, Ni, Co, and probably Ti, V, etc.). The well-known Fischer–Tropsch process[90]

$$n\,CO + \left(n + \frac{x}{2}\right) H_2 \rightarrow C_n H_x + n H_2 O$$

works very well under these conditions.

Analysis of hydrocarbons and nitrogen compounds found in several meteorites have suggested to several workers that this process has occurred on meteorites in the solar nebula.[90]

Fischer–Tropsch experiments show a distribution among the hydrocarbons that are similar to that for the meteorites. However, the relative abundance of aromatic compounds, at the expense of the lower-weight hydrocarbons suggests that reheating has transformed aliphatics to aromatics. The prevention of the condensation of graphite is considered essential to the generation of the amino acids and other compounds. This is termed a metastable equilibrium. Reactive molecules like CO, HCN and (CN)$_2$ form biologically important compounds at moderate temperatures. If CO and NH$_3$ are available, HCN can be produced at 900°K temperature. HCN is an important intermediate for many nitrogen compounds of biological interest, as discussed below.

5.2. Large interstellar molecules

Laboratory experiments (Miller[91]) have produced amino acids in a liquid above which there is a discharge containing ammonia, methane, hydrogen, and water vapour. The important intermediate molecule HCN was produced and is believed to be essential to the formation of nitriles which then are hydrolyzed in the liquid water to form amino acids.

Hydrogen cyanide can form directly from methane and ammonia or ammonia and carbon monoxide[92] (formed from methane and water vapor)

$$CH_4 + NH_3 \rightarrow HCN + 3H_2$$
$$CO + NH_3 \rightarrow HCN + H_2O$$

Hydrocarbons like $H_2C = CH_2$ (ethylene) and $HC \equiv CH$ (acetylene) are formed as well so that

$$C_2H_4 + HCN \rightarrow CH_3CH_2CN \text{ (propionitrile)}$$
$$C_2H_2 + 2HCN \rightarrow NCCH_2CH_2CN \text{ (succinodinitrile)}$$

Hydrolysis of the nitrile in the water yields,

$$CH_3CH_2CN + 2H_2O \rightarrow CH_3CH_2COOH \text{ (propionic acid)} + NH_3$$
$$NCCH_2CH_2CN + 4H_2O \rightarrow HOOCCH_2CH_2COOH \text{ (succinic acid)} + 2NH_3$$

Aldehydes form hydroxynitriles with HCN. These hydrolyze to acids,

$$CH_3CHO \text{ (acetaldehyde)} + HCN \rightarrow CH_3CH(OH)CN \xrightarrow{2H_2O}$$
$$CH_3CH(OH)COOH \text{ (lactic acid)} + NH_3$$

$$CH_3CH_2CHO \text{ (propionaldehyde)} + HCN \rightarrow CH_3CH_2CH(OH)CN \xrightarrow{2H_2O}$$
$$CH_3CH_2CH(OH)COOH \text{ (α–hydroxybutyric acid)} + NH_3.$$

With NH_3, the hydroxynitriles can form aminonitriles from which amino acids may be obtained by hydrolysis:

$$CH_3CH(OH)CN \xrightarrow{NH_3} CH_3CH(NH_2)CN + H_2O$$
$$\downarrow 2H_2O$$
$$CH_3CH_2CH(NH_2)COOH \text{ (alanine)} + NH_3.$$

$$CH_3CH_2CH(OH)CN \xrightarrow{NH_3} CH_3CH_2CH(NH_2)CN$$
$$\downarrow 2H_2O$$
$$CH_3CH_2CH(NH_2)COOH$$
$$\text{(α–aminobutyric acid)} + NH_3.$$

To form β-alanine, NH_3 makes a nucleophilic attack on the β carbon of acrylonitrile or acrylamide. Hydrolysis yields the acid.

$$CH_2CHCN \text{ (acrylonitrile)} \xrightarrow{NH_3} CH_2(NH_2)CH_2CN \text{ (β-aminopropionitrile)}$$

$$\xrightarrow{2H_2O} CH_2NH_2CH_2COOH \text{ (β-alanine)} + NH_3$$

$$CH_2CHCONH_2 \text{ (acrylamide)} \xrightarrow{NH_3} CH_2-CH-\underset{\underset{OH}{|}}{\overset{\overset{NH_2\ NH_2}{|\ \ \ \ |}}{C}} \text{ (β-aminopropionamide)}$$

$$\xrightarrow{H_2O} CH_2(NH_2)CH_2COOH \text{ (β-alanine)} + NH_3.$$

Temperature histories of heating followed by rapid cooling could be beneficial for forming amino acids followed by their polymerization into protein-like polymers, and the formation from these of microscopic structures (microspheres) which might have life-like characteristics, such as division and replication (Fox et al.).[92]

Oró and Kimball[93] have suggested that a mixture of HCN, NH_3 and H_2O yields adenine

With ammonia as a catalyst, the overall reaction is 5HCN → adenine.

$$2HC\equiv N \rightarrow N\equiv C - \underset{\underset{H}{|}}{C} = NH \xrightarrow{HCN} N\equiv C - \underset{\underset{H}{|}}{\overset{\overset{NH_2}{|}}{C}} - C\equiv N$$

$$\downarrow 2NH_3$$

$$\underset{HN}{\overset{H_2N}{\diagdown}}C = \underset{\underset{H}{|}}{\overset{\overset{NH_2}{|}}{C}} - \underset{NH_2}{\overset{NH}{\diagup\!\!\diagup}}C$$

↓ formamidine

adenine + 2NH$_3$ ←——formamidine—— $\underset{H_2N}{\overset{HN}{\diagup\!\!\diagup}}C\underset{\diagdown}{\overset{\overset{\overset{NH_2}{|}}{C}}{}}\!\!\underset{\underset{H}{N}}{\overset{N}{\diagdown}}$ + 2NH$_3$.

$$HCN + NH_3 \rightarrow HN = \underset{\underset{H}{|}}{C} - NH_2 \text{ (formamidine)}. \qquad AICAI$$

AICAI, 4-aminoimidazole-5-carboxamidine, gives rise to AICA, 4-aminoimidazole-5-carboxamide, which then yields the purines: guanine, xanthine, and hypoxanthine by means of reactions with formamidine, urea

$$\left(\underset{H_2N}{\overset{H_2N}{\diagdown\!\!\diagup}}C = O\right) \text{ or guanidine } \left(\underset{H_2N}{\overset{H_2N}{\diagdown\!\!\diagup}}C = NH\right).$$

An active dimer of HCN, aminocyanocarbene, could lead to proteins and purines.[94] The dimer is formed as follows:

$$2HCN \rightarrow \underset{HN}{\overset{H}{\diagdown\!\!\diagup}}C - C\equiv N \xrightarrow{\text{tautomerization}} H_2N - C^+ = C = N^-, (HCN)_2$$

(HCN)$_2$ rapidly polymerizes into polyaminoketenimine which, after tautomerization, reacts with HCN to form polyaminomalononitrile (PAMN):

$$n(\text{HCN})_2 \to \left(\begin{array}{c} -\text{C}=\text{C}=\text{N}- \\ | \\ \text{NH}_2 \end{array} \right)_n \to \left(\begin{array}{c} -\text{C}-\text{C}=\text{N}- \\ \| \ \ | \\ \text{NH} \ \text{H} \end{array} \right)_n \xrightarrow{\text{HCN}} \left(\begin{array}{c} \text{CN} \\ | \\ -\text{C}-\text{C}-\text{NH}- \\ \| \ \ | \\ \text{NH} \ \text{H} \end{array} \right)_n.$$

PAMN

PAMN would tend to form alpha-helices of both handedness. The projecting CN groups in PAMN would react with HCN to form various precursors of the alpha-amino acids. Water reacting with the projecting imino and cyano groups, followed by removal of CO_2, would then yield peptide structures (protein).[92]

An interesting possibility might be the formation of protein-like polymers without aqueous solutions, but with chemisorbed NH_3, H_2O, HCN and hydrocarbons on the interstellar grains, considering the liquid-like state of the adsorbed hydrogen.

5.3. Chemical reactions in shock waves

Formaldehyde in a few dark dust clouds appears in absorption against the microwave 2·7°K cosmic background. In order to explain this, one suggestion was that infrared fluorescence from formaldehyde molecules formed in a shock front could pump and anti-invert the microwave populations of the cooler molecules left downstream of the shock.[28] The details of this pumping are discussed above. However, the rates of formation of formaldehyde in the shock front need to be examined.

The general situation is that of rapid heating at the shock jump followed by gradual cooling by radiation. The species most important for cooling at these temperatures are H_2 and O. The grains become important for cooling at fairly high densities. If radicals can be formed in the heating zone, then chemical reactions can persist even into the cool regions. Recent shock tube measurements[95] have yielded the following rates for the oxidation of methane and the formation and destruction of formaldehyde:

$O + CH_4 \to CH_3 + OH$ $\quad k_1 = 2 \cdot 6 \times 10^{-10} \exp(-4000°K/T_k)$

$O + CH_3 \to CH_2O + H$ $\quad k_2 = 1 \times 10^{-10} \exp(-1600°K/T_k)$

$O + CH_2O \to H + CO + OH$ $\quad k_3 = 6 \times 10^{-10} \exp(-2500°K/T_k)$

The k's are rate constants in cm^3-sec^{-1}. The forward reactions are exothermic. The rate equations for the various species are (neglecting the back reactions)

$$\frac{d[CH_3]}{dt} \approx (k_1 [CH_4] - k_2 [CH_3]) [O]$$

$$\frac{d[CH_2O]}{dt} \approx (k_2 [CH_3] - k_3 [CH_2O]) [O].$$

The cooling rate is given by

$$\tfrac{3}{2} nk \frac{dT_k}{dt} \approx -\sum_{i,j} n_{ji} g_j E_{ij} \exp(-E_{ij}/kT_k) \left[W_{ij}^{-1} + \frac{1 - \exp(-E_{ij}/kT_k)}{A_{ij}/Q_{ij}} \right]^{-1}$$

where the sum is over all levels of O and H_2, with i the upper state and j the lower state. E_{ij} is the energy difference between the levels, A_{ij} is the Einstein-coefficient for spontaneous decay, Q_{ij} is the radiation trapping factor, W_{ij} is the collisional de-excitation rate, and n_{ji} is the population difference $(n_j/g_j) - (n_i/g_i)$. This complicated expression allows for the two cases that usually arise when collision rates are comparable to the radiative rates. When the W is large, the cooling rate is that for local thermodynamic equilibrium (LTE) at the kinetic temperature. When the A is large enough, the cooling rate is determined by the rate of exciting the fluorescent level by collisions. In either case, the rate of the slower process is important. When intensity I_0 from an external source of radiation is present, then the above cooling rate is modified by an approximate factor

$$1 - (I_0 L Q_{ij}/2I_{BB})$$

where L is the transmission function for the external flux at the resonant frequency, and I_{BB} is the black-body intensity for temperature T_k. It is assumed that the cloud is illuminated from one side only. These expressions are derived from the usual two-stream integral equation approach to radiative transfer described above, with the inclusion of collisional excitation and de-excitation, all with a two-level scheme. The multi-level, interlocked case is also simple to do by following the rules laid out in the discussion of collisional and radiative rates and steady-state populations.

The left-hand side of the above equation is simplified by the omission of cooling by volume expansion, given by $-kT_k \, dn/dt \, n$, with n being the total density. Contributions to the specific heat other than translational degrees of freedom were also neglected. An estimate of the characteristic cooling time τ_c appropriate to densities approaching 10^5 cm^{-3} is as follows

$$\frac{1}{\tau_c} = \frac{-1}{T_k}\frac{dT_k}{dt} \approx \frac{10}{27}\frac{[O]}{n} A E \exp(-E/kT_k)(kT_k)^{-1}$$

where $E/k \approx 228°K$, $W \simeq 3 \times 10^{-9}$ n sec^{-1} and $A \approx 9 \times 10^{-5}$ sec^{-1}. Therefore,

$$\tau_c \simeq \frac{3 \times 10^5 \text{ secs}}{[O]/n}$$

provided $n > 10^4$ cm^{-3} and $T_k \approx 1000°K$. Lower densities require the A-dominated rate. The first rate equation, when run to completion ($d[CH_3]/dt \approx 0$), gives

$$[CH_3]/[CH_4] \approx k_1/k_2.$$

Similarly, from the second rate equation, for chemical completion, $[CH_2O]/CH_3] \approx k_2/k_3$. Overall then

$$[CH_2O]/[CH_4] \approx k_1/k_3 \approx \tfrac{1}{2}\exp(-1500°K/T_k).$$

However, the cooling time may become shorter than the reaction times,

$$\tau_r \approx k_2^{-1}[O]^{-1} \quad \text{or} \quad k_3^{-1}[O]^{-1} \approx 10^{10} \text{ sec}/[O] \text{ cm}^3.$$

The cooling time in the lower-density optically-thin regime is given by

$$\tau_c^{-1} \approx \frac{2}{3}\frac{[O]}{n} W E \exp(-E/kT_k)(kT_k)^{-1} \approx \frac{[O]}{3 \times 10^9} \text{ sec}^{-1}.$$

Then, using the rate equations by neglecting the negative terms and assuming the CH$_4$ concentration is nearly constant, one obtains

$$[CH_3] \approx k_1 [CH_4][O]\tau_c$$
$$[CH_2O] \approx k_2 [CH_3][O]\tau_c/2 \approx k_1 k_2 [CH_4]^2 [O]^2 \tau_c^2/2.$$

Upon substituting the low-density τ_c into the above expressions, one has

$$[CH_3]/[CH_4] \approx \exp(-4000°K/T_k)$$
$$[CH_2O]/[CH_4] \approx \exp(-5600°K/T_k) \times 10^{-1}[CH_4].$$

The CH$_4$ concentration is unknown, but perhaps it is reasonable to assume that some fraction of the free carbon has been converted into methane by collision with hydrogen-coated grains in the shock front. With gas temperatures around 1000°K and $[CH_4] \simeq 10^{-5} [H_2]$, we obtain $[CH_2O]/[H_2] \approx 10^{-13} [H_2]$ for the lower density case and $[CH_2O]/[H_2] \approx 10^{-6}$ for the higher density case ($[H_2] > 10^4$ cm^{-3}). The requirements for adequate pumping by hot CH_2O was that $[CH_2O]/[H_2] \gtrsim 10^{-7}$, and the surveys of CH_2O 6-cm absorption throughout the galaxy give the ratio as $\sim 10^{-8}$. The frozen concentrations, left behind the rapidly cooling shock front, appear in the right range of values, provided the methane concentration is approximately $10^{-5} [H_2]$. If we equate the effective rate of collisions of C with the grains to the rate of destruction of CH_4 by O (given in the first chemical reaction), we obtain in steady state

$$\frac{d}{dt}[CH_4] \approx k_g[C][H_2] - k_1[O][CH_4] \approx 0$$

so that

$$[CH_4] \approx k_g[C][H_2]/k_1[O]$$

$$\approx 10^{-7} \exp(4000°K/T_k) \frac{[C]}{[O]}[H_2]$$

or

$$[CH_4] \approx 10^{-5} [H_2] \text{ for } T_k \approx 1000°K \text{ and } [C]/[O] \approx 1.$$

The coefficient k_g is approximately the usual grain cross-section per hydrogen atom 3×10^{-22} cm^2, times a sticking efficiency of about 1/3 times a relative velocity of 10^4 cm–sec^{-1}.

The rate of formation of H_2, OH and H_2O in the shock front is also of some interest. If, again, O is the more important element than C for reaction with hydrogen, the following reactions are to be considered:

$$O + H_2 \underset{-4}{\overset{4}{\rightleftarrows}} OH + H$$

$$H_2 + OH \underset{-5}{\overset{5}{\rightleftarrows}} H_2O + H.$$

The rate equations are

$$\frac{d[OH]}{dt} \approx k_4[O][H_2] - k_{-4}[H][OH] - k_5[H_2][OH] + k_{-5}[H_2O][H]$$

$$\frac{d[H_2O]}{dt} \approx k_5[H_2][OH] - k_{-5}[H][H_2O]$$

Steady-state conditions for $[H_2] \gg [H]$ give

$$[OH] \approx (k_4[H_2][O] + k_{-5}[H_2O][H])/(k_{-4}[H] + k_5[H_2]) \approx k_4[O]/k_5.$$

The second reaction probably will not go to completion, so that

$$[H_2O] \approx \int dT_k\, k_5[H_2][OH]\,\tau_c/T_k \approx k_4[H_2][O]\,\tau_c$$

otherwise $[OH] \approx k_4[O][H_2]/k_{-4}[H]$ and

$$[H_2O] \approx k_5[H_2][OH]/k_{-5}[H].$$

With [27]

$$k_4 \approx 10^{-12}\, T_k^{1/2} \exp(-4150°K/T_k)\, cm^3\text{-}sec^{-1}$$

$$k_{-4} \approx 10^{-11} \exp(-3700°K/T_k)\, cm^3\text{-}sec^{-1}$$

$$k_5 \approx 10^{-10} \exp(-3000°K/T_k)\, cm^3\text{-}sec^{-1}$$

and

$$k_{-5} \approx 10^{-10} \exp(-11000°K/T_k)\, cm^3\text{-}sec^{-1}$$

we obtain

$$[OH] \approx 10^{-2}[O]\, T_k^{1/2} \exp(-1200°K/T_k) \approx 10^{-1}[O]$$

$$[H_2O] \approx 3 \times 10^{-10}\, T_k^{3/2} \exp(-3900°K/T_k)[H_2]^2 \approx 10^{-7}[H_2]^2.$$

If the depletion of free O and H is not drastic, we would expect a large OH-to-hydrogen ratio like 10^{-4}. However, the observed value is approximately 10^{-8} in dark clouds. The destructive mechanism $O + OH \xrightarrow{6} O_2 + H$ proceeds at low temperatures with the rate constant $k_6 = 10^{-10} \exp(-600°K/T_k)\, cm^3\text{-}sec^{-1}$. Applying this rate for the lower-density cooling time, $\tau_c \approx T_k \times 10^7 \exp(230°K/T_k)[O]^{-1}$, gives

$$[OH]_f \approx [OH]_i \exp \int dT_k\, k_6[O]\, \tau_c/T_k \approx [OH]_i\, e^{-1}$$

where $[OH]_i$ is the OH concentration in the 1000°K region and $[OH]_f$ is the resulting concentration for the 200°K (or lower) region. The quenching of reaction 4 by cooling seems also to be unimportant.

The depletion of $[O]$ by a factor of about 10^{-3} in the hot region from its usual cosmic abundance $[O] \simeq 10^{-3}n$ allows H_2 to be the important cooling mechanism there, for which

$$\tau_c \approx 3 \times 10^4\, T_k^2 \exp(1680°K/T_k) \approx 3 \times 10^{11}\, \text{sec (for } T_k \approx 10^3\,°K)$$

Since $[O] < 10^{-6}[H_2]$, we have $[OH] < 10^{-7}[H_2]$ and $[H_2O] \approx 10^{-3}[H_2]$. The formation of H_2O accounts for most of the O-depletion. We have neglected the depletion of O and H_2O by means of condensation on or combination with the grains.

5.4. Protostars

If the gravitational potential exceeds about twice the average kinetic energy of a cloud, then there is gravitational instability.[96] If a major part of the kinetic energy is thermal then a uniform cloud is unstable if

$$\frac{kT_k}{m} < \frac{GM}{2R}.$$

A saturated maser implies that twice the induced transition rate W_m exceeds the combined net rates of emptying both states by radiation or collisions. Whether there is collisional or radiative pumping, we use the collision rate W_c across the microwave transition as a lower bound on the induced rate for a saturated maser, i.e.,

$$W_m > W_c.$$

This is rewritten as

$$(T_B/10^{10}\,°\text{K})(l_0/3 \times 10^{14}\,\text{cm})^2 > (n/10^5\,\text{cm}^{-3})(l/10^{17}\,\text{cm})^2$$

where nl^2 is nearly a constant determined by the velocity of collapse near the outer boundary, according to a variety of recent calculations.[28,96,97] The diameter l_0 is the apparent one determined by interferometry, and l is whatever diameter of the cloud for saturated maser action or in terms of the "hot spot" analysis given above, wherever $F(l/2) > 1$. Data on several sources give a value of approximately 10^3 for the left side of the above inequality. However, for gravitational collapse, the right side of the inequality must be numerically greater than about T_k which is estimated from millimeter and microwave molecular work to be between 5 and 50°K, depending on the type of cloud. Thus, it seems likely that the maser clouds are collapsing, yet with a large store of energy to account for the large microwave photon output as well as for the "hot spots."

According to calculations by Larson,[96] in approximately 10^{13} seconds a cloud about 10^{17} cm in diameter with a mass of about $1M_\odot$ and with a density of about 10^5 cm^{-3} will develop a small condensation within it having a central density of approximately 10^{11} cm^{-3}, while the temperature everywhere remains between 10° and 100°K. The diameter of this central region is about 3×10^{14} cm. A dense opaque core of $10^{-2}\,M_\odot$ and radius $\lesssim 10^{14}$ cm then develops inside this in about 300 years with an initial temperature of approximately 200°K and a density of about 10^{14} cm^{-3}. The core is bounded by a stationary shock which slows the infall of mass from outside, thereby producing hydrostatic equilibrium inside. The core rapidly heats up to about 2000°K, doubles its mass by continuing infall of mass and reaches a density of 10^{17} cm^{-3} in about a year. A second stellar core, with approximately $1R_\odot$ radius, 10^{-3} $M_\odot$, 10^{22} cm^{-3} and 20,000°K at its centre, then forms inside the first core. In a year, the parts of the cloud outside this core but within 10^{13} cm may be heated above 1000°K, while the stellar core has expanded to $10R_\odot$ and has acquired a surface temperature of 4000°K. Until this stage is reached the remainder of the cloud has a density profile that is inverse-square of the radius, from 10^{11} cm^{-3} at 10^{14} cm to 10^5 cm^{-3} at 10^{17} cm. This outer radius is kept fixed in the calculation.

From the known properties of the molecular emission and absorption we can conclude that only the part of the cloud outside the cores might be responsible for maser action, that the second core is too hot and too dense, and that the first core is too dense. The spatially-separated emitting points, which are supposed to be separate protostars, have intensities which differ by less than a factor of ten, though they may differ in polarization and bandwidth. This suggests a large and general source of pumping such as a shock wave covering the region ($100M_\odot$) and providing the outer regions of the protostars with sufficient energy in the form of molecular resonance lines for strong (but saturated) maser emission. For the strongest masers, the shock front area would be $\approx 10^{32}$ cm^2, with densities ahead of the front of $\approx 10^{11}$ cm^{-3}. There would be an additional pump source within each sphere (to assure "hot spots"), perhaps due to a small-diameter shockwave, perhaps to the stationary shock of $\approx 3 \times 10^{14}$ cm diameter that bounds the first core, or to a collapsing shockwave arising from an overpressure or a stellar wind acting on the outer boundary of a gravitationally-unstable cloudlet, or to an expanding shockwave due to reflection of the earlier one that collapsed all the way to the center (perhaps initiating a rapid series of events leading to a stellar core). In addition, for those cases of a neutral region that is imbedded in an H II region, a very slowly converging ionization front, at relatively large radii, would separate the neutral and the photo-ionized regions.[98]

5.5. Prediction of new microwave lines

The lowest rotational levels of a rigid asymmetric molecule (without internal rotation of one part with respect to the rest) have energies with respect to the ground state that are given by the rotational constants A, B, and C, as shown in Fig. 14.

These energies (divided by h) are easily derived from the matrix elements of the Hamiltonian for a rigid asymmetric rotor.[99] The selection rules for a-, b-, or c-type transitions (depending upon the electric dipole moment having a component along the a-, b-, or c-axis, respectively) are given in terms of the subscripts K_- and K_+ in the usual notation J_{K_-, K_+}. Table XI

Table XI

		$J' = J+1$	J	$J-1$
a-type ($\kappa \simeq -1$) $K_-' = K_-$		$\dfrac{(J+1)^2 - K_-^2}{J+1}$	$\dfrac{K_-^2(2J+1)}{J(J+1)}$	$\dfrac{J^2 - K_-^2}{J}$
b-type* ($\kappa \simeq -1$) $K_-' = K_- \pm 1$	ε_{K_-}	$\dfrac{(J \pm K_- + 1)(J \pm K_- + 2)}{2(J+1)}$	$\varepsilon_{K_-} \dfrac{(J \mp K_-)(J \pm K_- + 1)(2J+1)}{2J(J+1)}$	$\varepsilon_{K_-} \dfrac{(J \mp K_-)(J \mp K_- - 1)}{2J}$
b-type** ($\kappa \simeq +1$) $K_+' = K_+ \pm 1$	ε_{K_+}	$\dfrac{(J \pm K_+ + 1)(J \pm K_+ + 2)}{2(J+1)}$	$\varepsilon_{K_+} \dfrac{(J \mp K_+)(J \pm K_+ + 1)(2J+1)}{2J(J+1)}$	$\varepsilon_{K_+} \dfrac{(J \mp K_+)(J \mp K_+ - 1)}{2J}$
c-type ($\kappa \simeq +1$) $K_+' = K_+$		$\dfrac{(J+1)^2 - K_+^2}{J+1}$	$\dfrac{K_+^2(2J+1)}{J(J+1)}$	$\dfrac{J^2 - K_+^2}{J}$

* also c-type ($\kappa \simeq -1$).
** also a-type ($\kappa \simeq +1$).

$J_{K_- K_+} \leftrightarrow J'_{K_-' K_+'}$
a-type: $ee \leftrightarrow eo$
          $oo \leftrightarrow oe$

b-type $eo \leftrightarrow oe$
         $ee \leftrightarrow oo$
c-type $ee \leftrightarrow oe$
         $oo \leftrightarrow eo$

e means *even* integer, o means *odd*.

$$\kappa = \frac{2B - A - C}{A - C}$$

Note: $\varepsilon_{K_\pm} = \begin{cases} 1/2 \text{ if both } K_\pm \text{ and } K_\pm' \neq 0 \\ 1 \text{ if either } K_\pm \text{ or } K_\pm' = 0. \end{cases}$

gives these rules and expressions for the linestrengths if the asymmetry of the molecule is neglected. Corrections to the linestrengths due to asymmetry are easily made by using perturbation theory. If some frequency data (or the constants A, B, and C) are available but certain microwave

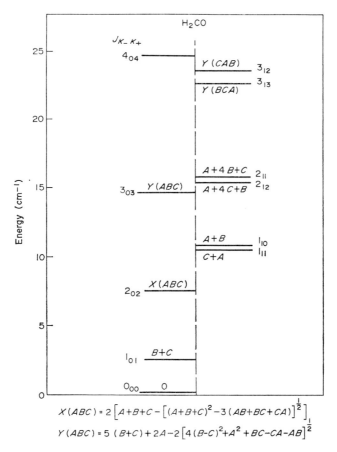

FIGURE 14

transitions like the $K = 1$ doublets have not been measured, then the frequencies can be calculated, with fairly good accuracy, from $(1/2)J(J+1) \times (B - C)$ for a prolate molecule. When internal rotation of one part of the molecule with respect to another (as OH with respect to CH_3 in methanol) occurs, a simple multiplying factor on $(B - C)$ often suffices for calculating the frequencies for the $K = 1$ doublets, for both high and low potential barriers for internal rotation.[100]

REFERENCES

1. Weinreb, S., Barrett, A. H., Meeks, M. L. and Henry, J. C., *Nature* **200**, 829 (1963).
2. Weinreb, S., *Technical Report* 412, *Research Laboratory of Electronics, M.I.T.* (1963).
3. Weaver, H., Williams, D. R. W., Dieter, N. H. and Lum, W. T., *Nature* **208**, 440 (1965); Gunderman, E., Ph.D. Thesis, Department of Astronomy, Harvard University (1965).
4. McGee, R. X., Robinson, B. J., Gardner, F. F. and Bolton, J. G., *Nature* **208**, 1193 (1965).
5. Cheung, A. C., Rank, D. M., Townes, C. H., Thornton, D. D. and Welch, W. J., *Nature* **221**, 626 (1969).
6. Knowles, S. H., Mayer, C. H., Cheung, A. C., Rank, D. M. and Townes, C. H., *Science* **163**, 1055 (1969).
7. Synder, L. E., Buhl, D., Zuckerman, B. and Palmer, P., *Phys. Rev. Letters* **22**, 679 (1969).
8. Barrett, A. H., *Scientific American* **219**, No. 6, 36 (1968).
9. Sullivan, W. T., *Ap. J.* **166**, 321 (1971).
10. Robinson, B. J. and McGee, R. X., *Ann. Rev. Astr. and Ap.* **5**, 183 (1967).
11. Litvak, M. M., McWhorter, A. L., Meeks, M. L. and Zeiger, H. J., *Phys. Rev. Letters* **17**, 821 (1966).
12. Palmer, P., Zuckerman, B., Buhl, D. and Snyder, L. E., *Ap. J. (Letters)* **156**, L147 (1969).
13. Turner, B. E., *Ap. J. (Letters)* **163**, L35 (1971).
14. Ball, J. A., Gottlieb, C. A., Lilley, A. E. and Radford, H. E., *Ap. J. (Letters)* **162**, L203 (1970).
15. Jefferts, K. B., Penzias, A. A. and Wilson, R. W., *Ap. J. (Letters)* **161**, L87 (1970).
16. Wilson, R. W., Jefferts, K. B. and Penzias, A. A., *Ap. J. (Letters)* **161**, L43 (1970).
17. Snyder, L. E. and Buhl, D., *Ap. J. (Letters)* **163**, L47 (1971).
18. Penzias, A. A., Solomon, P. M., Wilson, R. W. and Jefferts, K. B., *Ap. J. (Letters)* **168**, L53 (1971).
19. Wilson, R. W., Penzias, A. A., Jefferts, K. B., Kutner, M. and Thaddeus, P., *Ap. J. (Letters)* **167**, L97 (1971).
20. Kutner, M., Thaddeus, P., Jefferts, K. B., Penzias, A. A. and Wilson, R. W., *Ap. J. (Letters)* **164**, L49 (1971); Thaddeus, P., Wilson, R. W., Kutner, M., Penzias, A. A. and Jefferts, K, B.. *ibid.* **168**, L59 (1971).
21. Wilson, W. J. and Barrett, A. H., *Science* **161**, 778 (1968).
22. Wilson, W. J., Barrett, A. H. and Moran, J. M., *Ap. J.* **160**, 545 (1970).
23. Schwartz, P. R. and Barrett, A. H., *Ap. J. (Letters)* **159**, L123 (1970).
24. Weaver, H., Dieter, N. H. and Williams, D. R. W., *Ap. J. Suppl.* **16**, 219 (1968).
25. Goss, W. M., *Ap. J. Suppl.* **15**, 131 (1968).
26. Ball, J. A. and Staelin, D. H., *Ap. J. (Letters)* **153**, L41 (1968).
27. Litvak, M. M., *Science* **165**, 855 (1969).
28. Litvak, M. M., *Ap. J. (Letters)* **160**, L133 (1970).
29. Solomon, P. M. and Thaddeus, P. in preparation (cf. ref. 28).
30. Zuckerman, B., Palmer, P., Penfield, H. and Lilley, A. E., *Ap. J. (Letters)* **153** L69 (1968).

31. Yen, J. L., Zuckerman, B., Palmer, P. and Penfield, H. *Ap. J. (Letters)* **156**, L27 (1969).
32. Zuckerman, B. and Palmer, P., *Ap. J. (Letters)* **159**, L197 (1970).
33. Thacker, D. L., Wilson, W. J. and Barrett, A. H., *Ap. J. (Letters)* **161**, L191 (1970); Palmer, P. and Zuckerman, B., *Ap. J. (Letters)* **161**, L199 (1970).
34. Turner, B. E., Palmer, P. and Zuckerman, B., *Ap. J. (Letters)* **160**, L125 (1970).
35. Rydbeck, O. E. H., Kollberg, E. and Elldér, J., *Ap. J. (Letters)* **161**, L25 (1970).
36. Ball. J. A., Dickinson, D. F., Gottlieb, C. A. and Radford, H. E., *Astron. J.* **75**, 762 (1970); Ball, J. A., Gottlieb, C. A., Meeks, M. L. and Radford, H. E., *Ap. J. (Letters)* **163**, L33 (1971).
37. Churg, A. and Levy, D. H., *Ap. J. (Letters)* **162**, L161 (1970).
38. Dickinson, D. F., Litvak, M. M. and Zuckerman, B. M., *Sky and Telescope* **39**, 1 (1970).
39. Holtz, J. Z., *Ap. J. (Letters)* **153**, L117 (1968).
40. Townes, C. H. and Gwinn, W. D., unpublished (cf. ref. 27).
41. Moran, J. M., Burke, B. F., Barrett, A. H., Rogers, A. E. E., Ball, J. A., Carter, J. C. and Cudaback, D. D., *Ap. J. (Letters)* **152**, L97 (1968).
42. Burke, B. F., Papa, D. C., Papadopoulous, G. D., G. D., Schwartz, P. R., Knowles, S. H., Sullivan, W. T., Meeks, M. L. and Moran, J. M., *Ap. J. (Letters)* **160**, L63 (1970).
43. Litvak, M. M., *Ap. J.* **170**, 71 (1971); Highlights of Astronomy, **2** (Dordrecht–Holland: D. Reidel, 1971).
44. Landau, L. D. and Lifshitz, E. M., "Electrodynamics of Continuous Media" (Reading, Mass,: Addison Wesley, 1960). pp. 225–6.
45. Hyland, A. R., Becklin, E. E., Neugebauer, G. and Wallerstein, G., *Ap. J.* **158**, 619 (1969).
46. Herbig, G. H., *Ap. J.* **162**, 557 (1970); Wallerstein, G., *Ap. J.* **166**, 725 (1971).
47. Litvak, M. M., *Ap. J.* **156**, 471 (1969).
48. Heiles, C. E., *Ap. J.* **151**, 919 (1968).
49. Heiles, C. E., *Ap. J.* **157**, 123 (1969).
50. Cudaback, D. D. and Heiles, C. E., *Ap. J. (Letters)* **155**, L21 (1969).
51. Turner, B. E. and Heiles, C. E., *Ap. J.* **170**, 453 (1971).
52. Zuckerman, B., Palmer, P., Synder, L. E. and Buhl, D., *Ap. J. (Letters)* **157**, L167 (1969).
53. Scoville, N. Z., Solomon, P. M. and Thaddeus, P., *Ap. J.* **172**, 335 (1971); Kutner, M. and Thaddeus, P., *Ap. J. (Letters)* **168**, L67 (1971).
54. Heer, C. V., *Phys. Rev. Letters* **17**, 774 (1966).
55. Heer, C. V. and Settles, R. A. 1967, *J. Molec. Spectros.* **23**, 448 (1967).
56. Bender, P. L., *Phys. Rev. Letters* **18**, 562 (1967).
57. Litvak, M. M., *Phys. Rev. A.* **2**, 937 (1970).
58. Zuckerman, B., Ball, J. A. and Gottlieb, C. A., *Ap. J. (Letters)* **163**, L41 (1971).
59. Kraus, J. D., "Radio Astronomy" (New York: McGraw–Hill Co., 1966).
60. Herzberg, G. "Molecular Spectra and Molecular Structure II. Infrared and Raman Spectra of Polyatomic Molecules" (Princeton: D. van Nostrand Co., 1964). p. 300.
61. Evans, N. J., Cheung, A. C. and Sloanaker, R. M., *Ap. J. (Letters)* **159**, L9 (1970).
62. Field, G. B., Rather, J. D. G., Aanestaad, P. A. and Orszag, S. A., *Ap. J.* **151**, 953 (1969).
63. Oka, T., *J. Chem. Phys.* **47**, 13 (1967).
64. Lees, R. and Oka, T., *J. Chem. Phys.* **51**, 3027 (1969).
65. Oka, T. *J. Chem. Phys.* **48**, 4919 (1968).
66. Daly, P. W. and Oka, T., *J. Chem. Phys.* **53**, 3272 (1970).

67. Oka, T. and Shimizu, T., *Phys. Rev. A.* **2**, 587 (1970).
68. Rogers, A. E. E. and Barrett, A. H., *Ap. J.* **151**, 163 (1968); Goss, W. M. and Field, G. B., *Ap. J.* **151**, 177 (1968).
69. Hollenbach, D., Werner, M. W. and Salpeter, E. E., *Ap. J.* **163**, 165 (1971); Werner, M. W., *Ap. J. (Letters)* **6**, 81 (1970).
70. Townes, C. H. and Cheung, A. C., *Ap. J. (Letters)* **157**, L103 (1969).
71. Thaddeus, P., *Ap. J.* **173**, 317 (1972).
72. Litvak, M. M., "Interstellar Ionized Hydrogen", ed., Y. Terzian (New York: W. A. Benjamin, Inc., 1968). pp. 713–745.
73. Solomon, P. M., *Nature* **217**, 334 (1968).
74. Litvak, M. M., Zuckerman, B. and Dickinson, D. F., *Ap. J.* **156**, 875 (1969).
75. de Zafra, R. L., Marshall, A. and Metcalf, H., *Phys. Rev. A* **3**, 1557 (1971).
76. Low, F. J. and Aumann, H. H., *Ap. J. (Letters)* **162**, L79 (1971).
77. Herzberg, G. "Molecular Spectra and Molecular Structure, III. Electronic Spectra and Electronic Structure of Polyatomic Molecules" (Princeton, N. J.: Van Nostrand, 1967). pp. 518–522
78. Stief, L. J., Donn, B., Glicker, S., Gentieu, E. P. and Mentall, J. E., *Ap. J.* **171**, 21 (1972).
79. Aanestaad, P. A., unpublished.
80. Sobolev, V. V., "Treatise on Radiative Transfer" (Princeton, N. J.: D. van Nostrand Co., 1963).
81. Rubin, R. H., Swenson, G. W., Benson, R. C., Tigelaar, H. L. and Flygare, W. H., *Ap. J. (Letters)* **169**, L39 (1971); Gottlieb, C. A., Palmer, P., Rickard, L. J., and Zuckerman, B., *Ap. J.* (1972).
82. Barrett, A. H., Schwartz, P. R., and Waters, J. W., *Ap. J.(Letters)***168**, L101(1971).
83. Goldriech, P., unpublished.
84. Icsevgi, A. and Lamb, W. E., unpublished.
85. Litvak, M. M., *Phys. Rev. A* **2**, 2107 (1970).
86. Schlossberg, H. R. and Javan, A., *Phys. Rev.* **150**, 267 (1966).
87. Bloembergen, N., "Nonlinear Optics" (New York: Benjamin, 1965). pp. 96–100.
88. Stringer, T. E., *J. Nucl. Energy, Pt.* **C5**, 89 (1963).
89. Hollenbach, D. and Salpeter, E. E., *Ap. J.* **163**, 155 (1971); Hollenbach, D. and Salpeter, E. E., *J. Chem. Phys.* **53**, 79 (1971).
90. Studier, M. H., Hayatsu, R. and Anders, E., *Geochim. Cosmochim. Acta* **32**, 151 (1968); Hayatsu, R., Studier, M. H., Oda, A., Fuse, K. and Anders, E., *ibid.* **32**, 175 (1968).
91. Miller, S. L., *Biochim, Biophys. Acta* **23**, 484 (1957).
92. Keosian, J. "The Origin of Life" (New York: Reinhold Corp., 1968).
93. Oró, J. and Kimball, A. P., *Arch. Biochem. Biophys.* **96**, 293 (1962).
94. Kliss, R. M. and Matthews, C. N., *Proc. Natl. Acad. Sci.* **48**, 1300 (1962); Matthews, C. N., and Moser, R. E., *Proc. Natl. Acad. Sci.* **56**, 1087 (1966); *Nature* **215**, 1230 (1967).
95. Dean, A. M. and Kistiakowsky, G. B., *J. Chem. Phys.* **54**, 1718 (1971); Izod, T. P., Kistiakowsky, G. B., and Matsuda, S. *ibid.* **55**, 4425 (1971); Clark, T. C., Izod, T. P. J., and Matsuda, S. *ibid.* **55**, 4644 (1971).
96. Larson, R. B., *M.N.R.A.S.* **145**, 271 (1969).
97. Penston, M. V., *M.N.R.A.S.* **144**, 425 (1969).
98. Kahn, F. D., *Physica* **41**, 172 (1969).
99. Landau, L. D. and Lifshitz, E. M., "Quantum Mechanics" (Reading, Mass.: Addison-Wesley Co., 1958). pp. 373–8
100. Townes, C. H. and Schawlow, A. L., "Microwave Spectroscopy" (New York: McGraw-Hill Co., 1955). pp. 324–333

Radio Recombination Lines: An Observer's Point of View

E. CHURCHWELL

Max-Planck-Institute für Radioastronomie, Bonn, Germany

I. GENERAL CONSIDERATIONS

The range of principal quantum numbers over which radio recombination lines of hydrogen have been observed to the present time is:

$$56 \leqslant \eta \leqslant 253$$

which corresponds in frequency to a range from 37·5 to 0·40 GHz. In addition to those of hydrogen, recombination lines of helium and carbon have also been observed. In principle it should also be possible to observe radio recombination lines of the H_2 molecule, however no such detection has been made to the present time. Since we are dealing with hydrogenic atoms, the corresponding transitions of He and C are shifted, by the reduced mass effect, to progressively higher frequencies relative to those of hydrogen, and therefore it is possible to observe H, He, and C lines simultaneously.

To the present time such transitions (for hydrogen) have been observed in a large number of galactic HII regions, one extra-galactic HII region (30 Doradus in the Large Magellanic Cloud), two planetary nebulae (NGC 7027 and IC 418), and in the diffuse partially ionized interstellar gas. No radio recombination line has been observed in absorption.

As Professor Seaton has shown, the line intensity relative to the continuum is given by:

$$\Delta v_L \, T_L/T_C \propto v^{2 \cdot 1} \, T_e^{-1 \cdot 15}$$

if the emitting gas is in thermodynamic equilibrium, where Δv_L is the line full width at half maximum intensity, T_L is the line brightness temperature, T_C is the continuum brightness temperature, v is the line rest frequency, and T_e is the electron temperature of the emitting gas. Schematically the behaviour of the line intensity relative to the continuum is shown in Fig. 1.

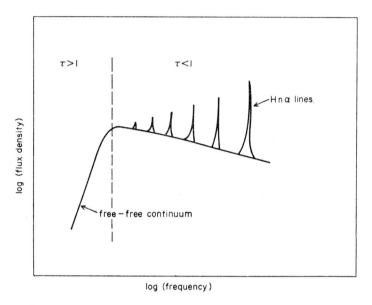

FIG. 1. Schematic illustration of the hydrogen-alpha line intensity relative to the free-free continuum emission of a typical HII region. τ is the continuum optical depth.

II. IMPORTANCE OF RADIO RECOMBINATION LINES IN ASTRONOMY

2.1. Probe of the Interstellar Medium

a. HII Distribution in the Galaxy.

By measuring the observed frequency of a radio recombination line relative to its rest frequency, one can get directly the radial velocity of an HII region. The radial velocity in conjunction with a rotation model of the galaxy gives the kinematic distance to the HII region. Therefore the distribution of HII regions in the galaxy can be determined. The importance of such a determination lies in the fact that HII regions are very young objects. They have life times of the order of 10^5 years which is quite short relative to the age of the galaxy (10^9 years). Therefore the distribution of HII regions give directly the region of large scale star formation within the galaxy. Figure 2 (after Mezger[1] and Churchwell and Mezger[2]) shows a histogram of the HI and HII regions as a function of distance from the galactic centre.

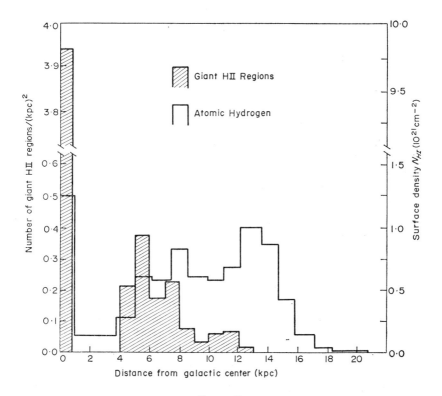

FIGURE 2.

Clearly the most active regions of star formation in our galaxy are in the nuclear disc and in a region between 4 kpc and 6 kpc from the galactic centre. It is also apparent that the HI gas is concentrated at larger distances from galactic centre than the HII gas. Recently measurements of hydrogen lines in the diffuse partially ionized interstellar gas have been made, from which one can get a direct observational handle on the ionization equilibrium in these regions. Also the "carbon" radio recombination lines give evidence of originating from very small cold clouds which some observers think may be protostars.

b. Determination of the Physical Conditions in the Emitting Gas.

Since Professor Seaton has discussed in detail the determination of the electron temperature T_e, the electron density N_e, and the emission measure $E = \int_0^s N_e^2 ds'$ of the emitting gas, I will make my discussion of these very brief. Basically, since the quantities which enter the line intensity equation have the following functional dependence

$$b_n = f(N_e, T_e), \tau_L = f(b_n, T_e, E), \tau_C = f(T_e, E)$$

where b_n is the ratio of the actual population of level n to what it should be in LTE at a given N_e and T_e, and τ_L and τ_C are respectively the optical depths in the line and in the continuum. Consequently, by using enough observed line intensities, one can get a weighted average of N_e, T_e and E. However, the interpretation of the meaning of these values must be considered with caution. Typical values which one obtains from a non-LTE analysis of radio recombination lines from galactic HII regions are: $T_e \simeq 10^4 \, K, N_e \simeq 10^3$–$10^4 \, cm^{-3}$, and $E \simeq 10^6$–$10^7 \, pc \, cm^{-6}$. The electron temperature can be determined independently of non-LTE considerations in several ways. One important way is to use the measured line width of H and He. Assuming that H and He are well mixed, and that each line has the same turbulent component then:

$$T_e \propto (\Delta v_L(H))^2 - (\Delta v_L(He))^2.$$

I point out in this regard that the Brocklehurst and Seaton model would predict low T_e values from this method, however our data at $\lambda 3 \, cm$ imply slightly high T_e values.

2.2. Atomic Processes Under Interstellar Conditions: Collisional Broadening in a High Temperature Low Density Gas

It was thought for many years that collisional broadening of radio recombination lines would make such lines unobservable. Therefore these transitions were only of academic interest until they were detected in 1964 by Dravskikh and Dravskikh,[3] and Sorochenko and Borodzich.[4] Ever since their first detection, observers have been trying to find collisional broadening effects in these lines without success. This is all the more puzzling in the light of recent theories regarding collisional broadening of radio recombination lines. Griem,[5] and more recently Brocklehurst and Leeman,[6] indicate that the ratio of the electron impact width to the Doppler width should go essentially as:

$$\frac{\Delta v_e}{\Delta v_D} \propto \frac{n^7 N_e}{(T_e T_D)^{1/2}}$$

where n is the principal quantum number, N_e the electron density, and T_e and T_D the electron and Doppler temperature respectively. From this one can see that:

$$\Delta v_e \propto n^4 \, N_e / T_e^{1/2}.$$

With such a strong dependence on n, it should be easily possible to detect collisional broadening in these lines.

I have just completed an analysis in which very strong evidence is given for electron impact broadening in radio recombination lines in Orion A and M17. The H109α line (at λ6 cm) and adjacent higher order lines H137β, H157γ, H172δ (all near the frequency of the H109α line) were observed with good signal-to-noise ratios. The fact that each line has nearly the same frequency means that each line also has the same continuum optical depth, the same Doppler component of line broadening, and the same turbulent width. Thus the only other variable is the extra contribution of collisional broadening with increasing n.

When the Doppler width (determined from the high frequency lines) is removed from the observed line width, it is found that they then follow an n^4 law rather accurately (Fig. 3). When Griem's theory is fitted to the data an electron density $N_e \simeq 3 \times 10^3$ cm^{-3} is implied. This value of N_e is low relative to that determined by optical methods and by a non-LTE analysis of the same lines. Perhaps this is a result of density fluctuations in HII regions as proposed by Brocklehurst and Seaton[7] or perhaps it is an indication that something is still lacking in the theoretical formulation.

c. Helium Abundances as a Function of Galactic Position.

There are no effective processes known which form or release helium into the general interstellar medium and by the same token once helium is in the interstellar gas there are no known processes which can effectively destroy it. These considerations in conjunction with observational data regarding the brightness of galaxies led Hoyle and Tayler[8] to the hypothesis that the helium which we observe in the interstellar medium and in the atmosphere of early type stars must have been formed during the first few seconds of the birth of the universe. The exact value of the $N(\text{He})/N(\text{H})$ ratio and its variation with galactic position (if any) has a direct bearing on the formation process of the universe in this picture.

With these assumptions the intensity ratio of corresponding transitions in hydrogen and helium gives directly the abundance ratio. These are: (1) He and H must be well mixed; (2) departures from LTE must be the same in both H and He; and (3) the Strömgren radii of both H and He must coincide. Fairly conclusive evidence has been given in support of the first two assumptions, however the third may be called into question.

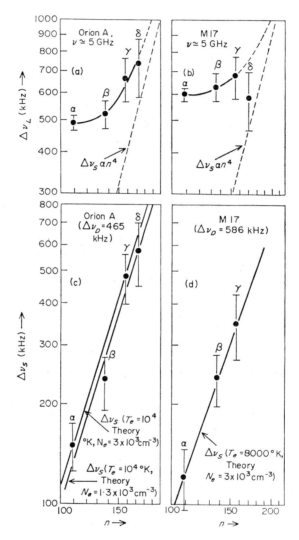

FIG. 3. (a) and (b): Observed line width, $\Delta\nu_L$, as a function of n at $\nu \simeq 5$ GHz. (c) and (d): The "stark width", $\Delta\nu_S$, as a function of n at $\nu \simeq 5$ GHz for a Doppler width of 465 kHz and 586 kHz respectively in Orion A and M17. The solid lines represent Griem's theoretical prediction for the indicated T_e and N_e values.

Over the past three years Dr. Mezger and I have been amassing helium abundance data using radio recombination lines. The results of our measurements can be catagorized into three groups. These are:

Group I: (12 sources) $< N(\text{He}^+)/N(\text{H}^+) > = 0.089 \pm 0.009$. Each source has three or more independent determinations and in the case of about three sources doubly ionized helium has been searched for with negative results.

Group II: (4 sources) $< N(\text{He}^+)/N(\text{H}^+) > = 0.059 \pm 0.007$. Each source independently measured three to four times at different frequencies. No doubly ionized Helium detectable in any of these sources.

Group III: No helium lines detected. All sources in the galactic centre region.

Source	$N(\text{He}^+)/N(\text{H}^+)$	$N(\text{He}^{++})/N(\text{H}^+)$
G0.2–0.0	< 0.02	< 0.07
G0.5–0.0	< 0.04*	—
G0.7–0.0	< 0.02	< 0.04

* Wilson et al.[9]

It is clear that it is still possible for these sources to have a normal abundance, however, it would require a large fraction of the helium to be in the doubly ionized state.

REFERENCES

1. Mezger, P. G. ,*Proc. IAU Symposium* No. 38 (W. Becker and G. Contopoulos, eds.), D. Reidel Pub. Co., p.107 (1970).
2. Churchwell, E. and Mezger, P. G., *Astr. and Ap.* in press (1971).
3. Dravskikh, Z. V. and Davskikh, A. F., *Astron. Tsirk.* No. 282 (English translation: *Sov. Ast. A. J.* **11**, 27 (1967)).
4. Sorochenko, R. L. and Borodzich, E. V., paper presented by V. V. Vitkevitch at the 12th General Assembly of the IAU, Hamburg (1964).
5. Griem, H. R., *Ap. J.,* **148**, 547 (1967).
6. Brocklehurst, M. and Leeman, S., *Astrophys. Lett.* **9**, 35 (1971).
7. Brocklehurst, M. and Seaton, M. J., *Astrophys. Lett.* in press (1971).
8. Hoyle, F. and Tayler, R. J., *Nature,* **203**, 1108 (1964).
9. Wilson, T. L., Mezger, P. G., Gardner, F. F. and Milne, D. K., *Astr. and Ap.* **6**, 364 (1970).

Some Recent Aspects of Spectroscopy at UV and X-ray Wavelengths

R. J. Speer

Imperial College. London. England.

I. INTRODUCTION

Spectroscopy may be defined as our concern with the observation and interpretation of light emitted and absorbed by collections of atoms. Within this definition spectroscopy at ultraviolet ($\lambda < 3000$ Å) and X-ray (say, $\lambda < 300$ Å) wavelengths becomes a matter of emphasis, for the atomic transitions responsible have just increased in energy from their visible counterparts resulting in a shift to shorter wavelengths. In general terms such transitions are characteristic of (a) higher temperatures, (b) more tightly bound electrons ("inner shell" electrons) and (c) "optical" or "valence" electrons that are under the influence of a higher nuclear charge than that of the neutral atom. In this last case the production and study of such ionized atomic systems is seen as being of major importance in the recent development of the subject.

In practice, spectroscopy below 3000 Å presents an impressive array of obstacles, not least of which are:

(i) (a) atmospheric absorption cut-off (astronomical), $\lambda \approx 3000$ Å.
    (b) air absorption cut-off (laboratory), $\lambda \approx 2000$ Å.
(ii)   last known material fails in transmission, $\lambda\, 1040$ Å (LiF).
(iii)   last known material fails in normal reflection $\lambda \approx 300$ Å.

The combination of these three particular constraints at $\lambda < 300$ Å to this day makes investigations technically very difficult. It remains an achievement to determine, within a factor of 2, the flux of photons from a specified transition anywhere in the soft X-ray region. In practice this is just the region where many current research interests are focussed. For example, by $\lambda \approx 200$ Å the astrophysically abundant elements H and He are completely ionized, and the resulting spectra from naturally occurring objects are dominated by transitions in ions from certain higher Z elements of lower abundance, e.g., C, N, Fe and Ni. Under certain circumstances a precise measure of the photon flux can result in an accurate element abundance determination.

Bearing in mind that (a) the interstellar medium again becomes transparent at $\lambda \approx 50$ Å and (b) the displacement law requires thermal emission sources to exceed say 10^6 K at these wavelengths, it is easy to understand the excitement and astrophysical implications of the recent observation of X-ray emitting sources in the celestial sphere. In this new discipline, X-ray astronomy, all the previously mentioned difficulties are compounded with an extremely low intrinsic flux level (≈ 10 photons cm^{-2} sec^{-1}), and a remote and hostile environment for the experimental hardware (stabilized rockets or satellites). It is of some topical interest that this particular branch of astronomical spectroscopy is currently undergoing a critical advance in knowledge as the first stabilized X-ray satellites become operational. In this respect, with all previous discoveries based upon rocket flights each of only several minutes duration (total several hours observation), the situation is reminiscent of when the first UV solar experiments were placed in orbit in 1962 (Orbiting Solar Observatory-1).

In the following notes we examine some aspects that are common to the seemingly widely different new sciences of Solar UV and X-ray Spectroscopy and X-ray Stellar Astronomy, and then relate these aspects to our knowledge gained from UV and XUV laboratory spectroscopy. Some very recent new developments in Eclipse Astronomy and Diffraction Grating Physics are included as they graphically illustrate the interdependence and progress between widely separated disciplines.

1.1. Historical Perspective

Although the foundations of ultraviolet spectroscopy were established over a century ago, a variety of factors has led to a remarkable upsurge of interest in recent years. In particular the extreme ultraviolet and soft X-ray region

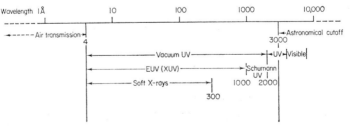

FIG. 1. 1Å–3000 Å. The sub-regions most commonly referred to in the current terminology.

can be seen to have progressed from a largely qualitative discipline to a mature and quantitative one. Before proceeding to trace these developments it is useful to distinguish the several wavelength regions that appear in the current terminology. These are shown in Fig. 1.

At 3000 Å the Hartley absorption band of telluric ozone results in the complete attenuation of UV photons from celestial objects, and by 1850 Å ground based laboratory spectroscopy becomes impossible without air evacuation from the spectrographic system. This wavelength marks the beginning of the vacuum ultraviolet region that extends without remission to X-ray wavelengths of several angstroms. It is useful to distinguish several subregions within the vacuum ultraviolet itself and the most important of these are (a) the Schumann region $\lambda\,2000\,\text{Å} \sim \lambda\,1000\,\text{Å}$ (b) the EUV or XUV $\lambda\,1000\,\text{Å} \sim 1\,\text{Å}$ and (c) the soft X-ray region $\lambda\,300\,\text{Å} \sim 1\,\text{Å}$. These wavelength limits are by no means fixed, and serve to indicate the approximate regions at which changes in the optical transmission and reflection characteristics of materials cause a change in instrumental technique.

(a) *The Schumann Region*

This name honours the pioneer of far ultraviolet spectroscopy, Victor Schumann. Apart from developing photographic emulsions of low gelatine content that still bear his name today, he constructed the first vacuum ultraviolet spectrograph using fluorite optics permitting an extension of the then known spectrum to 1300 Å. Today the region is characterised by the routine use of crystalline materials for windows and simple optics. The sequence, synthetic fused quartz, sapphire, SrF_2, CaF_2, MgF_2, progressively shifts this short wavelength transmission limit deeper into the UV. These limits are, respectively, 1600 Å, 1410 Å, 1280 Å, 1220 Å and 1120 Å. At 1040 Å we reach the transmission limit for the cleaved crystal LiF. This material, which does not occur in nature, is often used to the exclusion of all other. It demonstrates the shortest transmission cut-off wavelength known and marks the beginning of the "extreme ultraviolet" or "XUV" region.

(b) *The EUV or XUV*

We may characterize this region by the absence of transmitting optics. Historically the region was pioneered by Theodore Lyman who constructed the first concave grating vacuum spectrograph. By 1914 he had discovered the ground state resonance series of hydrogen that bears his name. The next few years up to 1924 saw a steady advance until a new intractable short wavelength limit was reached at about 150 Å. The steady progress up to this limit had come about by the use of special lightly ruled gratings and the introduction of the "vacuum" or "hot" spark. This period was particularly important in that exact wavelengths could be determined from the concave diffraction grating geometry alone, thus providing important confirmation of the Bohr theory of simple atomic systems. One can note in this period the fundamental observation that ortho and parahelium were not completely different atomic species (the 591 Å $1s^2\,{}^1S_0 - 1s2p\,{}^3P_1$ intersystem line of He I was observed by Lyman in 1924) and the discovery by Bowen and Millikan that the emission spectra of ionized atoms could be arranged systematically (iso-electronic series). Excellent reviews of this formative period in the exploration of this region can be found in articles by Flemberg[1] and Tousey.[2]

(c) *The Soft X-ray Region*

Further progress met with very great difficulty and revolved about the fact that the pioneers Lyman and Millikan allowed the incident radiation from the source to strike their concave reflection gratings at, or close to, normal incidence. In modern terms one can recognise that they had reached a short wavelength limit for normal incidence set by absorption of the radiation in the material of the grating surface. The reflection efficiency E_n (fraction of incident radiation returned into spectral order n, see Fig. 11(a)) falls effectively to zero at these angles and can only be retained by operating grating spectrographs at the large angles of incidence that satisfy the requirement for total external reflection.

This severe requirement appears to have been recognised first by Compton and Doan[3] who, in an experiment that is impressive to this day, demonstrated the characteristic X-ray spectrum of Cu (1·5 Å) using a lightly ruled grating at grazing incidence. This arrangement was subsequently exploited by Hoag[4] and Osgood[5] in the EUV vacuum region and was soon being applied to close the gap between the X-ray and UV region of the spectrum. This region, although it still presents the greatest technical difficulty, is currently undergoing intensive investigation due to the combined stimuli of high temperature plasma physics, solar and stellar X-ray and soft X-ray astronomy and solid state physics.

The development of atomic spectroscopy in the soft X-ray region is dominated by the contributions of Siegbahn and his collaborators and follows in logical order from the pioneering effort of Schumann and Lyman. Although recounted many times, it is interesting to trace the major steps and relate these in turn to some modern developments.

1.2. Period up to 1942

The fifteen years up to 1942 saw a remarkable advance in the investigation of atomic spectra in the EUV and soft X-ray region led by the work of the Physics Laboratory of the University of Uppsala. One can follow a steady sophistication in technique reflected in the production, diffraction and recording of the emission spectra of ions of higher and higher ionization potential, for example, Li II, Be III, B IV, C V (Edlen[6]). B IV, C V, N VI, O VII, F VIII (Tyren[7]) and Mg XI, Al XII (Flemberg[1]). The helium-like ion can be seen to play a particularly important role in this progression due to a common absence or weakness of hydrogen-like lines in the time integrated spectra recorded from the source plasma. This is due, in the case where sources are being pushed to their limit, to a marginal total plasma confinement time compared to the ionization relaxation time. The helium-like ion also shows the highest ionization potential among all atoms at the same stage of ionization. Cl XVII, A XVII, K XVII ... Fe XVII is such a sequence with A XVII showing the shortest wavelength spectra among the group. The resonance line $1s^2\ ^1S_0 - 1s2p\ ^1P_1$ of such helium-like ions has thus indicated the progress to shorter and shorter wavelengths. Figure 2 sketches this progression as a function of Z with an indication of the approximate timing for related instrumental developments.

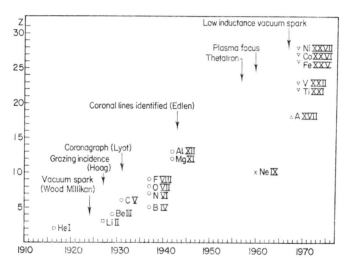

FIG. 2. The general run of progress for the laboratory excitation of spectra from ions in the configuration $1s^2$. The approximate timing for important instrumental developments has also been indicated.

A key step in the advance of astrophysical theory occurred in this period and concerns the proof by Edlen[8] of the high thermal temperature of the solar corona. Early in the development of solar physics it had become apparent that a unique group of solar emission lines defied interpretation. These lines, although seen in the visible, could only be recorded during the brief period of a total solar eclipse. The difficulties of eclipse spectroscopy are well known and it remained to Lyot[9] to demonstrate with his coronograph that the most intense of these lines could be observed at times other than total solar eclipse. In 1939 Grotrian remarked upon the possibility that two of the observed coronal emission line wavelengths corresponded in wavenumber with term differences derived from soft X-ray spectra at $\lambda \approx 100$ Å. The spectra in question were of highly ionized iron in the configuration $3p^n$ and had been recorded using a vacuum spark. In a classic piece of work Edlen[8] was able to account for nearly all of the observed solar lines. Thus 5303 Å and 6375 Å, which under typical solar conditions account for over half of the visible coronal line intensity, were assigned to the forbidden transitions [Fe XIV] $3p\ ^2P_{3/2}^0 - \ ^2P_{\frac{1}{2}}$ and [Fe X] $3p^5\ ^2P_{\frac{1}{2}} - \ ^2P_{3/2}$. This discovery shifted the research emphasis to what caused the necessarily high temperature ($T_e \approx 10^6\ °K$). In modern terms this temperature is seen as the result of the ineffectiveness of radiative cooling in a low density hydrogen plasma in the presence of energy dissipation by shock waves. The term structure of Fe X is reproduced in Fig. 3 with a calculation of the wavelength of the red coronal line as illustrated in Table I.

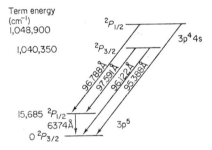

FIG. 3. The term structure of Fe X.

1.3. Period up to the present

The post war period saw renewed interest in the far ultraviolet that has continued progressively to this day. One can trace the main developments to the combined stimuli of space astronomy and high temperature plasma research. The increasing availability and sophistication of rocket and satellite platforms has led to a steady advance in our knowledge of the solar spectrum below $\lambda\ 3000$ Å, even though this region accounts for only about 5% of the total photon flux. The spectrum exhibits a remarkable range of phenomena and its study and interpretation has contributed greatly to our

Table I

Experimental data for FeX coronal line identification (Edlen [8])

Fe X $3s^23p^5 - 3s^23p^44s$	Å	cm^{-1}	Term difference cm^{-1}	Wave number, coronal line cm^{-1}
$^2P_{3/2} - {}^2P_{1/2}$	95·338	1,048,900		
			15,714	
$^2P_{1/2} - {}^2P_{1/2}$	96·788	1,033,186		
$^2P_{3/2} - {}^2P_{3/2}$	96·122	1.040,345		
			15,660	
$^2P_{1/2} - {}^2P_{3/2}$	97·591	1,024,685		
			15,687	15,685

understanding of the structure of the solar atmosphere. Probably the most significant observation is that a fundamental change occurs in the spectrum at $\lambda \approx 1600$ Å. At this wavelength the spectrum is observed to change from absorption to emission. This changeover is due to a reversal in temperature gradient in the outer atmosphere of the sun. Satellite data accumulated in the last few years shows that the temperature increases monotonically outwards from $T_e \approx 4300$ K to $T_e \approx 2 \times 10^6$ K over a very narrow region (see Fig. 4). Thus an orbiting spectrograph, capable of observing say from $\lambda = 361$ Å (Fe XVI $3s^2S_{\frac{1}{2}} - 3p^2P_{\frac{1}{2}}$, $T_e \approx 2 \times 10^6$ K) to 1600 Å is effectively probing the entire three dimensional structure of the solar atmosphere, from the temperature minimum to the coronal temperature maximum. Interpretation of UV data of this type, together with visible IR observations, appears to be leading to a common description of the average physical conditions in the solar atmosphere for the first time.

Figure 4 shows the general run of temperature T_e and density N_H as a function of height in the solar atmosphere, reproduced with permission from a recent review by Noyes.[10] The transition zone chromosphere/corona is characterized by an extremely steep temperature gradient and illustrates the importance of EUV observations. The approximate locations are marked for the emission from several abundant ions mentioned in the text. Observations involving the highest spatial resolution have also been conducted at all the major recent eclipses and are discussed in a subsequent section.

At soft X-ray and X-ray wavelengths, $\lambda < 50$ Å an intriguing new stimulus has occurred with the observation of X-ray emitting objects on the celestial sphere. At the time of writing approximately fifty such sources have been detected with about one dozen identifications with optical or radio counterparts. This list, although modest in numbers, incorporates the most dramatic events in our galaxy, for reliable X-ray identifications now exist for six supernovae remnants: the Crab Nebula, Cas A, the Cygnus Loop, Vela X, Puppis A and Tycho Brahe's Nova. Within this list the Crab Nebula appears to be unique following the recent observations of correlated X-ray, optical and radio pulsations of period 33 milliseconds (Fritz et al.,[11] Bradt et al.[12]). Although in the case of the Crab, relativistic electrons of 10^{13}–10^{14} eV are believed to cause the X-ray emission via the synchrotron process, considerable effort has been made during 1970 to observe optical line emission (e.g., Fe XXV $1s - 2p$, 1·9 Å) from the most intense X-ray emitting object Sco X-1, a thirteneth magnitude flickering blue star. The situation is not entirely clear, although opinion tends towards a thermal origin for the observed emission (≈ 40 photons cm^{-2} sec^{-1}). Despite the very low flux observed, with consequent difficulties for rocket and satellite observations, it is quite clear that X-ray emission is important on an absolute scale, in

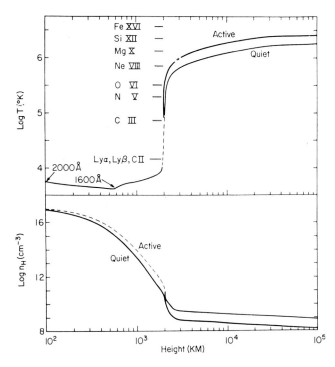

FIG. 4. The general run of electron temperature T_e and density N_H as a function of height in the solar atmosphere. (Noyes[10])

some cases exceeding all other forms of radiation output. This new astronomical discipline may offer the most intriguing new prospects for the spectroscopy of highly ionized systems.

II. ULTRAVIOLET SPECTROSCOPY AT TOTAL SOLAR ECLIPSE

The total solar eclipse offers two important and well known advantages over instrumental techniques for the study of the solar chromosphere and corona.

(a) *Observations emphasizing the intrinsic coronal spectrum*

The first advantage concerns the suppression of photospheric light which may typically exceed the intrinsic coronal flux level by a factor of a million or more. This suppression is so effective that if the eclipse combines with conditions of good seeing (low sky radiance) coronal structure may be observed out to 5 solar radii at 3×10^{-10} the disc intensity. Figure 5 shows the corona recorded by Newkirk and Lacey[13] under such ideal conditions at the recent total eclipse of March 7th 1970. The two most important

FIG. 5. Coronal structure observed by Newkirk and Lacey[13] at the total solar eclipse of March 7th, 1970.

components of this radiation in the visible are a Thomson scattered photospheric continuum flux from the electron population (K corona) and an intrinsic emission line spectrum consisting of forbidden (magnetic dipole) transitions in configurations of p electrons. Although this spectrum has been recorded and catalogued since photographic eclipse spectroscopy got under way in earnest in 1869, many lines remain to be assigned and there has been no comprehensive discussion of the excitation of the important coronal ions, although individual cases have been treated, e.g., Fe XV (Bely and Blaha[14]).

The advantages offered by suppression of the disc flux at total eclipse extends to all wavelengths showing significant continuum emission, i.e., approximately 0·1 μ to 100 μ. Only very recently has it proven technically possible to recover coronal eclipse spectra outside the visible. The infrared has been investigated with Fourier transform spectrometers in high flying aircraft in eclipse totality. In the UV use has been made of three-axis gyro-

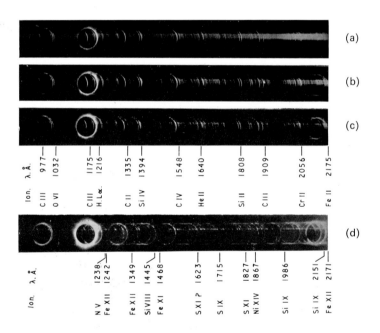

FIG. 6. Four plates reproduced from a sequence of fifty recorded by a rocket-borne VU spectrograph that passed through the imbra at the North American eclipse of March 7th, 1970. The leading edge of the moon untersects the solar image at 70° from the vertical (approximately 2 o'clock) and, at this point shows projected heights of (a) −2000 km, (b) −200 km, (c) +1800 km and (d) +8800 km respectively in relation to the solar limb (Speer et al.[22]).

stabilized rockets. The North America eclipse of March 7th 1970 was of outstanding importance in this respect as infra-red and ultra-violet experiments were successful for the first time between 1 μ–3 μ and 850 Å–2185 Å respectively. Interpretation of the infra-red data indicates that forbidden transitions in highly ionized Mg, A, Si, S and Cr have been observed for the first time (Olsen et al.[15]). In the UV, the photographic data offers the added advantage of being stigmatic (spatially resolved) in each wavelength. Figure 6(d) reproduces a plate from the sequence of 50 recorded during the eclipse (Gabriel et al.[16]) that is dominated by a purely coronal spectrum. Twenty-eight coronal lines are observed of which seventeen have been identified by Jordan.[17] Each line, of excitation potential χ eV, is characteristic of the electron temperature T_e in the solar atmosphere at which the calculable excitation function $g(T_e) = T_e^{-\frac{1}{2}} e^{-\chi/kT_e}$ (fractional ion abundance) has its maximum value (Pottasch[18]). The coronal lines may thus be ordered in temperature sequence and are reproduced from Jordan's paper in this form in Table II. The iso-electronic sequence is also indicated.

Table II

Newly observed UV coronal emission lines recorded at the total solar eclipse of March 7th 1970.

T_e ($\times 10^{6}$ °K)	Ion	Sequence		Transition	Obs.Å
0·69	Mg VII	C	$2p^2$	$^3P_1 - {}^1S_0$	1190·2
0·69	Si VII	O	$2p^4$	$^3P_2 - {}^1D_2$	2147·4
0·93	Si VIII	N	$2p^3$	$^4S_{3/2} - {}^2D_{3/2}$	1446·0
1·2	Si IX	C	$2p^2$	$^3P_2 - {}^1D_2$	1715·3
1·2	Si IX	C	$2p^2$	$^3P_1 - {}^1D_2$	1985·0
1·2	Si IX	C	$2p^2$	$^3P_2 - {}^1D_2$	2149·5
1·5	Fe XI	S	$3p^4$	$^3P_1 - {}^1S_0$	1467·0
1·7	Fe XII	P	$3p^3$	$^4S_{3/2} - {}^2P_{3/2}$	1242·2
1·7	Fe XII	P	$3p^3$	$^4S_{3/2} - {}^2P_{1/2}$	1349·6
1·7	Fe XII	P	$3p^3$	$^4S_{3/2} - {}^2D_{5/2}$	2169·7
1·9	Fe XIII	Si	$3p^2$	$^3P_1 - {}^1S_0$	1213·0
2·0	O VII	He	$1s^2$	$^3S_1 - {}^3P_2$	1624·0
2·0	S XI	C	$2p^2$	$^3P_1 - {}^1D_2$	1614·6
2·0	S XI	C	$2p^2$	$^3P_2 - {}^1D_2$	1826·0
2·0	Ni XIII	S	$3p^4$	$^3P_2 - {}^1D_2$	2126·0
2·0	Ni XIV	P	$3p^3$	$^4S_{3/2} - {}^2D_{5/2}$	1866·9
2·2	Ni XIV	P	$3p^3$	$^4S_{3/2} - {}^2D_{3/2}$	2185·1
2·5	Ni XV	Si	$3p^2$	$^3P_1 - {}^1D_2$	2085·7

It is apparent from this table that these newly observed coronal radiations are again from configurations $2p^n$ and $3p^n$. The lines are forbidden transitions within the ground term and in this respect complement and visible and infra-red spectra of the corona. At shorter wavelengths the coronal spectrum is dominated by allowed transitions in coronal ions and has been extensively reviewed (see, for example, Tousey[19]). At wavelengths longer than 2085 Å the coronal spectrum remains unknown until we reach the first line observed with ground based instruments at $\lambda\,3021$ Å. An operational failure of a rocket-borne spectrograph flown into the umbra of the March 7th, 1970 eclipse prevented recording this region. Recovery of this spectrum, which contains the well studied chromospheric Mg II resonance doublet, is one objective of coronal spectroscopy at forthcoming eclipse opportunities.

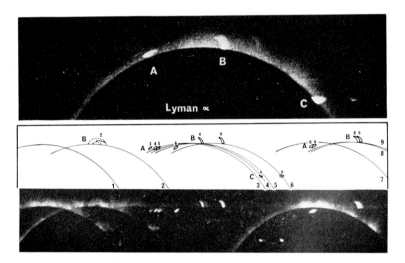

Fig. 7. UV eclipse limb spectroheliogram 1330Å–1470Å. The key to the observed features appears in Table 3.

In the case of collisionally excited optically thin lines the eclipse plate intensities for stigmatic images reflect the variation of $N_e^2\,dz$ along the line of sight. This is a sufficiently good approximation in many cases to permit a structural analysis of coronal limb activity, provided some assumption is made about the structure in the z, or line of sight, direction. Figure 7 shows a section of a UV limb spectroheliogram taken at the March 7th 1970 eclipse (Jones et al.[20]). The spectral range is 1330 Å to 1470 Å and conveniently illustrates the form of the emission for three active regions present on the day of the eclipse. Reference to Table III shows that the images can be

Table III

Identifications and temperature characteristics for the nine limb spectroheliograms presented in Fig. 7.

Temperature °K	Ion	Spectro-heliogram	Wavelength Å	Transition	Prominence image A	B	C
2×10^4	C II	8, 9	1335·8 1334·6	$^2P-^2D$	$A_{8,9}$	$B_{8,9}$	Off figure
$6·5 \times 10^4$	Si IV	4, 6	1402·8 1393·8	$^2S-^2P$	$A_{4,6}$	$B_{4,6}$	$C_{4,6}$
$1·2 \times 10^5$	O IV	3, 5	1404·8 1401·1	$^2P-^4P$	$A_{3,5}$	Weak	Weak
$9·3 \times 10^5$	Si VIII	2	1446·0	$^4S-^2D$	Absent	Loop structure	Absent
$1·5 \times 10^6$	Fe XI	1	1467·0	$^3P-^1S$	Absent	Diffuse coronal activity	Absent
$1·7 \times 10^6$	Fe XII	7	1349·6	$^4S-^2P$	Absent	Intense coronal activity	Absent

associated with a range of excitation temperatures from $2 \times 10^4\,°K$ to $1·7 \times 10^6\,°K$.

(b) *Observations Emphasizing Spatial Resolution*

Reference to Fig. 4 indicates the intriguingly narrow height range for the transition zone chromosphere/corona, with theoretically deduced gradients reaching values as high as 5000°K/km. Observational evidence remains to be obtained for the precise location and structure of this region. Expressed in angular terms the transition occurs in a fraction of an arc second (1 arc sec $\approx$ 720 km at the solar limb), an angular distance that remains well below the limit of optical imaging techniques in the ultraviolet. In principle, observations at eclipse can tie down the heights of emission to an arc-second or less as a consequence of this exceptionally small angular range over which the transition region ions exist. A stationary ground-based spectrograph observes an angular rate of the moon across the sun's limb of $\approx 0·47$ arc-sec per second of time, an impressively small rate that is exploited routinely in visible eclipse data. In the ultra-violet, however, a rocket platform will observe an apparent angular rate ranging somewhere between 0·2 and 5 arc seconds per second depending on the detailed relationship between the ballistic motion of the vehicle and the umbra. Although there is good evidence that the transition zone contours the rough spicular structure of the chromosphere, ultra-violet stigmatic eclipse spectra can yield the relative volume of transition zone material at each height and locate with precision the height at which emission commences. (Typically 1 arc-sec in movement of the lunar limb is equivalent to 1·7 km translational movement of the rocket spectrograph in the rest frame of the umbra. A rocket can be located with a considerably greater precision than this). Two experiments have been performed on three-axis gyro-stabilized Aerobee rockets to exploit this situation at eclipse (Brueckner et al.,[21] Speer et al.[22]) and each obtained ultra-violet data for the first time. Three adjacent frames that span second contact are shown in Fig. 6 to illustrate the very rapid modifications in character of the emission with arc second changes in the position of the moon's limb (the moon's limb progresses from left to right in the sequence (a), (b), (c), at projected heights of -2000 km, -200 km and $+1800$ km in the solar atmosphere at the centre of the crescent images). Initial data reduction indicates that emission scale heights are indeed averages over geometrical distances comparable to spicular structure scales, i.e., $\approx 10{,}000$ km.

III. LIGHT SOURCES

There have been many previous discussions of light sources for use in the ultra-violet (see, for example, Samson,[23] Chapter 5). In this section we limit ourselves to some relatively new aspect of source development for the production of very highly ionized systems and, in consequence, emphasize spectra recently obtained at X-ray wavelengths. References to Fig. 8 shows that the 1s–2p transition in He-like ions has been observed almost to Z = 30, taking the study of "optical" transitions into the calcite crystal region of Bragg spectroscopy and back to where air again becomes transparent at $\lambda \approx 4$ Å. Figures 2 and 8 sketch the progress made in observing these ions and mark some points of particular experimental significance. Order of magnitude estimates for the source electron temperatures are also indicated.

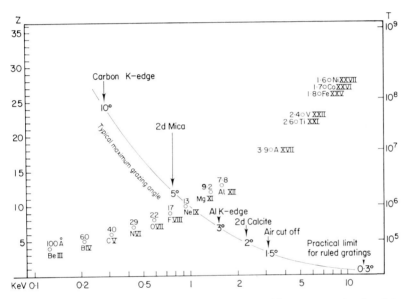

FIG. 8. Observed 1s–2p transitions in helium-like ions with some associated points of particular experimental significance.

(a) *The Vacuum Spark*

Lie and Elton[24–25] have recently exploited a modern version of the low pressure vacuum spark. In this device vaporized electrode material, constructed from alloys of elements in the iron group, can be compressed to a highly concentrated point plasma typically $\approx 15\,\mu$ in diameter and with electron density $N_e \approx 10^{20}$ cm^{-3}.

By virtue of this high electron density, the transient ion stripping time τ_z (assuming stepwise ionization), is comparable to the plasma lifetime and results in X-ray line emission originating from optical (dipole and intercombination) transitions in Helium-like Ti XXI, Fe XXV and Cu XXIII. In

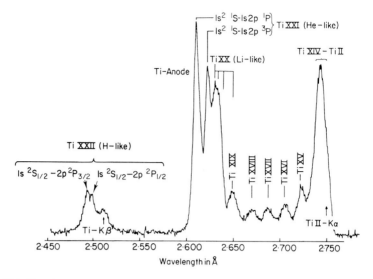

Fig. 9(a). Titanium spectrum recorded with a flat LiF Bragg crystal spectrograph using a low pressure vacuum spark as light source (Lie and Elton [24]).

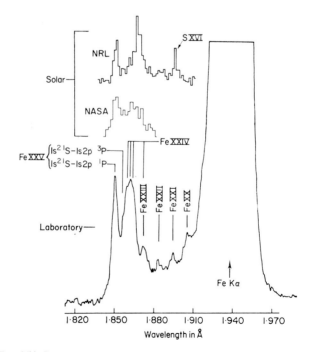

Fig. 9(b). Iron spectrum compared to solar data (Lie and Elton [25]).

addition emission from inner-shell X-ray transitions in Fe XIV–XXIV and Ti XV–XX ions have been identified. Figure 9(a) reproduces one of Lie and Elton's Ti spectra recorded with a flat LiF Bragg crystal spectrometer in which spectra from successive ion stages are grouped and can be clearly identified. Such features in Fe had not been observed previously in laboratory plasmas and compliment recent rocket and orbiting satellite data (Neupert and Swartz,[26] Doschek et al.[27]) on the emission spectra of solar flares, see Fig. 9(b). Both the solar and laboratory data await higher resolution studies to resolve the many blended components that are suggested in the available data.

(b) *Plasma Focus*

The metal walled coaxial gun device "plasma focus" originally described by Fillipov et al.,[28] has been investigated recently as a source of highly stripped ions. The device operates most successfully via the compression of cold hydrogen gas seeded with rare gas impurities at the 1% level. Densities are of the order of 10^{19} cm^{-3} with the mean thermal energies of $\approx$ 2keV. Spectra of neon and argon have been studied with both grating and crystal X-ray instrumentation (Peacock et al.[29–31]). Recent observation of helium-like argon (Peacock et al.[31])' are shown in Fig. 10.

The structure of the lithium-like satellites ($1s^22s - 1s2s2p, 1s2p^2$) to the helium-like parent resonance line $1s^2\ {}^1S_0 - 1s2p\ {}^1P_1$ is interpreted (Peacock

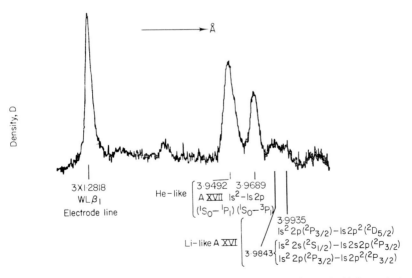

FIG. 10. Concave curved mica spectrum of a Plasma Focus discharge in high resolution (Peacock et al.[31]).

et al.[31]) as providing direct confirmation that dielectronic recombination is the dominant recombination process in the ion balance. These satellites are observed in both neon and argon at about 20% of the helium-like resonance line intensity. In the alternative case that the $1s2s2p$, $1s2p^2$ levels are populated directly from the lithium-like ion—only one component would be in evidence, the other level leading preferentially to autoionization.

IV. DIFFRACTION GRATINGS

A considerable body of data exists throughout the literature on the properties of the diffraction grating, the *sine qua non* of ultraviolet spectroscopy. A good current review has been published by Loewen.[32] Although very great advances have been made in the production of high reflection coatings, the reduction of scattered light levels, and raising the ruling frequency, one notices a distinct absence of systematic research in the "grazing incidence" or soft X-ray domain. This is due to the very great difficulties involved in the quantitative evaluation of research rulings in terms that are of real use to the practising spectroscopist, e.g., resolution, absolute efficiency and scattered light levels. Some new results however are outlined below and are regarded as very encouraging. It is useful to distinguish several classes of gratings, limiting our remarks to the context of soft X-ray performance.

4.1. Grating types

(a) *Amplitude Gratings* (Fig. 11(a))

These constitute the very lightly ruled gratings of the type first developed by Siegbahn and still available. Ruling is adjusted to maximize the function

$$E_n = \frac{1}{n^2\pi^2} \sin^2(n\pi\sigma) \quad \text{for} \quad n = +1.$$

This function, where $\sigma = a/(a+b) = a/d$, expresses the absolute fractional content of the diffracted order n and is maximized at $E_n = 1/\pi^2 \approx 10\%$, for the first diffracted order, by making $a = b$. Under these conditions 50% of the radiation is lost and even orders have zero intensity. In the limit of short wavelength use, i.e., extreme grazing angle, scattering from the uncontrollable ruling burr can further detract from the usable performance. Despite these shortcomings gratings of this type have played a fundamental role in the development of short wavelength spectroscopy. They appear in the literature under the alternative names, Siegbahn, scratch, or X-ray gratings.

(b) *Blazed Gratings* (Fig. 11(b))

The natural low limit in efficiency encountered above has tended to divert interest to the echellette type of ruling. In this case the angle β is adjusted during manufacture to satisfy the requirement of specular reflection into the desired diffracted image. The control of this process has reached the point where blaze angles of the order of $1°$ can be imparted to the groove profile at ruling frequencies of 2400 l/mm. The majority of gratings currently in use are of this type. The restriction on the angle of illumination set by the total external reflection condition at short wavelengths results in an unavoidable foreshortening of the illuminated part of the blaze profile and, by 10 Å, illumination is limited almost entirely to the tips of the grooves.

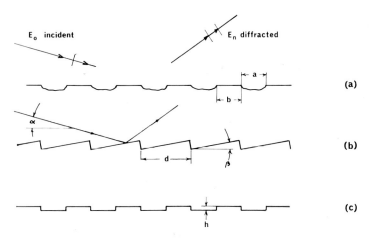

FIG. 11. Schematic representation of the principal types of reflection grating used at grazing incidence (soft X-ray domain) (a) amplitude (b) shallow blazed (c) laminar.

(c) *Laminar Gratings* (Fig. 11(c))

This type of grating may be regarded as a modification of (a) in which the grooves have now been incorporated as further diffracting elements. A simple calculation yields $(2/\pi)^2$, or 40%, as the maximum first order diffraction efficiency. In general terms, for a fixed step height h, diffraction efficiency is a function of α, λ and R_λ the reflection coefficient. Gold coated gratings of this type are just becoming available for soft X-ray spectroscopy, and some preliminary results from recent performance studies are included below (Franks and Sayce,[33] Speer,[34] Bennett[35]). Unlike type (a) or (b), performance is periodic, reflecting the increasing orders of interference between radiation diffracted from the higher and lower grooves. The generalized relation between the primary efficiency maximum and λ is given by:

$$\frac{\lambda}{d} = \frac{2\cos\alpha + (d/h)\sin\alpha}{1 + (d/2h)^2}$$

where α is the grazing angle and h and d are as indicated in Fig. 11(c).

For $d \gg h$, and in the limit of small α

$$h \simeq \frac{1}{2}\left(\frac{\lambda d}{2}\right)^{\frac{1}{2}}$$

which requires, for $\lambda \simeq 1$ Å and $d \approx 4\,\mu$ (300 l/mm), that $h \approx 100$ Å, or of the order of 30 atomic dimensions (the lattice spacing for gold ≈ 3 Å). It is remarkable that the practicable realization of such profiles over cm square areas is now possible (Franks[36]). Figure 12 shows test spectra taken by the author showing the behaviour of the short wavelength grating cut-off in an X-ray continuum using a 5m curved 300 l/mm X-ray grating of this type, where the limits on performance have been set by the optical constants and absorption edges of the diffracting surface.

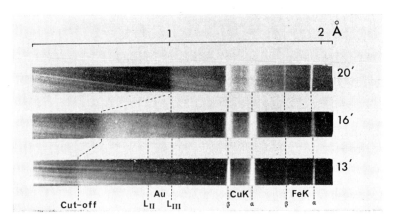

FIG. 12. First order short wavelength cut-off observed in continua using 300 l/mm Au coated NPL X-ray grating.

(d) *Holograpings*

The profile of this type of grating is intermediate between the blazed (triangular) type and sinusoidal. This shape derives from the processes used in transforming the optical fringes, recorded in photopolymer coated optical surfaces, into high reflectance metallized surfaces. They are included in this brief survey because of their potential use as soft X-ray stigmatic gratings in space astronomy (Labeyrie and Flamand,[37] Cordelle et al.,[38] Cordelle et al.[39]).

4.2. Grating performance at soft X-ray wavelengths

Absolute performance data on gratings that have been made under controlled conditions have been practically non existent in the grazing incidence domain and it is encouraging that this situation is beginning to improve with the advent of apparatus of sufficient engineering sophistication. The evidence to date strongly suggests that systematic progress depends upon determining quantitative values for the reflected, diffracted, scattered and absorbed fractions of (ideally) monochromatic soft X-ray radiation incident upon the grating, and then relating these data (a) to the surface profile and finish and (b) to the geometrical configuration of use. This requirement

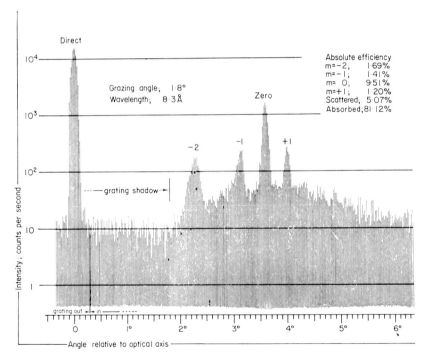

FIG. 13. Performance analysis of a typical soft X-ray ruling showing reflected, diffracted and scattered fractions at 8·4 Å.

poses severe technical problems and it is only recently that data of this type has become available in the range 1 Å < λ < 50 Å (Speer,[34] Bennett[35]). Figure 13 illustrates typical data where the reflected, diffracted and scattered fractions can be clearly identified. The ordinate is proportional to the number of photons detected as the sensor scans across the diffracted beam and is normalized in turn to the flux of photons incident upon the grating. The incident flux is displayed at the far left of the scan at D. It is thus possible to select the optimum conditions of use for the wavelength and order of diffraction desired for any specified astronomical or laboratory spectroscopic application, for example, Fig. 14(a) and (b) shows very recent data of this type for a 2 meter radius of curvature shallow blaze gold coated replica grating (d ≈ 1·7 μ). The quantity plotted is the absolute percentage of the incident monochromatic photon flux at 8·4 Å and 44 Å returned into orders 0, ±1 and +2.

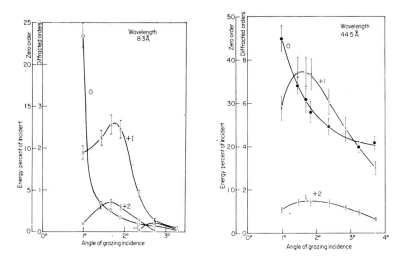

FIG. 14(a,b). The behaviour of the reflected ($n = 0$) and diffracted beams ($n = \pm 1, +2$) for a typical shallow blaze gold coated replica grating at 8·4 Å and 44 Å ($d = 1·7\mu$).

Laminar X-ray gratings have recently undergone intensive study under controlled conditions in view of their interesting short-wavelength performance. Typical data is shown in Fig. 15. The absolute distribution among the orders of the incident 8·4 Å flux (D) is shown plotted logarithmically; the scattered level can also be plainly seen. The sequence is one of increasing

SOME RECENT ASPECTS OF SPECTROSCOPY AT UV AND X-RAY WAVELENGTHS 307

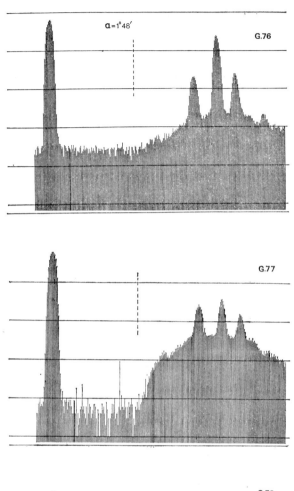

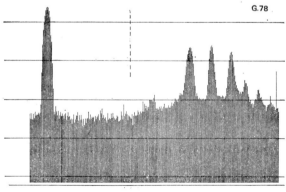

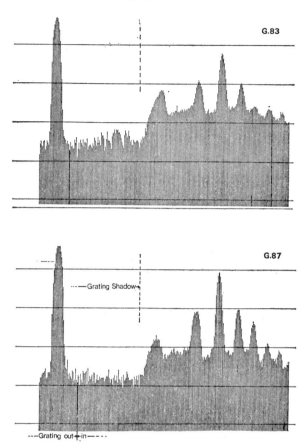

FIG. 15. The variation in total performance at one soft X-ray wavelength for a controlled sequence of laminar rulings. The groove depths are $h \approx 30$ Å, 70 Å, 100 Å, 200 Å and 350 Å respectively (Bennett[35]). Data format of Fig. 13.

groove depths with $h \simeq 30$, 70, 100, 200 and 350 Å respectively (Bennett[35]) but at a constant angle of grazing incidence ($\alpha = 1° 48'$). These results are sufficiently encouraging to mention the practical possibility of incorporating gratings of the laminar or holographic type as an integral part of the optical elements of the grazing incidence telescopes currently used non-dispersively in X-ray astronomy.

REFERENCES

1. Flemberg, H., *Arkiv for Matematic Astronomi och Fysik* **28**, 1 (1942).
2. Tousey, R. *Applied Optics* **1**, 679 (1962).
3. Compton, A. H. and Doan, R. L., *Proc. Nat. Acad. Sci. Am.* **11**, 598 (1925).
4. Hoag, J. B., *Ap. J.* **66**, 225 (1927).
5. Osgood, T. H., *Phys. Rev.* **30**, 567 (1927).
6. Edlen, B., *Nova. Acta. Reg. Soc. Sci. Uppsala.* **9**, No. 6 (1934).
7. Tyren, F., *Nova. Acta. Reg. Soc. Sci. Uppsala.* **12**, No. 1 (1940).
8. Edlen, B., *Z. Astrophys.* **20**, 30 (1942).
9. Lyot, B., *M.N.R.A.S.* **99**, 580 (1939).
10. Noyes, R., *Annual Review of Astronomy and Astrophysics* **9**, 209 (1971).
11. Fritz, G., Henry, R. C., Meekins, J. F., Chubb, T. A. and Friedman, H., *Science* **164**, 709 (1969).
12. Bradt, H. Rappaport, S., Mayer, W., Nather, R. E., Warner, B., McFarlane, M. and Kristian J., *Nature* **222**, 728 (1969).
13. Newkirk, G., and Lacey L., *Nature* **226**, 1098 (1970).
14. Bely, O. and Blaha, M., *Solar Physics* **3**, 563 (1968).
15. Olsen, K, H., Anderson, C. R. and Stewart, J. N. *Solar Physics,* **21**, 360 (1971).
16. Gabriel, A. H., Garton, W. R. S., Goldberg, L., Jones, T. J. L., Jordan, C., Morgan, F. J., Nicholls, R. W., Parkinson, W. H., Paxton, H. J. B., Reeves, E. M., Shenton, D. B., Speer, R. J., and Wilson, R., *Ap. J.* **169**, 595 (1971).
17. Jorden, C., *Solar Physics,* **21**, 381 (1971).
18. Pottasch, S. R., *Ap. J.* **137**, 945 (1963).
19. Tousey, R., "Beam-Foil Spectroscopy" **2**, 485 (Gordon and Breach, 1967).
20. Jones, T. J. L., Parkinson, W., Speer, R. J. and Yang, C., 1971, *Solar Physics,* **21**, 372 (1971).
21. Brueckner, G. E., Bartoe, J. F., Nicolas, K. R. and Tousey, R., *Nature* **226**, 1132 (1970).
22. Speer, R. J., Garton, W. R. S., Goldberg, L., Parkinson, W. H., Reeves, E. M., Morgan, J. F., Nicholls, R. W., Jones, T. J. L., Paxton, H. J. B., Shenton, D. B. and Wilson, R., *Nature* **226**, 249 (1970).
23. Samson, J. A. R., "Techniques of Vacuum Ultraviolet Spectroscopy" (Wiley, 1967).
24. Lie, T. N. and Elton, R. C., *Phys. Rev. A.* **3**, 865 (1971).
25. Lie, T. N. and Elton, R. C., *Space Science Reviews,* to be published (1971).
26. Neupert, W. M. and Swartz, M., *Ap. J. (Letters)* **160**, L189 (1970).
27. Doschek, G. A., Meekins, J. F., Kreplin, R. W., Chubb, T. B. and Friedman, H., *Ap. J.,* in press (1971).
28. Filippov, N. V., Filippova, T. I. and Vinograsov, V. P., *Nucl. Fusion. Suppl.* Part 2, 577 (1962).
29. Peacock, N. J., Wilcock, P. D., Speer, R. J. and Morgan, P. D., *Plasma Physics and Controlled Nuclear Fusion Research* **2**, 51 (1969).
30. Peacock, N. J., Speer, R. J. and Hobby, M. G., *J. Phys. B.* **2**, 798 (1969).
31. Peacock, N. J., Hobby, M. G. and Morgan, P. D., *Proc. 4th Conf. on Plasma Physics and Controlled Nuclear Fusion Research,* CN-28/D-3 (1971).
32. Loewen, E. G., *J. Phys. B.* **3**, 953 (1970).
33. Franks, A. and Sayce, L. A., *Proc. Roy. Soc. A.* **282**, 353 (1964).

34. Speer, R. J., *Advances in X-ray Analysis* **13,** 382 (1970).
35. Bennett, J. M., *Ph .D. Thesis* (London University, 1971).
36. Franks, A., "Applied X-rays, Electrons and Ions" (New Delhi Press, 1970).
37. Labeyrie, A. and Flamand, J., *Optics Communications* **1,** 5 (1969).
38. Cordelle, J., Flamand, J. and Pieuchard, G., "Optical Instruments and Techniques" (Oriel Press, 1969).
39. Cordelle, J., Laude, J. P., Petit, R. and Pieuchard, G., *Nouv. Rev. d'Optique Appliquée* **1,** 149 (1970).

Spectral Intensities from Helium-like Ions

A. H. GABRIEL

Astrophysics Research Unit, Culham Laboratory, Berkshire, England.

I. INTRODUCTION

Spectral intensities are influenced primarily by two plasma parameters, electron temperature and electron density. An ideal diagnostic measurement would be sensitive to one of these and independent of the other. Furthermore it would be based upon line intensity ratios rather than the technically difficult measurement of absolute intensities. After considering briefly the principles of the two types of measurement, we shall examine the particular advantages of helium-like ions for density measurement. For a further development of many of the points raised, the reader is referred to a recent review article by Gabriel and Jordan.[1]

Helium-like ion spectra offer a unique opportunity for the measurement of plasma electron density. This arises through the fact that the four $n = 2$ terms are relatively close in energy and that three of these are long-lived or metastable levels. It is shown that the relative intensities of the three strongest emission lines from such an ion will be sensitive to density over a range of nearly ten orders of magnitude. Moreover these three lines are close together in wavelength, thereby simplifying the experimental proceedures. The value of this approach will be greatly strengthened when confidence in the values of the collision rates used in the interpretation can be improved.

II. TEMPERATURE AND DENSITY MEASUREMENT

For highly-charged ions, over the entire density range considered in this paper, excited level populations are not populated according to Local Thermodynamic Equilibrium or Boltzmann statistics. They approximate rather to the so-called "coronal" population. In this case excited levels are populated by electron impact excitation from the ground, and, except for metastable levels, decay by spontaneous radiation. In such a model virtually all the population of an ion is in the ground state, with only a very small fraction in excited states. If we consider a ground state g, and an excited state i, then in the steady state

$$N_g N_e C_{gi} = N_i A_{ig} + N_e N_i C_i^* \qquad (1)$$

where N_e, N_g are populations of electrons, state g, etc, C_{gi} is the collisional excitation rate coefficient $g \to i$, A_{ig} is the spontaneous decay rate $i \to g$, and C_i^* is the collisional loss rate from the level i.

Rearranging we get

$$N_g N_e C_{gi} = N_i (A_{ig} + N_e C_i^*) \qquad (2)$$

$$\simeq N_i A_{ig} \propto I_{ig}, \qquad (3)$$

since the second term in the right-hand side of (2) is small compared with the first, where I_{ig} is the photon intensity of $i \to g$ radiation.

For two excited levels i and j we then have

$$\frac{I_{ig}}{I_{jg}} = \frac{C_{gi}}{C_{gj}} \qquad (4)$$

which is independent of the density terms.

The coefficients C can be expressed as $\langle vQ \rangle$, that is the product of the relative velocity of the colliding particles and their collision cross-section, averaged over the Maxwellian distribution of velocities. C depends in detail on the behaviour of Q, but always contains terms of the type $F(T) \exp(-E/kT)$ where E is the threshold excitation energy and $F(T)$ is some slowly varying function of the temperature T. As a result, the principal temperature dependence of the intensity ratio of Eqn (4) arises from a term

$$\exp[-(E_i - E_j)/kT].$$

This is most sensitive to temperature changes when $E_i - E_j > kT$ and least sensitive when $E_i - E_j < kT$. The principal requirements, therefore, for using two lines g–i and g–j for temperature measurement can be summarized: (a) They must both be strongly allowed transitions to justify the approximation of Eqn (3), and (b) The upper levels must be separated by an energy interval larger than the temperature being measured. This type of reasoning has led to the choice of the line ratio $(2s-2p)/(2s-3p)$ in lithium-like ions as being ideal for such a measurement (Heroux[2]).

Suppose now that of the two levels i and j, j is metastable; that is that A_{jg} is very much smaller than it would have been for an allowed level. Both terms of Eqn (2) may now be important for the level j and we have

$$\frac{I_{ig}}{I_{jg}} = \frac{C_{gi}}{C_{gj}}\left(1 + \frac{N_e C_j^*}{A_{jg}}\right) \tag{5}$$

showing that the intensity ratio now becomes density dependent when N_e begins to approach the value A_{jg}/C_j^*. For N_e much larger than A_{jg}/C_j^*, the intensity ratio of Eqn (5) will become so large that I_{jg} may be too weak to measure. The condition for sensitive density measurement therefore becomes

$$N_e C_j \approx A_{jg}. \tag{6}$$

In Eqn (5) it was assumed that the population transferred from j by the process C_j^* was lost to the levels considered. In some systems, it will be transferred to the level i, and the equation must then be slightly modified.

$$\frac{I_{ig}}{I_{jg}} = \frac{C_{gi}}{C_{gj}}\left[1 + \frac{N_e C_{ji}}{A_{jg}}\left(1 + \frac{C_{gj}}{C_{gi}}\right)\right]. \tag{7}$$

This correction is not sufficient to affect the requirement (6). If the intensity ratio of Eqn (7) becomes very large due to a large value of N_e, then a simplification is possible in which we consider the sum of the two excitation rates, resulting in emission from the level i only. That is

$$I_{ig} = N_e N_g (C_{gi} + C_{gj}) \tag{8}$$

so that i and j can in effect be treated as one combined level.

Referring to Eqn (5) or (6) we can now define the ideal requirements for density measurement from the ratio g–i to g–j. These are (a) The levels i and j must be close together, so that C_{gi}/C_{gj} is not too dependent on temperature, and (b) one level, say j, must have a spontaneous decay rate A_{jg} much smaller than the other A_{ig}, or in effect one level must be metastable. The temperature dependence of the coefficient C_{ji} (or C_j^*) will be small if the excitation is transferred from j to a nearby level.

The above considerations leading to ideal conditions for temperature and density measurement will suffice for the following discussion on helium-like ions. However it should be realized that these are not necessary conditions, and that other more complex term structures can lead to further methods for measuring these parameters.

III. THE HELIUM-LIKE ION

We define a parameter Z to determine the position in the isoelectronic sequence, such that $Z - 1$ is the total charge on the ion. Z in then the effective charge seen by an excited electron. For the helium-like sequence, the first excitation potential is over 75% of the ionization energy. It is therefore not possible to find two excited levels so that $E_i - E_j > kT$, and such ions are not suited to this method of temperature measurement.

On the other hand, the four $n = 2$ levels provide three pairs of levels ideal for density measurement, each covering a different density regime. The $n = 2$ levels are close together, covering in total only $\approx 1/(8Z)$ of the ionization energy. The temperature dependence of the excitation rate ratios then derive entirely from the slowly varying terms $F(T)$ (see above) and can normally be neglected.

Figure 1 is a simplified term diagram of a helium-like ion, showing the four $n = 2$ terms and their decay processes. These terms can be abbreviated as 2^1P, 2^3P, 2^1S and 2^3S. All four are excited from the ground, with excitation rates varying between them by not more than a factor of 5. The decay process are as follows, in decreasing order of transition probability.

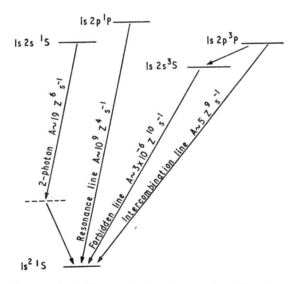

FIG. 1. Schematic energy level diagram showing the ground and first four excited states of a helium-like ion. Radiative decay mechanisms are shown with their approximate Z-scaling.

2^1P decays by normal allowed electric dipole through the resonance line with a rate of $\approx 10^9 Z^4 \text{ sec}^{-1}$. An alternative transition to 2^1S has such a small rate in comparison that it can be neglected.

2^3P consists of three levels which must in some cases by treated separately. As a simplification at present we can take it as a single level. Mixing through spin–orbit interaction with 2^1P introduces an intercombination line with an effective rate of $\approx 5Z^9 \sec^{-1}$ (Drake and Dalgarno[3]). An alternative allowed transition to 2^3S has a larger rate below carbon V but a smaller one above. This must be allowed for in the theory, but does not affect the general principles followed.

2^1S decays by two-photon electric dipole radiation to the ground, with a rate of $\approx 19\, Z^6 \sec^{-1}$ (Drake, Victor and Dalgarno[4]). This results in a continuum rather than a line, and cannot readily be measured.

2^3S decays by relativistic magnetic dipole radiation with a rate of $\approx 3 \times 10^{-6}\, Z^{10} \sec^{-1}$ (Drake[5]). This will be referred to as the forbidden line.

These last three transitions each satisfy the relation of Eqn (6) in a different density regime. We can thus identify three separate regimes as follows:

1. *Low Density.* 2^3S is progressively transferred by collisions to 2^3P. The intensity ratio

$$\frac{\text{Intercombination line}}{\text{Forbidden line}}$$

increases with N_e.

2. *Intermediate Density.* All the triplet excitation decays through the intercombination line. 2^1S is transferred by collisions to 2^1P. The intensity ratio.

$$\frac{\text{Resonance line}}{\text{2-photon continuum}}$$

increases with N_e. This cannot be measured, but due to the increase in the resonance line, the ratio

$$\frac{\text{Resonance line}}{\text{Intercombination line}}$$

also increases with N_e.

3. *High Density.* Only the resonance and incombination lines now exist. Collisions progressively transfer $(2^3S + 2^3P)$ population to the 2^1S and 2^1P levels. Consequently, the ratio

$$\frac{\text{Resonance line}}{\text{Intercombination line}}$$

increases with N_e.

There is of course an even higher density where LTE conditions apply, but since under these conditions only the resonance line is seen, it is of purely academic interest and offers no diagnostic possibilities for density measurement.

It should be noted that, in all that follows, a collision rate C_{ig} to a level i with $n = 2$ includes not only direct excitation from the ground, but a small contribution by cascade from higher n levels plus recombination into excited levels.

IV. LOW DENSITY

As recently as three years ago, the 2^3S level was believed to decay primarily by two-photon emission. Only when solar soft X-ray spectra became available from Rugge and Walker[6] and Jones, Freeman and Wilson,[7] was the forbidden line first seen in oxygen VII and neighbouring ions. Gabriel and Jordan[8] identified this as due to a single photon magnetic dipole transition, and went on to develop a theory for the density dependence of the line intensity ratios (Gabriel and Jordan[9]). This showed that the ratio of the forbidden to intercombination line intensities R is related to the ratio R_0 at zero density by the expression

$$\frac{R_0}{R} = 1 + \frac{(N_e C^*_{2^3S} + \Phi)}{A_{2^3S}} (1 + F_3) \qquad (9)$$

where F_3 is the ratio of collisional excitation rates to 2^3S and 2^3P, and $C^*_{2^3S}$ is the transfer rate $2^3S \to 2^3P$. Φ is a photo-excitation rate $2^3S \to 2^3P$ important in the sun for carbon V only. This is directly analogous to Eqn (7). R_0 is given by

$$R_0 = \frac{1 + F_3}{B} - 1 \qquad (10)$$

a complication deriving from the alternative decay $2^3P \to 2^3S$. B is here the effective branching ratio for decay to 1^1S rather than 2^3S, and includes allowance for the three 2^3P levels.

Solar observations of R were interpreted in terms of electron density at the source. Early work was based upon values of the forbidden line transition probability A_{2^3S} calculated by Griem.[10] These were subsequently shown to be in error (Freeman et al.[11]) and the densities re-derived using a semi-empirical expression for A_{2^3S}. This expression was supported later by new calculations by Drake.[5] The densities derived by Freeman et al. range up to 10^{13} cm^{-3} for active regions. The criteria used for deciding that R departs from its low density value is that R should vary between different observations by an amount which is larger than the observational errors claimed. If one now re-examines the data with a more critical estimate of the observational errors, it is possible to dismiss many of the observations as being indistinguishable from R_0. However, measurements in neon IX and magnesium XI remain as strongly indicating departures from R_0 and thus densities of over 10^{12} cm^{-3}. Recent improved calculations of $C^*_{2^3S}$ by Blaha[12] have the effect of reducing the densities derived by Freeman et al. by a factor of ≈ 2.

V. INTERMEDIATE DENSITY

The expression for the intermediate density regime can readily be obtained by replacing triplets by singlets in Eqns (9) and (10). This is further simplified by the fact that $B = 1$ for the singlets. Then the intensity ratio

$$\frac{\text{Resonance line}}{\text{2-Photon Continuum}} = \frac{1}{F_1}\left[1 + \frac{N_e C^*_{2^1S}}{A_{2^1S}}(1 + F_1)\right] \quad (11)$$

$$= \text{say } \alpha$$

from which the intensity ratio

$$\frac{\text{Resonance line}}{\text{Intercombination line}} = \frac{\alpha}{1 + \alpha} \times \frac{\text{Total singlet excititaon}}{\text{Total triplet excitation}}. \quad (12)$$

This density regime has not been studied experimentally and there is some doubt about the value of the excitation rate ratio F_1. There has been one reported observation of the two-photon continuum from neon IX in a laboratory plasma (Elton, Palumbo and Griem[13]), but such continua have not been positively identified in astrophysical sources. The continuum has a threshold at the $1s^2\,{}^1S - 1s2s\,{}^1S$ energy interval and has a maximum on a wavelength scale at 1·4 times the threshold wavelength.

VI. HIGH DENSITY

For this regime, 2^1S and 2^1P behave as one level decaying through the resonance line, while 2^3S and 2^3P behave as one level decaying through the intercombination line. As the density increases, the combined triplet levels lose population by (a) transfer to 2^1S and 2^1P at a rate C_{31}, and (b) by ionization at a rate S. Equations (5) and (7) then lead to the intensity ratio

$$\frac{\text{Resonance line}}{\text{Intercombination line}} = \frac{C_1}{C_3}\left[1 + \frac{N_e}{A}\left(S + C_{31}\left(1 + \frac{C_3}{C_1}\right)\right)\right] \quad (13)$$

$$= \frac{C_1}{C_3}\left[1 + \frac{N_e}{A}P\right] \text{ say} \quad (14)$$

Here C_1 and C_3 represent total singlet and triplet excitation rates respectively, while A is in the intercombination line decay rate suitably averaged over the combined triplet levels.

This region has been studied experimentally in laboratory plasmas by Kunze, Gabriel and Griem[14] and Gabriel, Paget and Kunze.[15] The combined coefficient P can be determined by measuring the line ratio as a function of density. This has been carried out for carbon V, nitrogen VI and oxygen VII using theta-pinch plasmas. Whereas the value of P can be well determined, separation into its components C_{31} and S is subject to less direct arguments. The above experiments indicate also that a third process might be important; i.e. collisional transfer from the $n = 2$ triplet levels to singlet levels of $n = 3$ and higher. However, values of the combined rate P (see Eqn 14) are all that are required to use the line ratios as a measure of density and these are now known.

VII. CONCLUSIONS

The theory for the three regions can be combined to give the relative intensities of the resonance, intercombination and forbidden lines as a function of density over a wide range. This has been done for carbon V in Fig. 2. It is necessary to assume values for the relative excitation rates from the ground, and these have been taken as follows

2^1S	1·0
2^1P	1·0
2^3S	0·3
2^3P	0·8

These are the values that derive from combining existing observational data, but it is thought likely that the 2^1S rate is in reality much smaller.

It is seen in Fig. 2 that the line ratios are sensitive to changes in N_e over a density range of nearly ten orders of magnitude. The use of other ions in the sequence can further extend this density range. Unfortunately the theory used to construct Fig. 2 is based upon collision rates which are not at present known with sufficient confidence. It is therefore clear that improved calculations of the excitation rate coefficients to the four $n = 2$ levels for helium-like ions would be of considerable value for the use of these ions for density measurement.

It is of particular importance to examine closley the low density regime since this has been used for solar density measurement. In this connection, the precise value of the excitation rate ratio F_3 is of less importance than the assumption made that it is independent of temperature. An attempt has been made recently by Blumenthal, Drake and Tucker[16] to compute the effect of temperature variation on this measurement. Unfortunately some of the data available to them is inadequate in reliability for this purpose. The results they obtain would appear to justify the earler assumption that temperature variations can be neglected. However, more precise studies of this regime would be an advantage when improved collision rates become available.

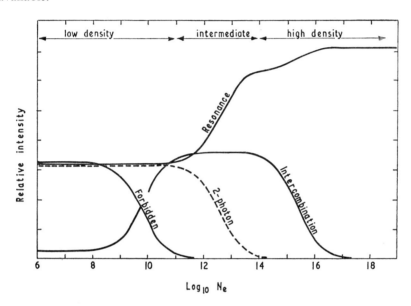

FIG. 2. The predicted variation in relative intensities of the resonance, intercombination and forbidden lines, plotted as a function of electron density for carbon V, at a temperature of 10^6 °K.

ACKNOWLEDGEMENTS

I am pleased to acknowledge many helpful discussions with Dr Carole Jordan and Dr H. J. Kunze, relating to the Low and High Density Regimes respectively.

REFERENCES

1. Gabriel, A. H. and Jordan, C., *In* "Case Studies in Atomic Collision Physics" –II, eds, McDaniel and McDowell (North-Holland: Amsterdam, 1971).
2. Heroux, L., *Proc. Phys. Soc.* (*London*) **83**, 121 (1964).
3. Drake, G. W. F., and Dalgarno, A., *Ap. J.* **157**, 459 (1967).
4. Drake, G. W. F., Victor, G. A. and Dalgarno, A., *Phys. Rev.* **180**, 25 (1969).
5. Drake, G. W. F., *Phys. Rev.* **3**, 908 (1971).

6. Rugge, H. R. and Walker, A. B. C., *In* "Space Research" Vol. 8. p, 439 (North-Holland: Amsterdam, 1968).
7. Jones, B. B., Freeman, F. F. and Wilson, R., *Nature* **219,** 252 (1968).
8. Gabriel, A. H. and Jordan, C., *Nature* **221,** 947 (1969).
9. Gabriel, A. H. and Jordan, C., *M.N.R.A.S.* **145,** 241 (1969).
10. Griem, H. R., *Ap. J.* (*Letters*) **156,** L103 (1969).
11. Freeman, F. F., Gabriel, A. H., Jones, B. B. and Jordon, C., *Phil. Trans. Roy. Soc.* (*London*) *A* **270,** 127 (1971).
12. Bahla, M., *Bull. Amer. Phys. Soc.* **3,** 246 (1971).
13. Elton, R. C., Palumbo, L. J. and Griem, H. R., *Phys. Rev. Letters* **20,** 783 (1968).
14. Kunze, H. J., Gabriel, A. H. and Griem, H. R., *Phys. Rev.* **165,** 267 (1968).
15. Gabriel, A. H., Paget, T. M. and Kunze, H. J., *J. Phys. B.,* to be published (1972).
16. Blumenthal, G. R., Drake, G. W. F. and Tucker, W. H., *Ap. J.*, **172,** 205 (1972).

Abundances in the Solar Corona

H. OLTHOF

University of Groningen, Groningen, Holland.

The chemical composition of the solar corona as determined from the measurements of the forbidden line emission has been previously considered by Woolley and Allen,[1] Shklovskii,[2] Pottasch[3] and Jordan.[4] This article is a short description of a new analysis of the coronal line spectrum carried out at the Kapteyn Astronomical Laboratory in Groningen. A more detailed description is published elsewhere by de Boer, Olthof and Pottasch.[5]

The analysis is based on the observation of the visual line spectrum observed during the eclipse of May 1965 by Jefferies, Orrall and Zirker.[6] Most of these lines in the coronal line spectrum are magnetic dipole transitions in the ground configurations of highly ionized atoms. The line emission from a given volume element in the solar atmosphere is given by

$$E_\nu = \frac{h\nu}{4\pi} N_u A_{ul} \text{ (erg cm}^{-3} \text{ sec}^{-1} \text{ ster}^{-1})$$

where u and l refer to the upper and lower level of the transition involved. N_u is the population density in the upper level and A_{ul} is the radiative transition probability. The observed intensities

$$I_\nu \text{ (erg cm}^{-2} \text{ sec}^{-1} \text{ ster}^{-1})$$

can be written

$$I_\nu = \int E_\nu e^{-\tau_\nu} dl = \int E_\nu \, dl = \frac{h\nu}{4\pi} A_{ul} \int N_u \, dl$$

when the optical depth τ_ν is small.

From the observed line intensities $\int N_u \, dl$ can be calculated. This is related to the total ion density assuming statistical equilibrium. To obtain the ion densities we have to take into account all the populating and depopulating processes of the different levels of the ions, such as radiative excitation, direct collisional excitation by electrons, collisional excitation by electrons to higher levels followed by cascade to the upper level of the forbidden transition and proton excitation. Bahcall and Wolf[7] give asymptotic expressions for a few ions. Bely and Faucher[8] give numerical values of these rates only for the $2p$, $2p^5$, $3p$ and $3p^5$ isoelectronic sequences. It turns out that the proton collision rates are comparable with the direct electron collision rate at temperatures higher than 2×10^6 °K. Only for temperatures higher than 5×10^6 °K do the proton collisions influence the upper level population density by more than 30%. Solving the equations of statistical equilibrium we obtain N_u/N_{ion} as a function of N_e and T_e for several ions which we are going to apply to circumstances in the solar corona.

From the observations we have obtained $\int N_u \, dl$. This can be expanded as

$$\int N_u \, dl = \int \frac{N_u}{N_{\mathrm{ion}}} \frac{N_{\mathrm{ion}}}{N_{\mathrm{H}}} \frac{N_{\mathrm{H}}}{N_e} N_e \, dl$$

$$= \frac{N_{\mathrm{H}}}{N_e} \frac{N_{\mathrm{ion}}}{N_{\mathrm{H}}} \int \frac{N_u}{N_{\mathrm{ion}}} N_e \, dl.$$

So the ion abundance follows from

$$\frac{N_{\mathrm{ion}}}{N_{\mathrm{H}}} = \frac{\int N_u \, dl}{\frac{N_{\mathrm{H}}}{N_e} \int \frac{N_u}{N_{\mathrm{ion}}} N_e \, dl}.$$

We have to be somewhat careful now, because we do not yet know anything about the temperature and the electron density distribution along the line of sight. The value of $N_{\mathrm{ion}}/N_{\mathrm{H}}$ we obtain in this way is not the value of $N_{\mathrm{ion}}/N_{\mathrm{H}}$ at a certain place in the line of sight but more or less a mean value over the line of sight. Theoretical calculations on the ionization equilibrium indicate those ions we can expect in a coronal plasma which has a temperature between 1 to 5×10^6 °K. In the further calculations we assumed a temperature of 2×10^6 °K. The electron density distribution can be derived from the continuum emission of the solar corona, also measured by Jefferies, Orall and Zirker.[6] It can be shown that the continuum emission is mostly due to electron scattering. Scattering by dust grains contributes less than 5% and the free-free emission is negligible. From this continuum emission $\int N_e \, dl$ can be determined. For most of the observed lines the intensity distribution as a function of position angle on the solar corona is restricted to a small area, a condensation. From the equations of statistical equilibrium it can be shown that for most ions

N_u/N_{ion} is rather insensitive to temperature variations, but more sensitive to variations in the electron density as a function of the position angle. To obtain a model for the N_e distribution we assumed that the total line and continuum intensity was formed within this condensation. We split it up into two parts

$$N_{e1} = 10^9 \text{ cm}^{-3}$$

$$N_{e2} = 3 \times 10^8 \text{ cm}^{-3}$$

and choose the values of Δl in such a way that

$$\int N_e \, dl = N_{e1} \Delta l_1 + N_{e2} \Delta l_2.$$

Putting this in the quation for the ion abundance we obtain

$$\frac{N_{ion}}{N_H} = \frac{\int N_u \, dl}{\dfrac{N_H}{N_e} \left[\left(\dfrac{N_u}{N_{ion}}\right)_1 N_{e1} \Delta l_1 + \left(\dfrac{N_u}{N_{ion}}\right)_2 N_{e2} \Delta l_2 \right]}$$

where N_u/N_{ion1} is calculated with $N_e = 3 \times 10^8 \text{ cm}^{-3}$, $T_e = 2 \times 10^6$ °K and N_u/N_{ion2} is calculated with $N_e = 10^9 \text{ cm}^{-3}$, $T_e = 2 \times 10^6$ °K.
In this way we obtain an average value of the ion abundance in the condensation.

For only two elements, Fe and Ni, a sufficient number of ionization stages are observed so that the abundance can be found simply by adding the ionic values corrected with values for the "missing" ions. To obtain abundances of elements for which only one or two ionization stages are observed we followed the same procedure as Pottasch.[3] We plotted the obtained abundances of the Fe ions against the ionization potential of the next lower state of ionization. The Ni ions are added by multyplying their abundances by the ratio Fe–Ni. In this way we obtain an ionization curve (Fig. 1). The abundances of the other elements with similar ionization potentials can be found by scaling the ion abundance to this ionization curve, under the assumption that the ionization curves of the other elements resembles the Fe–Ni curve. For Ca and A we assumed a different ionization curve because only lines of Ca ions with an ionization potential of the next lower stage above 500 eV are observed. Furthermore it is likely that because of the large difference in ionization potential between CaX (211eV) and CaXI (592 eV) only ions of CaXI and higher are present. The same arguments are valid for the A ions. The Ca abundance is found from adding the ion abundances. The A abundance can be determined by scaling to this curve. The results are given in Table I.

Table I

relative coronal abundances
(logarithmic scale, $\log N_H = 12$).

element	this work	Pottasch[3]
H	12	12
A	5·5	7·1
Ca	6·4	6·8
Cr	6·7	6·0
Mn	5·3	5·7
Fe	8·0	7·9
Co	6·1	5·6
Ni	6·7	6·7
Cu	4·6	
Zn	7·0	

A discussion on the line identifications is given by Jefferies[9] and by de Boer, Olthof and Pottasch.[5] From the table it can be seen that the Fe and Ni abundances are in rather good agreement with earlier determinations. This is not true for most of the other species. The uncertainty in their abundancies is rather high because they are mostly based on one or two points on the ionization curve. Unfortunately other lines of these highly ionized atoms fall either in the UV below 3000Å or in the infrared above 8000Å. About 40% of the observed lines are still unidentified mostly because it is not possible to make a cross-check. Sometimes there is an

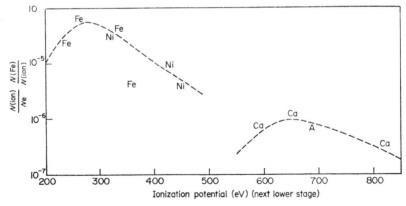

FIGURE 1

identification for a line but then it turns out that the ascribed ion should have a much stronger line at another wavelength which is not observed. More observations are necessary to extend and test the reliability of the two ionization curves. Lines of ions with lower ionization potentials below 200 eV and between 400 eV and 600 eV all fall in the infrared region of the spectrum. By turning around the whole procedure as sketched before we can make a prediction of the infrared lines based on this and other abundance determinations. By doing this we expect rather strong lines of the ions given in Table II.

Table II

expected strong infrared lines

ion	$\lambda(\mu)$
SIX	1·25
SiX	1·43
ScIX	2·59
MgVIII	3·03
AlVIII	3·7
SiIX	3·9
FeXI	6·1

The lines of MgVIII ($3\cdot03\mu$) and SiX ($1\cdot43\mu$) have already been detected by Münch, Neugebauer and McCammon.[10]

REFERENCES

1. Woolley, R. v. d. R. and Allen, C. W., *M.N.R.A.S.* **108**, 292 (1948).
2. Shklovskii, I. S., *Ap. J. Crimea* **6**, 105 (1951).
3. Pottasch, S. R., *M.N.R.A.S.* **128**, 73 (1964).
4. Jordan, C. *M.N.R.A.S.* **132**, 463, 515 (1966).
5. de Boer, K. S., Olthof, H. and Pottasch, S. R., *Astr. and Ap.* **16**, 417.
6. Jefferies, J. T., Orall, F. Q. and Zirker, J. B., *Sol. Phys.* **16**, 103 (1971).
7. Bachall, J. N. and Wolf, R. A., *Ap. J.* **152**, 701 (1968).
8. Bely, O. and Faucher, P., *Astr. and Ap.* **6**, 88 (1970).
9. Jefferies, J. T., *In* "Les transitions interdites dans les spectres des astres", Université de Liège, p. 213 (1969).
10. Münch, G. Neugebauer, G. and McCammon, D., *Ap. J.* **149**, 681 (1967).

The Formation of H_2 Molecules in Dark Interstellar Clouds

T. DE JONG

(*Sterrewacht, Leiden*)

I. INTRODUCTION

Although at the present time most astrophysicists agree that H_2 molecules are formed in interstellar space through surface recombination on grains, the theory of this process is very poor from a purely physical point of view and many uncertainties are involved (Hollenbach *et al*.[1]). Recently Schutte[2] reported preliminary results of an experiment to measure surface recombination of H_2 in the laboratory. Such an approach may lead to a more conclusive picture of the details of this process in the near future.

Another possibility is formation through associative detachment of the H^- ion: $H^- + H \rightarrow H_2 + e$, which was first suggested by McDowell.[3] In this process the electrons instead of the grains act as catalysts for H_2 formation. Now that ionization by low energy cosmic rays has been recognized to increase the earlier estimates of the electron density in the interstellar medium by about two orders of magnitude, it appears worthwhile to consider the problem of H_2 formation by this process anew.

We shall present calculations of the density of atoms (H), positive ions (H^+), negative ions (H^-), molecules (H_2), positive molecular ions (H_2^+, H_3^+) and electrons (e) in a dark cloud with kinetic temperature $T_k = 50°K$. The calculation is done for clouds with total hydrogen density $n = 100 \text{ cm}^{-3}$ and 1000 cm^{-3} and we let the low-energy cosmic ray flux inside the cloud vary over three decades. We consider two different cases. In case (A) both surface recombination and associative detachment are operative; in case (B) H_2 molecules are formed by associative detachment only. A similar calculation, limited to case (A) and including about half of the reactions considered here (see Table I), was carried out by Solomon and Werner.[4]

Table I

Reaction	Rate coefficient ($T_k = 50°K$)	Reference
1. $2H$ + grain → H_2 + grain	$k_1 = 1 \times 10^{-17}$ cm^{3}sec^{-1}	text
2. $H + H^-$ → $H_2 + e$	$k_2 = 1.3 \times 10^{-9}$ cm^{3}sec^{-1}	H
3. $H + H_2^+$ → $H^+ + H_2$	$k_3 = 5.8 \times 10^{-10}$ cm^{3}sec^{-1}	H
4. $H + e$ → $H^- + h\nu$	$k_4 = 5.3 \times 10^{-17}$ cm^{3}sec^{-1}	text
5. $H + p$ → $H^+ + e + p$	$k_5 = 0.6 \times \zeta_0$ sec^{-1}	SW
6. $H^+ + H^-$ → $2H$	$k_6 = 2.3 \times 10^{-7}$ cm^{3}sec^{-1}	H
7. $H^+ + e$ → $H + h\nu$	$k_7 = 1.1 \times 10^{-11}$ cm^{3}sec^{-1}	SW
8. $H^- + H_2^+$ → $H + H_2$	$k_8 \leqslant 2.3 \times 10^{-7}$ cm^{3}sec^{-1}	text
9. $H^- + h\nu$ → $H + e$	$k_9 = 2.4 \times 10^{-7}$ sec^{-1}	,,
10. $H^- + p$ → $H + e + p$	$k_{10} = 11 \times \zeta_0$ sec^{-1}	,,
11. $H_2 + p$ → $H_2^+ + e + p$	$k_{11} = 0.95 \times \zeta_0$ sec^{-1}	SW
12. $H_2 + p$ → $H^+ + H + e + p$	$k_{12} = 0.05 \times \zeta_0$ sec^{-1}	,,
13. $H_2 + p$ → $2H + p$	$k_{13} = 0.1 \times \zeta_0$ sec^{-1}	,,
14. $H_2^+ + e$ → $2H$	$k_{14} = 5 \times 10^{-9}$ cm^{3}sec^{-1}	,,
15. $H_2^+ + p$ → $H + H^+ + p$	$k_{15} = 1.4 \times \zeta_0$ sec^{-1}	text
16. $H_2^+ + p$ → $2H^+ + e + p$	$k_{16} = 0.16 \times \zeta_0$ sec^{-1}	text
17. $H_2^+ + H_2$ → $H_3^+ + H$	$k_{17} = 2.1 \times 10^{-9}$ cm^{3}sec^{-1}	SW
18. $H_3^+ + e$ → $H_2 + H$	$k_{18} = 5 \times 10^{-9}$ cm^{3}sec^{-1}	,,

SW = Solomon and Werner[4]
H = Hirasawa[5]

II. THE REACTIONS

In Table I we show the reactions and rate coefficients at $T_k = 50°K$ that are included in the calculations. Changing the temperature within a factor of two will not affect the results very much because the gas kinetic reactions in Table I are all exothermic and therefore not very temperature dependent. The rate of reaction $A + B \to C + D$ is given by $K_{AB} n_A n_B$. When B consists of photons or protons the reaction rate equals $K_A n_A$. We shall now discuss some of the reactions in Table I.

Following Spitzer and Tomasko[6] and Field et al.[7] we assume the low energy cosmic rays (LECR) to consist of monoenergetic 2 MeV protons that are ejected by supernovae into interstellar space. In Table I we distinguish between H^+ (thermal protons that are the result of H ionization) and p (MeV protons).

The energetic electrons produced upon LECR ionization of H and H_2 are very rapidly thermalized by collisions with the gas. Habing and Goldsmith[8] estimate that only a fraction 10^{-7} of all electrons will be in the high energy tail of the Maxwell velocity distribution; apart from secondary ionization their effect will be neglected.

Most reactions with photons are excluded. In a dense cloud with about half of the gas in molecular form, H_2 molecules in the outer layer absorb the dissociating UV photons of the interstellar radiation field around 1000 Å so that the molecules inside are shielded against photodissociation. The only photoreaction included is photodetachment of H^- because photons around 1μ may penetrate into the cloud.

Recombination of atomic and molecular hydrogen ionized by LECR may provide some UV photons in the interior of the cloud. In the case of atomic hydrogen this has been accounted for by only considering recombination to excited levels. In the case of molecular hydrogen we assume that only dissociative recombination, reaction (14), takes place (Dubrovskii and Ob'edkov[9]). The rate of surface recombination of H_2, reaction[1], is computed for grains with radius $v_{gr} = 0.15\mu$ assuming that one out of every four colliding H atoms contributes to molecule formation. Since the number density of grains is proportional to the total hydrogen density ($n_{gr} \simeq 1.2 \times 10^{-12} n$) it is convenient to define k_1 such that the reaction rate equals $k_1 n n_H$.

From the analytic expression for the photodetachment cross-section of H^-, reaction (9), given by Ohmura and Ohmura[10] together with an estimate of the radiation field inside a dark cloud (Greenberg[11]), characterized by a radiation temperature of 4000°K and a dilution factor of 3.9×10^{-13}, we calculate the value of k_9 in Table I. Using the principle of detailed balance (e.g., Rosseland[12]) the cross-section and reaction rate of reaction (4), which is the inverse of reaction (9), can then easily be obtained.

The rate coefficient of reaction (8), $H^- + H_2^+ \to H + H_2$, is probably similar to that of reaction (6), $H^- + H^+ \to 2H$. Because in all cases considered $n_{H_2^+} \ll n_{H^+}$ (see Fig. 1) the rate of this process will be negligibly small compared to the other destruction mechanisms of H^- and is therefore omitted in the calculations. In view of the very small H^- densities (see Fig. 1) the same holds for the destruction rate of H_2^+ by this reaction.

In order to estimate the cosmic ray detachment cross-section of H^- (reaction 10) we note that for proton energies around 2 MeV where binary encounter theory may be applied, the cross-section is roughly inverse proportional to the binding energy of the electron so that

$$k_{10} \approx 17 k_5 \approx 11 \zeta_0 \text{ sec}^{-1}$$

Cross-sections for high energy protons colliding with H_2^+ are given by

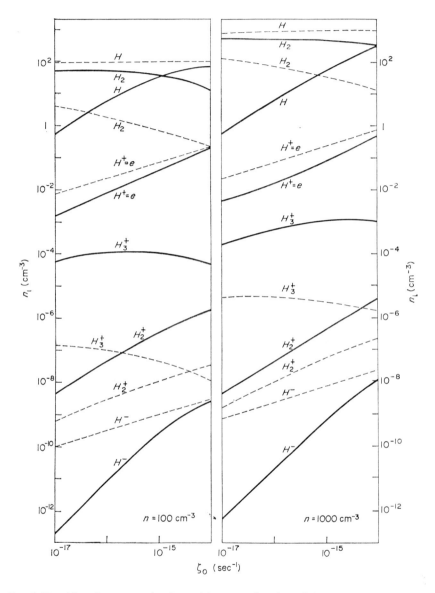

FIG. 1. Densities of atoms, molecules and ions as a function of the LECR flux in clouds with total hydrogen densities $n = 100 \text{ cm}^{-3}$ and $n = 1000 \text{ cm}^{-3}$. The magnitude of the LECR flux is given by the parameter ζ_0, the number of H_2 ionizations per second. The solid lines represent case (A) where all reactions of Table I are included; the dashed lines represent case (B) where surface recombination of H_2 is excluded

Bates and Holt.[13] More extensive calculations for energetic electrons on H_2^+ are presented by Peek.[14] When the energies are properly scaled, we find from Peek's data a total cross-section of $0.30\,\pi a_0^2$ for 2 MeV protons. The reaction rates k_{15} and k_{16} are derived, assuming a branching ratio of 10 percent for dissociative ionization (16) with respect to dissociation (15) and taking the effect of secondary electrons and protons into account by increasing the rates by about 70 percent.

III. RESULTS

Using the processes in Table I we can write down a steady state equation for every atom, molecule and ion. In addition, equations for particle and charge conservation have to be satisfied. The resulting set of non-linear equations has been solved by an iterative procedure.

The results of the calculation for total density $n = 100\,\text{cm}^{-3}$ and $1000\,\text{cm}^{-3}$ are displayed in Fig. 1 as a function of the parameter ζ_0. In order to investigate the effect of a varying LECR flux inside the cloud we let ζ_0, the rate coefficient of H_2 ionization by MeV protons, vary between 10^{-14} and 10^{-17} sec^{-1}. Using the estimates of Field et al.,[7] Solomon and Werner[4] calculated $\zeta_0 = 10^{-15}$ sec^{-1} in general interstellar space.

The solid lines in Fig. 1 represent the equilibrium values for case (A) in which all processes of Table I are included; the dashed lines represent case (B) in which H_2 formation on the surface of interstellar grains is left out ($k_1 = 0$). From the figure it can be read that 50 to 100 percent of the gas will be in molecular form if molecule formation on grains is included (case (A)). It turns out, however, that the cycle of gas-kinetic reactions (4), $H + e \rightarrow H^- + h\nu$, and (2), $H^- + H \rightarrow H_2 + e$, alone, is able to convert 1 to 25 percent of the gas into molecules. Electrons instead of grains then act as catalysts taking some of the formation energy of H_2 with them. The low energy cosmic rays are essential for the gas-kinetic formation of H_2 as they provide the catalyzing electrons.

Discussing Figure 1 in somewhat more detail we note that the density of electrons is equal to the density of H^+ ions in both case (A) *and* (B) because ionization and recombination of atomic hydrogen dominate the production and consumption of electrons.

Although the electron density is not very much affected by the absence of surface recombination, the H^- density is much higher in this case because more H^- ions are formed if a large fraction of the gas is atomic.

The behaviour of H_2^+ and H_3^+ is qualitatively the same in both cases. Without grain surface recombination $n_{H_2^+}$ is about one order of magnitude lower ($\propto n_{H_2}$) and $n_{H_3^+}$ two orders of magnitude ($\propto n_{H_2}^2$). The only importance of H_3^+ in this calculation is that it eats up H_2 molecules and H_2^+ ions;

neglecting H_3^+ by omitting reaction (17) and (18) would result in somewhat higher H_2 densities and much higher H_2^+ densities, of the same order of magnitude as H_3^+ in Fig. 1.

We note that the equation of particle conservation will not be upset by condensation of solid hydrogen on the surfaces of interstellar grains as Greenberg and de Jong[15] showed that this is impossible at realistic grain temperatures.

IV. DISCUSSION

Although at the present time it is thought that both diffuse X-rays and low-energy cosmic rays can explain the required degree of ionization of the interstellar gas, in the case of dark clouds ionization and dissociation by X-rays can be ruled out. Upon ionization X-rays give almost all their energy to the photo-electrons and are therefore absorbed in a thin outer layer of the cloud whereas low-energy cosmic rays loose only about 30 eV per ionization, a very small fraction of their total energy. Nevertheless the LECR flux inside the cloud will be smaller than in general interstellar space. The range of 2 MeV protons in hydrogen gas is about 5×10^{-1} gr cm^{-3} corresponding to path-lengths of 10 and 1 pc in gas of total densities $n = 100$ cm^{-3} and 1000 cm^{-3}, respectively.

For comparison with observations we shall introduce a dark cloud model, with $n = 1000$ cm^{-3}, a radius $r = 1$ pc, $\zeta_0 = 10^{-17}$ sec^{-1} and a visual extinction $\Delta m = 4$ along a path equal to the radius of the cloud, calculated on the basis of the gas–dust ratio given by Heiles.[16]

21-cm observations in the direction of dark clouds often show less neutral hydrogen than one would expect on the basis of a constant gas–dust ratio (Garzoli and Varsavsky[17,18]). More detailed measurements of the emission profiles even show absorption dips (Heiles,[16,19] Sancisi and Wesselius,[20] Sancisi[21]). This "anti-correlation" of gas and dust may be due to conversion of atomic into molecular hydrogen and is often interpreted as such. It has been emphasized, however, by Sancisi and Wesselius that such measurements can also be explained by cold H I gas if one drops the assumption of an optically thin cloud. It therefore appears premature to draw definite conclusions about molecular hydrogen in dark clouds from 21-cm observations alone. In this respect we do not agree with the final conclusion of Solomon and Werner[4] that calculations like theirs and ours can explain the observed low column densities of neutral hydrogen only if low energy cosmic rays are absent.

As an example we consider the dark cloud L 134 that has been studied by Sancisi.[21] The diameter of this cloud is $d = 1$ pc, the extinction in the centre amounts to 8 magnitudes and if one assumes a constant gas–dust ratio the total hydrogen density is about 2000 cm^{-3} so that it may be compared with our cloud model. The 21-cm observations of Sancisi show a peak brightness temperature $T_1 = 51°$K in the direction of the cloud as compared with $T_2 = 58°$K at nearby positions in the sky. If the brightness depression is caused by a foreground cloud of cool neutral hydrogen that absorbs 21-cm line radiation (Sancisi's model b) the following relation applies

$$T_1 = T_{cl}(1 - e^{-\tau}) - T_2 e^{-\tau}$$

yielding

$$\tau = \ln\left(\frac{T_2 - T_{cl}}{T_1 - T_{cl}}\right).$$

Furthermore the volume density of atomic hydrogen n_H is related to the opacity of the cloud τ by the expression

$$N_H = n_H d = 1\cdot 82 \times 10^{18} T_{cl} \tau \Delta v$$

where $\Delta v = 5$ km sec^{-1} is the velocity halfwidth of the observed absorption feature. Using these formulae we find that the brightness temperature in the direction of the cloud can be explained by the atomic hydrogen densities of our cloud model if the gas temperature in the cloud is as given in Table II. We note that the value of $1\cdot 7°$K calculated for case (A) is below the temperature of the 3°K background radiation. The coincidence with the $1\cdot 8°$K excitation temperature of formaldehyde, that has been derived from radio measurements of the 1_{11}–1_{10} rotational transition of this molecule in L 134 (Palmer et al.[22]), must be considered accidental. The gas temperature is determined by the heat balance in the interior of a dark cloud. We shall assume that the gas is heated by low-energy cosmic rays and cooled by rotational excitation of H_2 molecules and by accommodation of kinetic energy on the grain surfaces when atoms and molecules in the gas collide with the grains. Both cooling mechanisms convert thermal energy to far-infrared photons which escape from the cloud.

Table II

Cloud model fitting to Sancisi's[21] observations of L 134.

	n_H(cm^{-3})	n_H(cm^{-3})	τ	$T_{cl}(°K)$
case (A)	0·6	500	0·16	1·7
case (B)	740	130	5·0	50

The LECR heating rate can be written

$$R_{heat} \approx \zeta_0 n \Delta E \approx 5 \times 10^{-12} n\zeta_0 \text{ erg cm}^{-3} \text{ sec}^{-1}$$

where ΔE is the amount of energy available for heating of the gas upon ionization of H and ionization and dissociation of H_2. In the case of ionization of atomic hydrogen Spitzer and Tomasko[6] found $\Delta E = 3.4$ eV. We have put $\Delta E = 3$ eV.

At low temperatures the rate of cooling by H_2 molecules equals

$$R_{cool,mol} \approx \frac{g_2}{g_0} n_{H_2} E_{20} A_{20} \exp(-E_{20}/kT)$$

$$\approx 7 \times 10^{-24} n_{H_2} e^{-510/T} \text{ erg cm}^{-3} \text{ sec}^{-1}$$

where the indices refer to the $J = 0$ and $J = 2$ rotational levels of H_2. The transition between these two levels is characterized by $g_0 = 1$, $g_2 = 5$, $E_{20} = 7 \times 10^{-14}$ erg and $A_{20} = 2 \times 10^{-11}$ sec^{-1}.

Cooling by accommodation on the grain surface amounts to

$$R_{cool,gr} \approx \alpha n \sigma_{gr} v n_{gr} \tfrac{3}{2} k(T - T_{gr})$$

$$\approx 1 \times 10^{-33} n^2 T^{\frac{1}{2}} (T - T_{gr}) \text{ erg cm}^{-3} \text{ sec}^{-1}$$

where we have inserted for the accommodation coefficient $\alpha = 0.5$ (Schutte[2]), $\sigma_{gr} = 7 \times 10^{-10}$ cm^2, $v = 1.45 \times 10^4 T^{\frac{1}{2}}$ cm/sec and $n_{gr} = 1.2 \times 10^{-12} n$ cm^{-3}. For the cloud model under consideration ($n = 1000$ cm^{-3} and $\zeta_0 = 10^{-17}$ sec^{-1}) we find after equating heating and cooling rates and putting $T_{gr} = 10°$K a gas temperature $T = 20°$K. At this temperature cooling by grains is dominant.

Returning to the question of the H/H_2 ratio it is obvious from the set of values in Table II that one should be very cautious in interpreting 21-cm observations of dark clouds because any atomic hydrogen density can be explained in combination with some gas temperature smaller than or equal to the observed brightness temperature. Measurements with high frequency resolution like those carried out by Knapp[23] could help to resolve some of this ambiguity by providing restrictions on the range of possible temperatures.

V. CONCLUSION

We have shown that formation of H_2 molecules in dark clouds may also proceed through associative detachment of the H$^-$ ion. This is important since our understanding of surface recombination of H_2 on grains is very poor. On the basis of 21-cm observations of atomic hydrogen one cannot decide between these two formation mechanisms as long as no reliable estimate of the gas temperature in dark clouds is available.

If other negative ions like C^- and O^- are also present in dark clouds with sufficient high densities, and there is no reason why they should not be there, the interesting possibility suggests itself that some of the molecules discovered in interstellar space during the last few years could be formed via reactions with negative ions. This would not be surprising because in general reactions with ions are faster than those between neutral particles.

ACKNOWLEDGEMENT

This work was supported by the Netherlands Foundation for the Advancement of Pure Research (ZWO).

REFERENCES

1. Hollenbach, D. J., Werner, M. W. and Salpeter, E. E., *Ap. J.* **163**, 165 (1971).
2. Schutte, A., following paper.
3. McDowell, M. R., *Observatory* **81**, 240 (1961).
4. Solomon, P. M. and Werner, M. W., *Ap. J.* **165**, 41 (1971).
5. Hirasawa, T., *Prog. Theor. Phys.* **42**, 523 (1969).
6. Spitzer, L. and Tomasko, M. G., *Ap. J.* **152**, 971 (1968).
7. Field, G. B., Goldsmith, D. W. and Habing, H. J., *Ap. J.* **154**, L149 (1969).
8. Habing, H. J. and Goldsmith, D. W., *Ap. J.* **166**, 525 (1971).
9. Dubrovskii, G. V. and Ob'edkov, V. D., *Sov. Astr.* **11**, 305 (1967).
10. Ohmura, T. and Ohmura, H., *Phys. Rev.* **118**, 154 (1960).
11. Greenberg, J. M., *Astr. and Ap.* **12**, 240 (1971).
12. Rosseland, S., "Theoretical Astrophysics" University Press, Oxford (1936).
13. Bates, D. R. and Holt, A. R., *Proc. Phys. Soc.* **85**, 691 (1965).
14. Peek, J. M., *Phys. Rev.* **154**, 52 (1967).
15. Greenberg, J. M. and de Jong, T., *Nature* **224**, 251 (1969).
16. Heiles, C., *Ap. J.* **156**, 493 (1969).
17. Garzoli, S. L. and Varsavsky, C., *Ap. J.* **145**, 79 (1966).
18. Garzoli, S. L. and Varsavsky, C., *Ap. J.* **160**, 75 (1970).
19. Heiles, C., *Ap. J.* **160**, 51 (1970).
20. Sancisi, R. and Wesselius, P. R., *Astr. and Ap.* **7**, 341 (1970).
21. Sancisi, R., *Astr. and Ap.* **12**, 232 (1971).
22. Palmer, P., Zuckerman, B., Buhl, D. and Suyder, L. E., *Ap. J.* **156**, L147 (1969).
23. Knapp, G. R., paper presented at IAU General Assembly, Brighton (1970).

Formation of Molecular Hydrogen on Cold Surfaces

A. SCHUTTE*

The formation reaction $H + H \xrightarrow{surface} H_2$ has been studied quite extensively during the last ten years by astrophysicists.[1-2] This reaction, proposed in 1948 by Van de Hulst,[3] is thought to be the most efficient recombination process for H_2 in interstellar clouds. Surface reactions, however, are very often not well understood, and therefore it is useful to simulate some aspects of this problem in a laboratory experiment. An earlier approach has been made by Brackman and Fite.[4] However, their results were limited by rather poor vacuum conditions.

At the University of Genova, together with G. P. Marenco, F. Tommasini and G. Scoles, I studied this problem in a molecular beam apparatus from an experimental point of view. Recently we got our first results[5] and in this seminar I wish to report on them.

In a three stage molecular beam apparatus a H or H_2 beam is produced, collimated and detected, using a microcalorimetric technique: A He-cooled Ge-bolometer of $2 \cdot 5 \times 2 \cdot 5$ mm^2 serves both as detector and reaction surface. The surface is outgassed at room temperature for several hours and then cooled down. During cooling it can be expected that some (two) layers of background gas condense on the surface. During a run one extra layer is formed every three hours, thus ensuring a stable surface for shorter periods. Beam intensities used, were: $1-30 \times 10^{13}$ cm^{-2} sec^{-1} for H, $0 \cdot 5-15 \times 10^{13}$ cm^{-2} sec^{-1} for H_2. The minimum detectable H flux is 10^8 cm^{-2} sec^{-1}. The condensing background gas was equivalent to a flux of 10^{11} cm^{-2} sec^{-1}. In a typical interstellar cloud, $n = 15$ cm^{-3}, the flux is 10^6 cm^{-2} sec^{-1}, therefore the difficulty of overcoming scaling is not so much the detection of extreme small atomic hydrogen fluxes, but rather suppressing possible unknown reactions with background gas condensing simultaneously on the surface. Chemisorption processes, releasing a large amount of energy as well, have to show saturation, and can therefore be separated from chemical reactions of condensing particles (e.g., H_2 recombination).

* Deceased; on March 30, 1972, A. Schutte died in a tragic accident.

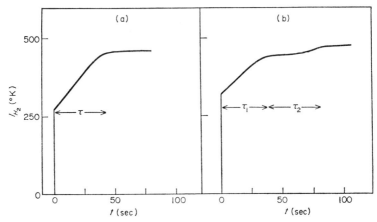

FIG. 1. Time evolution of the measured energy per H_2 molecule (a) 5°K, (b) 3°K.

In Figure 1 the energy measured per H_2 molecule as a function of time is shown. Due to formation of a mono-layer (Fig. 1a) the nature of the surface changes and therefore the measured signal. At lower temperatures the beam pressure is higher than the vapor pressure of H_2 and more layers are formed (Fig. 1b). With an *ad hoc* estimate of 10^{15} molecules cm^{-2} necessary for the formation of a layer, the sticking coefficient S can be calculated. It was found to be roughly 0·25. Because the bonding energy is

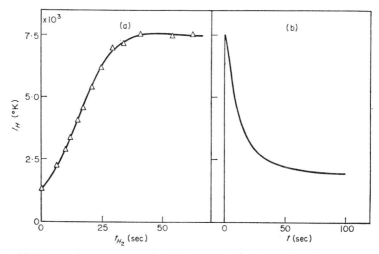

FIG. 2. (a) Measured energy per pair of H-atoms in function of surface contamination with H_2. (b) Time evolution of the atomic hydrogen signal after depositing one layer of H_2 on the surface. Due to clearing the signal decreases.

rather low, the major contribution of the signal I is kinetic energy of the beam particles: $I_{kin} = 2kT(\alpha(1-S) + S)$.

From the measurements on atomic hydrogen three conclusions can be drawn:

a) the H signal is much more sensitive to surface contamination. Indeed after deposition on the surface of a certain amount of H_2 (by exposure, during a time t_{H_2}, to the H_2 beam), the H signal is measured (Fig. 2a). An increase of almost one order of magnitude is observed.

b) Recombination takes place. In fact, in the case of the presence of H_2 on the surface the measured signal is much larger than what can be expected from available kinetic energy. Without H_2 on the surface the situation is less clear; sticking may be low, but it is quite possible that the formed species leave the surface with high internal and/or kinetic energy, without having the possibility of thermalizing.

c) Formed, and present H_2 molecules evaporate. Following the time evolution of the H signal (Fig. 2b) it appears that the signal decreases rapidly due to sputtering of H_2 molecules. The cleaning rate is rather high: for every molecule formed several extra molecules are sputtered when the surface is covered with one mono-layer H_2. This implies that the formation of a mono-layer of H_2 on grains will be slowed down by the simultaneous occurrence of the recombination reaction.

More detailed measurements on better specified surfaces (rare gases, H_2O etc.) and analysis of formed and reflected particles are in progress and will be reported soon.

ACKNOWLEDGEMENTS

I wish to thank Prof. J. J. M. Beenakker, Dr. H. F. P. Knaap, Dr. C. J. N. Van den Meydenberg and Drs. T. De Jong for useful discussions and the help to carry out this work.

The experiment is a joint program of the universities of Leiden, Holland and Genova, Italy. Work supported by the N.A.T.O. Office for Scientific Research under grant No. 419 and carried out with the facilities of the Gruppo Nazionale di Struttura della Materia (G.N.S.M.), a division of the Italian N.R.C.

This paper was written while the author was at the Istituto di Scienze Fisiche dell'Università, Genova, Italy, on leave of absence from Kamerlingh Onnes Laboratorium, University of Leiden, Leiden, The Netherlands.

REFERENCES

1. Knaap, H. F. P., Van den Meydenberg, C. J. N., Beenakker, J. J. M. and Van de Hulst, H. C., *Bull. Astr. Inst. Neth.* **18,** 256 (1966).
2. Hollenbach, D. and Salpeter, E. E., *Ap. J.* **163,** 155 (1971), and references quoted therein.
3. Van de Hulst, H. C., *Rech. Astr. Obs. Utrecht* **11,** (1948).
4. Brackmann, R. T. and Fite, W. L., *J. Chem. Phys.* **34,** 1572 (1961).
5. Marenco, G., Schutte, A., Scoles, G. and Tommasini F., Proceedings of the International Conference on Solid Surfaces—I.C.S.S. Boston (U.S.A.), October 1971.

Emission-Line Spectra as Probes of Dust Clouds

B. E. J. PAGEL

Royal Greenwich Observatory, Herstmonceux Castle, Hailsham, Sussex.

Many objects with emission-line spectra are strong infra-red sources, often because of the presence of dust in the object which can be detected in other ways such as excess reddening or anomalous intensity ratios of emission lines, or from the presence in extended objects of dark patches and reflection nebulae. Even for Seyfert galaxies, a dust model (Rees et al.[1]) has attractive features and cannot be ruled out until variations of infra-red flux have been more firmly established than is the case at present. I wish to talk about the use of emission lines to deduce some properties of dust clouds in three cases: M82, Seyfert galaxies and the remarkable object η Carinae in our own Galaxy.

I. M82

M82 is a relatively nearby galaxy of our Local Group classified as Irr II on the Hubble–Sandage System, highly flattened and seen nearly edge-on. It is very dusty and a moderate non-thermal radio source. Photographs taken by Sandage and Miller[2] show filaments in Hα and continum radiation along the minor axis, the continuum being polarised. Doppler observations of the Hα line by Lynds and Sandage[3] revealed that the line is sloping, corresponding to a line-of-sight velocity of the order of 100 km/sec (or about 1000 km/sec after allowing for the projection angle) at a distance of 1·5 kpc. This corresponds to an enormous explosion in which something like 10^{56} ergs (equivalent to a million supernova outbursts) were liberated $1\frac{1}{2}$ million years ago. However, new observations by Sandage and Visvanathan[4] reveal that the Hα line is highly polarised (implying emission by scattering) and less than 10 angstroms wide (implying scattering probably by dust and certainly not by free electrons). Scattering by dust was previously suggested by Elvius.[5]

Recently Sanders and Balamore[6] have pointed out that, for Hα light originating near the centre of the galaxy and scattered by material in the filaments, the full Doppler shift is observed with no projection factor. Hence the kinetic energy is reduced by a factor of 100 and the time scale increased by a factor of 10. The equivalent rate of energy generation is reduced by a factor of 1000 to about a million times the luminosity of the Sun and can be accounted for by a few young, hot stars expelling dust along the rotational axis.

According to this analysis, then, M82 is no longer a pin-up example of violent explosions in galaxies and we may conceivably also have a new mechanism for producing anomalous red shifts.

II. SEYFERT GALAXIES

Seyfert galaxies (Seyfert[7]) are galaxies having compact nuclei with a variable non-stellar continuum and emission lines resembling those of planetary nebulae but much broader. They are strong infra-red and X-ray sources and often strong radio sources, especially NGC 1275 = Perseus A. Interest in Seyfert galaxies greatly increased after the discovery of quasars, which resemble Seyfert nuclei on a more powerful scale (at least if the cosmological interpretation of the large red shifts is correct). A fairly detailed model of the line-emitting region of the quasar 3C 273 has been put forward by Bahcall and Kozlovsky,[8] who found an electron density of the order of 10^7 cm^{-3} and a size of a few parsecs.

The emission spectra of Seyfert galaxies have been studied several times, e.g. by Osterbrock and Parker[9] who noted that the Balmer decrement was anomalously steep. This raises the possibility that there is intrinsic reddening, as seems to be the case for the infra-red strong planetary nebula NGC 7027 (Krishna Swamy and O'Dell[10]).

A specific test for intrinsic reddening was suggested by Miller[11] and applied to Seyfert galaxies by Wampler.[12] This test is based on the measurement of relative intensities of infra-red and violet [S II] lines with common upper levels and known transition probabilities, and it gave substantial amounts of reddening in most cases studied. Similar amounts of reddening were deduced from a comparison of Paschen and Balmer lines. The question then arises as to whether the reddening applies to the non-stellar continuum which (in view of its variability) presumably arises from a smaller central region. If the dust is embedded in ionised hydrogen filaments, one may see a fairly undistorted continuum which then is non-thermal and somewhat like the radio continuum. Andrillat and Souffrin[13] have argued in favour of this hypothesis because the stellar continuum is unreddened even close to the nucleus. On the other hand, Kaneko[14] has pursued the hypothesis that the reddening also applies to the continuum. In this case it often approaches the Rayleigh–Jeans distribution for a hot thermal source and

Kaneko shows that this may account for the ionisation and the infra-red radiation quantitatively if one imagines the central source to be a supermassive star of the type suggested by Hoyle and Fowler[15] with a mass of about 10^6 solar masses, an effective temperature of 50,000 to 70,000 K and a radius of 10^{14} cm. A clear-cut decision between these two models of the continuum cannot yet be made.

III. η CARINAE

η Carinae is a compact object in our own Galaxy in the very young galactic cluster Trumpler 16, which underwent a nova or supernova-like outburst with a maximum ($m_v \simeq -1$) in 1843, followed by a quasi-linear decline that was much slower than that of any known supernova. Its properties have been discussed by Burbidge,[16] who shows that it is probably a star of large mass (~ 100 solar masses) losing material owing to vibrational instability. Westphal and Neugebauer[17] found an enormous infra-red excess indicating a total luminosity ($\sim 10^{40}$ ergs/sec) similar to its visual luminosity at maximum.

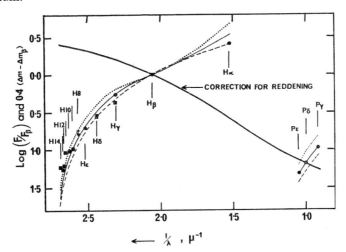

FIG. 1. Relative intensities of Paschen and Balmer lines in η Carinae (Hβ = 1) after correction according to a standard interstellar reddening curve with colour excess 1^m15 in (B–V) (shown by the thick solid line). Curves show theoretical line ratios computed for a recombination spectrum by Capriotti[21] with optical depths in the centre of Hα of 1 (broken line), 25 (thin solid line) and 63 (dotted line). Observational points are taken from Rodgers and Searle[18] (filled circles) and Aller and Dunham[22] (filled squares).

Rodgers and Searle[18] studied the emission lines and continuum and corrected for interstellar reddening corresponding to a colour excess of 0^m69 in (B–V), deduced from the surrounding H II region. The continuum appeared to be non-thermal, but here again the possibility of intrinsic reddening arises. Using an idea similar to that of Miller and Wampler, Pagel [19-20] studied the [Fe II] lines (which are especially strong in η Carinae) and deduced a total colour excess of $1^m\!.20$ in $(B-V)$, strikingly confirmed by the ratio of Paschen to Balmer lines (Fig. 1). If this reddening also applies to the continuum, the latter is thermal, corresponding either to a black-body distribution at 25,000 K or to two-photon emission from neutral hydrogen in the $2s$ state.

Rodgers and Searle noted that no actual star is visible and that the co-existence of Fe^+ and H^+ is most easily explained by a collisionally excited plasma with an electron temperature of about 20,000 K. In this case the conditions required for two-photon emission are difficult, but perhaps not impossible, to fulfill. Davidson,[23] on the other hand, proposes ionization by a central star (invisible because of scattering by dust) and an electron temperature of only 8,000 K, agreeing with the observed excitation temperature of [Fe II] (Thackeray[24]) and with an estimate of thermal balance. The difficulty with this model is the predominance of Fe^+ over Fe^{++}, which is hard to explain unless Fe^{++} has a remarkably high recombination coefficient; on the other hand, it explains the Balmer decrement very well (Fig. 1).

A further serious complication is the fact that Rodgers[25] has now applied the [S II] intrinsic reddening test with negative results! This seems to imply that [S II] lines must be formed in a different sort of region from hydrogen and Fe II lines, so that the usual methods for estimating electron temperatures, densities and abundances become very dubious. It is difficult to think of any model that will satisfy all of the observations.

REFERENCES

1. Rees, M. J., Silk, J. I., Werner, M. W. and Wickramasinghe, N. C., *Nature* **223**, 788 (1969).
2. Sandage, A. R. and Miller, W. C., *Science* **144**, 382 (1964).
3. Lynds, C. R., and Sandage, A. R., *Ap. J.* **137**, 1005 (1963).
4. Sandage, A. R. and Visvanathan, N., *Ap. J.* **157**, 1065 (1969).
5. Elvius, A., *Lowell Obs. Bull.* No. 19, 281 (1962).
6. Sanders, R. H. and Balamore, D. S., *Ap. J.* **166**, 7 (1971).
7. Seyfert, C. K., *Ap. J.* **97**, 28 (1943).
8. Bahcall, J. N. and Kozlovsky, B. Z., *Ap. J.* **155**, 1077 (1969).
9. Osterbrock, D. E. and Parker, R. A. R., *Ap. J.* **141**, 892 (1965).
10. Krishna Swamy, K. S. and O'Dell, C. R., *Ap. J.* **151**, L61 (1968).
11. Miller, J. S., *Ap. J.* **154**, L57 (1968).
12. Wampler, E. J., *Ap. J.* **154**, L53 (1968).
13. Andrillat, Y. and Souffrin, S., *Astr. and Ap.* **11**, 286 (1971).
14. Kaneko, N., *Pub. Astr. Soc. Japan* **24**, 145 (1972).
15. Hoyle, F. and Fowler, W. A., *M.N.R.A.S.* **125**, 169 (1963).

16. Burbidge, G. R., *Ap. J.* **136,** 304 (1962).
17. Westphal, J. A. and Neugebauer, G., *Ap. J.* **156,** L45 (1969).
18. Rodgers, A. W. and Searle, L., *M.N.R.A.S.* **135,** 99 (1967).
19. Pagel, B. E. J., *Nature* **221,** 325 (1969).
20. Pagel, B. E. J., *Astrophys. Lett.* **4,** 221 (1969).
21. Capriotti, E. R., *Ap. J.* **139,** 225 (1964).
22. Aller, L. H. and Dunham, T. Jr., *Ap. J.* **146,** 126 (1966).
23. Davidson, K., *M.N.R.A.S.* **154,** 415 (1971).
24. Thackeray, A. D., *M.N.R.A.S.* **135,** 51 (1967).
25. Rodgers, A. W., *Ap. J.* **165,** 665 (1971).

The Investigation of UV Oscillator Strengths in C, N and O Ions

JEFFRY V. MALLOW
University of Jerusalem, Israel.

I. INTRODUCTION

An accurate knowledge of atomic (and molecular) oscillator strengths is important for a large number of astrophysical problems. In the course of this summer school, we have seen the need for accurate f-values in determination of collisional cross sections, intensity ratios, and recombination rates, to mention only a few applications.

Reliable measurements of absolute f-values, and the related quantities, spontaneous transition probabilities and radiative lifetimes, are important not only to astrophysics, but also to atomic theory, which at present seems able to give us simple methods for calculating transition energies, but which becomes considerably more cumbersome and haphazard when trying to predict other observables, such as transition probabilities.

Professor J. Burns and I have measured a number of radiative lifetimes for resonance transitions in ions and neutral atoms of carbon, nitrogen, and oxygen.[1] We have modified a technique developed by Bennett and Dalby[2] for measuring molecular lifetimes in the range 20–100 nanoseconds, and in the visible part of the spectrum; our measurements are for vacuum ultraviolet resonance transitions with lifetimes typically less than 10 nsec.

II. RELEVANT THEORETICAL FORMULAE AND EXPERIMENTAL PROCEDURE

The line strength for a dipole transition between an atomic level m and some lower level n is given by

$$S_{mn} = |\langle \Psi_m |\mathbf{P}| \Psi_n \rangle|^2$$

where $\mathbf{P}$ is the electric dipole operator and m and n are the initial and final states of the atom, respectively. The oscillator strength is

$$f_{mn} = \frac{8\pi^2 m_e v_{mn}}{3e^2 h\omega_n} S_{mn}$$

where v_{mn} is the transition frequency, m_e is the electron mass, and ω_n is the statistical weight of the state n. The transition probability is

$$A_{mn} = \frac{64\pi^4 v_{mn}^3}{3hc^3 \omega_m} S_{mn}.$$

The radiative lifetime of the upper state m is just $\tau_m = A_{mn}^{-1}$ for transition to one lower state n, or $\tau_m = (\sum_n A_{mn})^{-1}$ for several lower states n. If a gas of atoms or ions is excited, say by electron collision, so that the atomic level m is well populated, and if the collision beam is abruptly cut off, then the state m decays according to

$$\frac{dN_m(t)}{dt} = -\frac{N_m(t)}{\tau_m}$$

where $N_m(t)$ is the instantaneous population of m at time t. The solution to this equation is $N_m(t) = N_m(0) \exp(-t/\tau_m)$. Since each decay emits a photon of energy hv_{mn}, one can measure τ_m by monitoring photon counts as a function of time after cut-off of the exciting beam. Figure 1 shows the

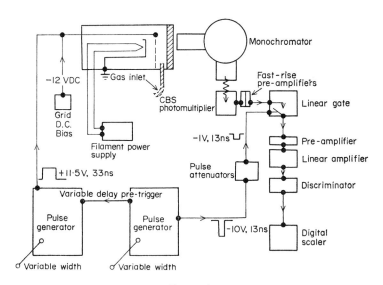

FIGURE 1

experimental set-up. The excitation chamber consists of a triode electron gun. The gas of interest is leaked into a field-free region within the anode of the gun, at a pressure of $2 \cdot 5 \times 10^{-3}$ Torr. Electrons from the hot cathode are prevented from reaching the gas by a negative DC bias applied to the grid of the gun. At a given instant, a positive square pulse from the first pulse generator cancels the DC grid bias, and the beam of electrons is accelerated to 200 eV and allowed to collide with the gas. The pulse rise and fall times are 0·7 nsec each. A variable delay pre-trigger connects the pulse generator to a second pulse generator of the same type, which turns on the photon detection apparatus. This last consists of a vacuum monochromator, a windowless uv sensitive photomultiplier, some very fast response preamplifiers, a linear gate which passes or blocks the amplified photo-multiplier pulses in response to the second pulse generator's signal, amplification and noise discrimination circuitry, and a digital pulse scaler.

Figure 2 shows the sequence of experimental steps in this "pulse decay" technique. Since counting statistics were very low, the excitation-decay-detection cycle was repeated at 4 kHz for each delay setting between the pulse generators, and we counted for typically 5 minutes at each setting. Delay settings were taken in random order, so that possible drifts in the system would blur the decay curve rather than shift its slope.

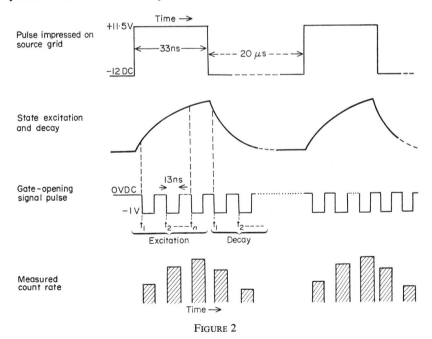

FIGURE 2

III. RESULTS

If each set of data were actually just the decay for the transition of interest, then a semilogarithmic plot of photon counts *vs* time would yield a straight line of slope $-1/\tau_m$. This is not the case, since each set of data contains

1. the actual decay
2. possible cascades into the upper state
3. the response time of the system.

The cascades, if undetected, give spuriously long decay times. The system response time, which was not a problem for the molecular lifetime measurements,[2] is a serious problem for our case, since it is comparable to the lifetimes being measured. It is, however, a very good approximation to assume that each of the above effects can be represented by an exponential, or sum of exponentials, with characteristic decay constants. The raw count rate $C(t)$, is then given by a sum of these contributions

$$C(t) = A_1 e^{-t/\tau_1} + A_2 e^{-t/\tau_2} + \ldots.$$

If one is fortunate, and the decay constants are considerably different one from the other, then in the asympotic (large t) region, the component of longest decay is all that one sees, and as we go to smaller t, the other components begin to contribute significantly to the curve. We can then use the well-known graphical technique from nuclear physics for "peeling off" the various terms, one by one. Using this procedure, we found that all of the data for the various transitions contained components with time constants 0·75, 1·6 and 3·1 ($\pm$ 0·1) nsec. These must be due to system response, and in fact, seem to be related to the 0·7 nsec risetime of the pulse generators. In addition, we found components with decay constants in the range 10–65 nsec. In many cases, we were able to identify these with known cascades, and we believe all of these to be cascade transitions, some as yet unidentified.

After all of these effects were "peeled off", what remained were the lifetimes of the transitions which we wished to measure. These are given in Table I, and compared with results of other experiments. Of particular significance is the comparison of our results with those of Hutchinson,[3] since his measurements were carried out using the same system, but exciting the states with an rf signal, and measuring the phase shift between this signal and the output rf signal from the photomultiplier. In the two cases where his results are significantly larger than ours, namely the C II 1335 Å and the N I 1135 Å lines, we were able to identify cascades which were apparently missed by the phase shift analysis, and which account for the difference

Table I

Experimental results and comparison with lifetimes by other investigators

Atom	Transition	λ(Å)	τ, nsec (this work)	τ (other work)	
C II	$2p\,^2P^0 - 2p^2\,^2D$	1335	$3\cdot2 \pm 0\cdot7$	$4\cdot3 \pm 0\cdot6$	(a)
				$3\cdot9 \pm 0\cdot4$	(b)
C I	$2p^2\,^3P - 2p^3\,^3D^0$	1560	$3\cdot6 \pm 0\cdot4$	$6\cdot6 \pm 3\cdot3$	(c)
C I	$2p^2\,^3P - 3s\,^3P^0$	1657	$3\cdot0 \pm 1\cdot2$	$2\cdot4 \pm 1\cdot2$	(c)
O I	$2p^4\,^3P - 3s\,^3S^0$	1304	$3\cdot0 \pm 1\cdot0$	$2\cdot6 \pm 1\cdot3$	(e)
N II	$2p^2\,^3P - 2p^3\,^3D^0$	1084	$2\cdot8 \pm 0\cdot3$	$2\cdot4 \pm 0\cdot5$	(a)
				$2\cdot7 \pm 0\cdot4$	(b)
N II	$2p^2\,^3P - 2p^3\,^3P^0$	916	$1\cdot2 \pm 0\cdot5$	$1\cdot8 \pm 0\cdot5$	(a)
N I	$2p^3\,^4S^0 - 2p^4\,^4P$	1135	$5\cdot5 \pm 1\cdot5$	$9\cdot9 \pm 1\cdot0$	(a)
				$7\cdot2 \pm 1\cdot1$	(b)
				$7\cdot0 \pm 0\cdot2$	(f)
				$4\cdot3 \pm 2\cdot2$	(d)
N I	$2p^4\,^3P - 3s^1\,^3D^0$	1243	$2\cdot5 \pm 0\cdot4$	$6\cdot9 \pm 0\cdot2$	(a)
				$2\cdot2 \pm 1\cdot1$	(d)
				$2\cdot6 \pm 0\cdot1$	(f)

Notes: (a) Phase shift—Hutchison[3]
(b) Phase shift—Lawrence and Savage[4]
(c) Stabilized emission are—Boldt[5]
(d) Stabilized emission are—Labuhn[6]
(e) Afterglow emission—Prag and Clark[7]
(f) Beam foil—Bickel et al.[8]

between our respective results. This identification is particularly important for the 1335 Å transition, since it allows us a greater confidence in the result, in spite of the fact that it is very close to one of the system response time constants, and therefore more difficult to determine by the graphical method. The lifeline of other transitions which were close to system time constants (e.g., 1657 Å, 1304 Å), were assigned large errors, in part because of this, in part for other reasons, discussed in reference 1.

IV. CONCLUSIONS

The advantage of this pulse decay technique over the comparable phase shift method are:

1. no calibration to a "known standard" lifetime is necessary
2. cascades are easily identified

3. noise levels for photon counting are lower than the corresponding levels in the phase shift experiment.

The disadvantages of the pulse decay technique are:

1. difficulty of extending the measurements to lifetimes below 1 ns:c
2. low counting rates.

A combination of the two techniques, whereby the pulse decay method is used to measure good standards in the range of 1 nsec, and the phase shift method is used, with these standards, in the range below 1 nsec, where it is applicable with present day equipment, may be a fruitful direction for this research to follow in the future.

REFERENCES

1. Mallow, J. V. and Burns, J., sumitted for publication.
2. Bennett, R. G. and Dalby, E. W., *J. Chem. Phys.* **31,** 2, 434 (1959).
3. Hutchison, R. B., *J. Quant. Spectrosc. Radiat. Transfer* **11,** 81 (1971).
4. Lawrence, G. M. and Savage, B. D., *Phys. Rev.* **141,** 67 (1966).
5. Boldt, G., *Z. Naturforsch.* **18A,** 1107 (1963).
6. Labuhn, F., *Z. Naturforsch.* **20A,** 998 (1965).
7. Prag, A. B. and Clark, K. C., *Phys. Rev. Letters* **12,** 34 (1964).
8. Bickel, W. S., Desesquelles, J., Berry, H. G. and Bashkin, S., *J. Opt. Soc. Amer.*

Author Index

The numbers in brackets are the reference numbers and those in italic refer to the Reference pages where the references are listed in full. Absence of page number indicates a general reference.

A

Aanestaad, P. A., 227 (62), 236 (79), *275, 276*
Abrines, R., 74 (6), 75 (10), 81 (6, 24, 25), 82 (10), *83*
Adams, W. S., 198 (7), *198*
Alder, K., 76 (13), *83*
Allen, C. W., 321 (1), *325*
Aller, L. H., 121 (7), 130 (20, 21), 131 (23), *152, 153*, 343 (22), *345*
Altenhoff, W., 147 (46), *153*
Anders, E., 259 (90), 261 (90), *276*
Anderson, C. R., 295 (15), *309*
Anderson, P. W., 109 (28), 116 (28), *118*
Andrick, D., 27 (51), *61*
Andrillat, Y., 342 (13), *344*
Arthurs, A. M., 50 (121), *63*
Asundi, R. K., 58 (142, 143), *63*
"Atomic and Molecular Processes", *60*
Aumann, H. H., 236 (76), *276*
Avila, C., 115 (46), *118*
Ayl, B., van, 329 (15), *335*

B

Bachall, J. N., 322 (7), *325*, 342 (8), *344*
Balamore, D. S., 341 (6), *344*
Balha, M., 317 (12), *320*
Ball, J. A., 148 (49), *153*, 198 (16, 27, 28), *199*, 201 (14), 205 (26), 206 (36), 209 (41), 219 (14, 58), 244 (14, 58), *274, 275*
Baranger, M., 85 (9), 97 (9, 17), 103 (17), 106 (17), 108 (17), *118*

Bardsley, J. N., 53 (135), 54 (136, 137), 55 (136, 137), 57 (135, 137), 58 (137), 59 (137), 60 (136), *60, 63*
Barnes, K., 106 (24). 107 (24), 112 (24), *118*
Barrett, A. H., 198 (22), *199*, 201 (1, 8), 205 (21, 22, 23), 206 (33), 209 (22, 41), 218 (21, 22, 23), 227 (68), 242 (82), 244 (82), *274, 275, 276*
Bartoe, J. F., 298 (21), *309*
Bashkin, S., 351 (7), *352*
Bassel, R. H., 22 (40), 23 (40), *61*
Bates, D. R., 48 (116), 49 (116, 117, 118, 119), 59 (145), *63*, 74 (7), *83*, 198 (8), *198*, 329 (13), *335*
Becklin, E. E., 218 (45), *275*
Bederson, B., 28 (60), *61*, 31 (67, 68), 32 (67), *62*
Beenakker, J. J. M., 337 (1), *340*
Behmenburg, W., 117 (57), *119*
Beigmann, I. L., 32 (73), *62*, 82 (28), *83*
Bely, O., 14 (20), 27 (55), 30 (66), 31 (71), *60, 61, 62*, 106 (23), 107 (23), 112 (23), *118*, 294 (14), *309*, 322 (8), *325*
Bender, P. L., 219 (56), 252 (56), *275*
Bennett, J. M., 303 (35), 306 (35), 308 (35), *310*
Bennett, R. G., 347 (2), 350 (2), *352*
Benson, R. C., 241 (81), 244 (81), *276*
Berkner, K. H., 83 (33), *83*
Berman, L., 130 (19), *152*
Bernstein, R. B., 76 (14), *83*
Berry, H. G., 351 (7), *352*
Bethe, H. A., 6 (12), 40 (93), *60, 62*
Bickel, W. S., 351 (7), *352*
Blaha, M., 294 (14), *309*

Blakenbecler, R., 18 (32), *61*
Blatt, J. M., 6 (11), *60*
Bloembergen, N., 252 (87), *276*
Blumenthal, G. R., 319 (16), *320*
Bohm, K. H., 85 (4), *117*
Boland, B. C., 30 (65), *61*
Boldt, G., 351 (5), *352*
Bolton, J. G., 201 (4), *274*
Born, M., 76 (12), 82 (12), *83*
Borodzich, E. V., 280 (4), *283*
Bowen, I. S., 125 (9), 130 (20), 131 (23), *152, 153*
Brackmann, R. T., 74 (9), *83*, 337 (4), *340*
Bradt, H., 292 (12), *309*
Bransden, B. H., *60*
Brattsev, V. F., 82 (26), *83*
Brechot, S., 106 (22), *118*
Breene, R. G., 85 (3, 8, 10), *117, 118*
Briglia, D. D., 59 (144), *63*
Brissaud, A., 116 (50), *118*
Brocklehurst, M., 116 (52), *119*, 130 (15, 16), 131 (16), 139 (36, 37), 145 (37), *152, 153*, 280 (6), 281 (7), *283*
Browne, J. C., 60 (146), *63*
Brueckner, G. E., 298 (21), *309*
Buhl, D., 198 (17, 23, 25), *199*, 201 (7, 17), 219 (12, 52), *274, 275*, 333 (25), *335*
Burbridge, G. R., 343 (16), *345*
Burch, D. S., 42 (99, 100), *62*
Burgess, A., 25 (47), 26 (47), 49 (120), *61, 63*, 73 (3), 81 (3), *83*
Burke, B. F., 147 (46), *153*, 209 (41), 213 (42), *275*
Burke, P. G., 20 (35, 36, 37), 22 (39), 25 (45), 26 (45, 48), 28 (58), 29 (63), 30 (63), 31 (69), 33 (74, 75), 34 (75, 79), 44 (105), 53 (129), 54 (138), *60, 61, 62, 63*
Burns, J., 347 (1), *352*

C

Capriotti, E. R., 131 (26), *153*, 343 (21), *345*
Carruthers, G. R., 198 (11), *198*
Carter, J. C., 198 (22), *199*, 209 (41), *275*
Cesarsky, D., 148 (49), *153*
Chandrasekhar, S., 42 (97), *62*, 113 (38), *118*
Chappelle, J., 112 (36), *118*
Chen, S., 85 (2), 116 (53), *117, 119*
Cheung, A. C., 198 (19, 20, 21, 24), *199*, 201 (5, 6), 226 (61), 229 (70), 231 (70), *274, 275, 276*
Chubb, T. A., 292 (11), *309*
Chubb, T. B., 301 (27), *309*
Churchwell, E., 139 (40), 145 (42), 147 (45), *153*, 279 (2), *283*
Churg, A., 206 (37), *275*
Clark, K. C., 351 (8), *352*
Codling, K., 42 (103), *62*
Collins, R. E., 31 (67, 68), 32 (67), *62*
Compton, A. H., 288 (3), *309*
Conneely, M. J., 46 (113), 47 (113), *63*
Cooper, J., 85 (15), 101 (20), 109 (20), 111 (29), 116 (20, 48), *118*
Cooper, J. W., 26 (48), 44 (104, 109), 45 (109), *61, 62, 63*
Cordelle, J., 305 (38, 39), *310*
Crawford, O. H., 53 (130), *63*
Crompton, R. W., 52 (128), *63*
Cudaback, D. D., 209 (41), 219 (50), *275*

D

Dalby, E. W., 347 (2), 350 (2), *352*
Dalgarno, A., 48 (116), 50 (121), 52 (126), 53 (130), 60 (146), *63*, 315 (3, 4), *319*
Daly, P. W., 227 (66), *275*
Dance, D. F., 25 (46), 26 (46), *61*
Danks, A. C., 127 (10), *152*
Davidson, K., 344 (23), *345*
Davies, R. D., 145 (41), *153*
Davis, J., 112 (33), *118*
Davis, R. D., 139 (38, 39), 144 (38), 145 (38), 147 (38), *153*
Dean, A. M., 265 (95), *276*
Desesquelles, J., 351 (7), *352*
Dickinson, D. F., 206 (36, 38), 233 (38), 234 (74), *275, 276*
Dieter, N. H., 201 (3), 205 (24), *274*
Doan, R. L., 288 (3), *309*
Donn, B., 236 (78), *276*
Doschek, G. A., 301 (27), *309*
Doughty, N. A., 42 (98), 43 (98), *62*

AUTHOR INDEX

Drake, G. W. A., 132 (30), *153*
Drake, G. W. F., 315 (3, 4, 5), 316 (5), 319 (16), *319*, *320*
Dravskikh, A. F., 280 (3), *283*
Dravskikh, Z. V., 280 (3), *283*
Dubrovskii, G. V., 329 (9), *335*
Dufay, J., 121 (6), *152*
Dunham, T., Jr., 343 (22), *345*
Dunn, G. H., 329 (15), *335*
Dupree, A. K., 148 (48, 49), *153*

E

Edlen, B., 289 (6, 8), 290 (8), 291 (8), *309*
Edmonds, A. R., 37 (83), *62*
Edrich, J., 139 (40), *153*
Ehrhardt, H., 27 (51), 53 (134), *61*, *63*
Eisner, W., 34 (80), 35 (80), 36 (80), 37 (80), *62*
Eliezer, I., 55 (140), *63*
Ellder, J., 206 (35), *275*
Elton, R. C., 27 (54), *61*, 299 (24, 25), 300 (24, 25), *309*, 317 (13), *320*
Elvius, A., 341 (5), *344*
Englander, P., 28 (60), *61*
Evans, N. J., 226 (61), *275*

F

Fano, V., 44 (104, 106, 109), 45 (109), *62*, *63*
Faucher, P., 322 (8), *325*
Ferguson, E., 114 (42), 115 (42), *118*
Feshbach, H., 18 (33), *61*
Field, G. B., 227 (62, 68), *275*, *276*, 328 (7), 331 (7), *335*
Filippov, N. V., 301 (28), *309*
Filippova, T. I., 301 (28), *309*
Fite, W. L., 22 (38), *61*, 74 (9), *83*, 337 (4), *340*
Flamand, J., 305 (37, 38), *310*
Flannery, M. R., 74 (5), *83*
Flather, E., 127 (11), *152*
Flemberg, H., 288 (1), 289 (1), *309*
Flower, D. R., 121 (4), 128 (12, 13, 14), *152*
Flygare, W. H., 241 (81), 244 (81), *276*

Fock, V., 72 (1), *83*
Fowler, W. A., 343 (15), *344*
Fox, R. E., 26 (49), 27 (49), *61*
Franco, V., 22 (40), 23 (40), *61*
Franks, A., 303 (33), 304 (36), *309*, *310*
Fraser, P. A., 42 (98), 43 (98), *62*
Freeman, F. F., 316 (7, 11), *320*
Friedman, H., 292 (11), 301 (27), *309*
Frish, U., 116 (50), *118*
Fritz, G., 292 (11), *309*
Fuse, K., 259 (90), *276*

G

Gabriel, A. H., 27 (53), *61*, 295 (16), *309*, 311 (1), 316 (8, 9, 11), 318 (14, 15), *319*, *320*
Gailitis, M., 15 (24), 18 (24), *61*
Gardner, F. F., 201 (4), *274*, 283 (9), *283*
Garton, W. R. S., 294 (22), 295 (16), 298 (22), *309*
Garzoli, S. L., 332 (20, 21), *335*
Geltman, S., 39 (89), 52 (125), 53 (125), *62*, *63*
Gentieu, E. P., 236 (78), *276*
Gerjouy, E., 22 (40), 23 (40), *61*, 52 (127), *63*
Gibson, D. K., 52 (128), *63*
Gilbody, H. B., 74 (8), *83*
Giomousis, G., 60 (148), *63*
Glauber, R. J., 7 (17), *60*
Glicker, S., 236 (78), *276*
Goldberg, L., 136 (33), 148 (48, 49), *153*, 294 (22), 295 (16), 298 (22), *309*
Goldriech, 245 (83), *276*
Goldsmith, D. W., 328 (7), 329 (8), 331 (7), *335*
Goldstein, M., 31 (67, 68), 32 (67), *62*
Gordon, M. A., 145 (43), *153*
Goss, W. M., 205 (25), 209 (25), 227 (68), *274*, *276*
Gottlieb, C. A., 198 (16, 27, 28), *199*, 201 (14), 206 (36), 219 (14, 58), 244 (14, 58), *274*, *275*
Greenberg, J. M., 329 (11), 332 (18), *335*
Griem, H., 132 (29), 139 (35), 147 (35), *153*

Griem, H. R., 27 (53), *61*, 85 (11), 106 (23, 27), 107 (23), 111 (31), 112 (23, 35, 37), 114 (37, 41), 115 (27, 44, 45), 116 (51), *118*, 280 (5), *283*, 316 (10), 317 (13), 318 (14), *320*
Griffing, G. W., 74 (7), *83*
Gryzinsky, M., 24 (43), *61*
Gwinn, W. D., 208 (40), 209 (40), 222 (40), *275*

H

Habing, H. J., 328 (7), 329 (8), 331 (7), *335*
Hann, Y., 18 (29, 30), *61*
Hammer, D. G., 74 (9), *83*
Hara, S., 52 (124), 53 (124), *63*
Harris, F. E., 60 (147), *63*
Harrison, M. F. A., 25 (46), 26 (46), *61*
Hayatsu, R., 259 (90), 261 (90), *276*
Heer, C. V., 219 (54, 55), 250 (54), *275*
Heiles, C., 332 (19, 22), *335*
Heiles, C. E., 219 (48, 49, 50, 51), 226 (51), *275*
Henery, J. C., 201 (1), *274*
Henry, R. C., 292 (11), *309*
Henry, R. J. W., 33 (74, 75), 34 (75), 40 (94), 46 (112), 52 (123), 53 (123, 133), *62*, *63*
Herbig, G. H., 218 (46), *275*
Heroux, L., 29 (61, 62), *61*, 312 (2), *319*
Herzberg, G., 155 (1, 2, 3), 165 (1), *198*, 223 (60), 236 (77), *273*, *276*
Herzenberg, A., 54 (136, 137), 55 (136, 137), 56 (141), 57 (137), 58 (137), 59 (137), 60 (136), *63*
Hibbert, A., 34 (79), *62*
Hindmarsh, W. R., 116 (55), 117 (56), *119*
Hirasawa, T., 328 (5), *335*
Hjelliming, R. M., 145 (41, 42, 43), *153*
Hoag, J. B., 288 (4), *309*
Hobby, M. G., 301 (30, 31), *309*
Hollenbach, D., 229 (69), 259 (89), 261 (69), *276*, 337 (2), *340*
Hollenbach, D. J., 327 (1), *335*
Holt, A. R., 329 (13), *335*
Holt, H. K., 26 (50), *61*
Holtz, J. Z., 206 (39), *275*
Hooper, C. F., 113 (39), *118*
Hougen, J. T., 172 (6), *198*
Hoyle, F., 281 (8), *283*, 343 (15), *344*
Hummer, D. G., 25 (47), 26 (47), *61*
Hutchison, R. B., 350 (3), 351 (3), *352*
Hyland, A. R., 218 (45), *275*

I

Icsevgi, A., 245 (84), *276*
Inglis, D. R., 115 (47), *118*
Ireland, J. V., 74 (8), *83*
Itikawa, Y., 52 (122), 53 (122, 131), *63*

J

Jackson, J. D., 6 (11), *60*
Jacobs, V., 46 (110), *63*
Jahoda, F. C., 30 (65), *61*
Javan, A., 250 (86), *276*
Jefferts, K. B., 198 (9, 10, 12, 13, 26, 29, 30), *198*, *199*, 201 (15, 16, 18, 19, 20), *274*
Jeffries, J. T., 321 (6), 322 (6), 324 (9), *325*
John, T. L., 42 (102), *62*
Jones, B. B., 316 (7, 11), *320*
Jones, T. J. L., 30 (65), *61*, 294 (22), 295 (16), 296 (20), 298 (22), *309*
Jong, T., de, 332 (18), *335*
Jordan, C., 295 (16, 17), *309*, 311 (1), 316 (8, 9, 11), *319*, *320*, 321 (4), *325*

K

Kahn, F. D., 271 (98), *276*
Kaneko, N., 342 (14), *344*
Karule, E. M., 28 (56, 57), 31 (56, 57), 32 (56, 57), *61*
Kauppila, W. E., 22 (38), *61*
Keck, J., 82 (27), *83*
Keller, J. B., 6 (13), *60*
Keosian, J., 262 (92), *276*
Kepple, P., 115 (44), *118*
Kimball, A. P., 263 (93), *276*
King, I. I., 83 (33), *83*
Kingston, A. E., 49 (117, 118, 119), *63*, 80 (18, 19), 82 (19), *83*

Kistiakowsky, G. B., 265 (95), *276*
Klarsfeld, S., 111 (30), *118*
Klein, A., 6 (10), *60*
Kliss, R. M., 264 (94), *276*
Knapp, G. R., 334 (26), *335*
Knapp, H. F. P., 337 (1), *340*
Knowles, S. H., 201 (6), 213 (42), *274, 275*
Kohn, W., 13 (18), *60*
Kolb, A., 112 (37), 114 (37), *118*
Kollberg, E., 206 (35), *275*
Koppendorfer, W. W., 27 (54), *61*
Kozlovsky, B. Z., 342 (8), *344*
Kramer, K. H., 76 (14), *83*
Kraus, J. D., 220 (59), *275*
Kreplin, R. W., 301 (27), *309*
Krishna Swamy, K. S., 342 (10), *344*
Kristian, J., 292 (12), *309*
Krotkov, R., 26 (50), *61*
Kunze, H. J., 27 (53), *61*, 318 (14, 15), *320*
Kutner, M., 198 (13, 26), *199*, 201 (19, 20), 221 (53), *274*

Lewis, E. L., 118 (59), *119*
Lewis, M., 85 (6), *118*
Lewis, N., 114 (40), 115 (40), *118*
Lie, T. N., 299 (24, 25), 300 (24, 25), *309*
Lifshitz, E. M., 214 (44), 271 (99), *275, 276*
Lilley, A. E., 147 (44, 47), 148 (49), 149 (44), *153*, 198 (28), *199*, 201 (14), 206 (30), 219 (14), 244 (14), *274*
Linder, F., 53 (134), *63*
Lipsky, L., 40 (94), 46 (113), 47 (113), *62, 63*
Litvak, M. M., 201 (11), 205 (27, 28), 206 (38), 214 (43), 219 (47, 57), 221 (28), 222 (28), 226 (28, 47), 230 (27), 233 (38, 72), 234 (11, 74), 237 (27), 239 (47), 240 (47), 244 (47), 245 (85), 250 (57), 252 (57), 265 (28), 270 (28), *274, 275, 276*
Loewen, E. G., 302 (32), *309*
Low, F. J., 236 (76), *276*
Lum, W. T., 201 (3), *274*
Lynds, C. R., 341 (3), *344*
Lyot, B., 290 (9), *309*

L

Labeyrie, A., 305 (37), *310*
Labuhn, F., 351 (6), *352*
Lacey, L., 293 (13), *309*
Lamb, W. E., 245 (84), *276*
Landau, L. D., 214 (44), 271 (99), *275, 276*
Lane, A. M., 14 (22), *60*
Lane, N. F., 52 (123, 125), 53 (123, 125), *63*
Langhans, L., 53 (134), *63*
Larson, R. B., 270 (96), *276*
Laude, J. P., 305 (39), *310*
Lauer, J. E., 80 (18, 19), 82 (19), *83*
Lawrence, G. M., 351 (4), *352*
Layzer, D., 34 (81), *62*
"Lectures in Theoretical Physics", *60*
Leeman, S., 116 (52), *119*, 139 (36), *153*, 280 (6), *283*
Lees, R., 227 (64), *275*
Lemaire, J. L., 117 (58), *119*
Levy, B. R., 6 (13), *60*
Levy, D. H., 206 (37), *275*
Lewis, B. A., 29 (63), 30 (63), *61*

M

McCammon, D., 325 (10), *325*
McCoyd, G. C., 83 (31), *83*
McDowell, M. R., 327 (3), *335*
McEachran, R. P., 42 (98), 43 (98), *62*
Macek, J. H., 18 (31), 42 (101), *61, 62*
McFarlane, M., 292 (12), *309*
McGee, R. X., 201 (4, 10), *274*
McIntosh, A. I., 52 (128), *63*
McKinley, W. A., 18 (31), *61*
McNamara, L. F., 118 (59), *119*
McVicar, D. D., 25 (45), 26 (45), 44 (105), *61, 62*
McWhirter, R. W. P., 30 (65), 49 (118), *61*
McWhistler, R. W. P., 49 (117), *63*
McWhorter, A. L., 201 (11), 234 (11), *274*
Madden, R. P., 42 (103), *62*
Mallow, J. V., 347 (1), *352*
Mandl, F., 53 (135), 54 (136, 137), 55 (136, 137), 56 (141), 57 (135, 137), 58 (137), 59 (137), 60 (136), *60, 63*

Mansbach, P., 82 (27), *83*
Marenco, G., 337 (5), *340*
Margenau, H., 85 (6), *118*
Marshall, A., 235 (75), *276*
Massey, H. S. W., *60*
Matthews, C. N., 264 (94), *276*
Mayer, C. H., 201 (6), *274*
Mayer, W., 292 (12), *309*
Mazing, M. A., 85 (7), *118*
Meekins, J. F., 292 (11), 301 (27), *309*
Meeks, M. L., 198 (16, 22), *199*, 201 (1), 206 (36), 213 (42), *274, 275*
Meels, M. L., 201 (11), 234 (11), *274*
Mentall, J. E., 236 (78), *276*
Menzel, D. H., 41 (96), *62*, 138 (34), *153*
Metcalf, H., 235 (75), *276*
Mezger, P. G., 147 (44, 45, 46, 47), 149 (44), *153*, 279 (1, 2), 283 (9), *283*
Michels, H., 118 (59), *119*
Miller, J. S., 131 (24), *153*, 342 (11), *344*
Miller, S. L., 261 (91), *276*
Miller, W. C., 341 (2), *344*
Milne, D. K., 283 (9), *283*
Milne, E. A., 48 (115), *63*
Minaeva, L. A., 83 (30), *83*
Minkowski, R., 130 (20, 21), *152*
Moffett, R. J., 52 (126), *63*
Moiseiwitsch, B. L., 332 (17), 335 (17), *60, 335*
Moores, D. L., 31 (69), *62*
Moran, J. M., 205 (22), 209 (22, 41), 213 (42), 218 (22), *274, 275*
Morgan, F. J., 295 (16), *309*
Morgan, J. F., 294 (22), 298 (22), *309*
Morgan, L. A., 34 (78), *62*
Morgan, P. D., 301 (29, 31), *309*
Moser, R. E., 264 (94), *276*
Mott, N. F., *60*
Münch, G., 325 (10), *325*

N

Nather, R. E., 292 (12), *309*
Neugebauer, G., 218 (45), *275*, 325 (10), *325*, 343 (17), *345*
Neupert, W. M., 301 (26), *309*
Newkirk, G., 293 (13), *309*
Newsom, G. H., 31 (70), *62*

Nicholls, R. W., 294 (22), 295 (16), 298 (22), *309*
Nicolas, K. R., 298 (21), *309*
Norcliffe, A., 72 (2), *83*
Norcross, D. W., 46 (111), *63*
Noyes, R., 291 (10), *309*
Nussbaumer, H., 34 (80), 35 (80), 36 (80), 37 (80), *62*

O

Ob'edkov, V. D., 329 (9), *335*
Ochkur, V. I., 14 (21), *60*, 82 (26), *83*
Oda, A., 259 (90), *276*
O'Dell, C. R., 342 (10), *344*
Oertel, G. K., 111 (29), *118*
Ohmura, H., 329 (10), *335*
Ohmura, T., 329 (10), *335*
Oka, T., 227 (63, 64, 65, 66, 67), *275, 276*
Olsen, K. H., 295 (15), *309*
Olthof, H., 321 (5), 324 (5), *325*
O'Malley, T. F., 6 (14), 18 (29, 30), *60, 61*
Omidvar, K., 83 (32), *83*
Orall, F. Q., 321 (6), 322 (6), *325*
Ormonde, S., 20 (36), 26 (48), *61*
Oro, J., 263 (93), *276*
Orszag, S. A., 227 (62), *275*
Osgood, T. H., 288 (5), *309*
Oster, L., 133 (31), *153*
Osterbrock, D. E., 32 (72), *62*, 121 (3), 127 (11), 131 (3, 25), *152, 153*, 342 (9), *344*
Ott, W. R., 22 (38), *61*

P

Pagel, B. E. J., 344 (19, 20), *345*
Paget, T. M., 318 (15), *320*
Palmer, P., 147 (44, 47), 149 (44, 50, 51), *153*, 198 (14, 15, 23, 25), *199*, 201 (7, 12), 206 (30, 31, 32, 33, 34), 219 (12, 52), *274, 275*, 333 (25), *335*
Palumbo, L. J., 317 (13), *320*
Papa, D. C., 213 (42), *275*
Papadopoulous, G. D., 213 (42), *275*
Parker, R. A. R., 342 (9), *344*
Parkinson, W., 296 (20), *309*

Parkinson, W. H., 294 (22), 295 (16), 298, (22) *309*
Paxton, H. J. B., 294 (22), 295 (16), 298 (22), *309*
Peach, G., 106 (24), 107 (24), 112 (24), *118*
Peacock, N. J., 301 (29, 30, 31), *309*
Pedlar, A., 139 (39), 149 (52), *153*
Peek, J. M., 329 (14), *335*
Pekeris, C. L., 41 (96), *62*
Penfield, H., 147 (44, 47), 149 (44), *153*, 206 (30, 31), *274, 275*
Penston, M. V., 270 (97), *276*
Penzias, A. A., 198 (9, 10, 12, 13, 26, 29, 30), *198, 199*, 201 (15, 16, 18, 19, 20), *274*
Percival, I. C., 24 (44), *61*, 72 (2), 73 (3), 74 (4, 6), 75 (10), 77 (15), 80 (15, 20, 21, 22), 81 (3, 6, 24, 25), 82 (10), *83*, 130 (17, 18), *152*
Perel, J., 28 (60), *61*
Peterkop, R. K., 28 (57), 31 (57), 32 (57), 38 (84, 85, 86, 87), 39 (90), *61, 62*
Petford, A. D., 117 (56), *119*
Petit, R., 305 (39), *310*
Pfennig, H., 115 (43), *118*
Pieuchard, G., 305 (38, 39), *310*
Pottasch, S. R., 295 (18), *309*, 321 (3, 5), 323 (3), 324 (3, 5), *325*
Prag, A. B., 351 (8), *352*
Prats, F., 44 (104), *62*
Presnyakov, L. P., 82 (29), *83*

R

Radford, H. E., 198 (16, 28), *199*, 201 (14), 206 (36), 219 (14), 244 (14), *274, 275*
Rank, D. M., 198 (19, 20, 21), *199*, 201 (5, 6), *274*
Rapp, D., 59 (144), *63*
Rappaport, S., 292 (12), *309*
Rather, J. D. G., 227 (62), *275*
Rau, A. R. P., 39 (91), *62*
Rees, M. J., 341 (1), *344*
Reeves, E. M., 294 (22), 295 (16), 298 (22), *309*
Reifenstein, E. C., III, 147 (46), *153*

Rich, J. C., 46 (114), 47 (114), *63*
Richards, D., 24 (44), *61*, 74 (4), 77 (15), 80 (15, 20, 21, 22), *83*, 130 (17, 18), *152*
Riviere, A. C., 83 (33), *83*
Robb, D. W., 34 (79), *62*
Robbins, R. R., 131 (27, 28), *153*
Roberts, D. E., 112 (33), *118*
Robinson, B. J., 201 (4, 10), *274*
Rodgers, A. W., 343 (18), 344 (25), *345*
Rogers, A. E. E., 209 (41), 227 (68), *275, 276*
Rosenberg, L., 6 (14), *60*
Ross, M. H., 15 (23), *60*
Rosseland, S., 329 (12), *335*
Rostas, F., 117 (58), *119*
Roueff, E., 117 (60, 61), *119*
Rubin, K., 31 (68), *62*
Rubin, R. H., 241 (81), 244 (81), *276*
Rudge, M. R. H., 38 (8, 88), 39 (89), *60, 62*
Rugge, H. R., 316 (6), *320*
Rydbeck, O. E. H., 206 (35), *275*

S

Sahal-Brechot, S., 106 (25, 26), 107 (26), 111 (25, 26), 112 (26, 36), *118*
Salpeter, E. E., 40 (93), *62*, 229 (69), 259 (89), 261 (69), *276*, 327 (1), *335*, 337 (2), *340*
Samson, J. A. R., 44 (107), *62*, 298 (23), *309*
Sancisi, R., 332 (23, 24), 333 (24), *335*
Sandage, A. R., 341 (2, 3, 4), *344*
Sanders, R. H., 341 (6), *344*
Saraph, H. E., 23 (42), 24 (42), 33 (76), 34 (80), 35 (76, 80), 36 (80, 82), 37 (76, 80), *61, 62*, 79 (17), 82 (17), *83*
Savage, B. D., 351 (4), *352*
Sayce, L. A., 303 (33), *309*
Schawlow, A., 156 (4), 174 (4), *198*
Schawlow, A. L., 273 (100), *276*
Schey, H. M., 20 (35), *61*
Schlossberg, H. R., 250 (86), *276*
Schlüter, H., 114 (42), 115 (42, 46), *118*
Schnopper, K., 44 (108), *62*
Schulz, G. J., 26 (49), *61*, 53 (132), 56 (132), 58 (142, 143), *63*

Schutte, A., 327 (2), *335*, 337 (5), *340*
Schwartz, C., 17 (27), 18 (27), *61*
Schwartz, P. R., 198 (22), *199*, 205 (23), 213 (42), 218 (23), 242 (82), 244 (82), *274*, *275*, *276*
Scoles, G., 337 (5), *340*
Scott, E. H., 331 (16), 334 (16), *335*
Scoville, N. Z., 219 (53), 221 (53), *275*
Searle, L., 343 (18), *345*
Seaton, M. J., 7 (15, 16), 14 (19), 15 (25), 23 (41), 30 (64), 31 (64), 32 (64), 33 (76, 77), 34 (80), 36 (80, 82), 37 (76, 80), 38 (88), 39 (89), 41 (95), *60*, *61*, *62*, 79 (16), *83*, 100 (19), 111 (19), *118*, 121 (1, 2, 4), 131 (22), 139 (37), 145 (37), *152*, *153*, 281 (7), *283*
Settles, R. A., 219 (55), *275*
Seyfert, C. K., 342 (7), *344*
Sharp, T. E., 59 (144), *63*
Shaw, G. L., 15 (23), *60*
Shemming, J., 33 (76), 35 (76), 37 (76), *62*
Shen, K. Y., 112 (37), 114 (37), *118*
Shenton, D. B., 294 (22), 295 (16), 298 (22), *309*
Shimizu, T., 227 (67), *276*
Shklovskii, I. S., 321 (2), *325*
Siegert, A. J. F., 55 (139), *63*
Silk, J. I., 341 (1), *344*
Sin Fai Lam, A. L., 33 (75), 34 (75), 53 (129), *62*, *63*
Sloanaker, R. M., 226 (61), *275*
Smith, A. C. H., 25 (46), 26 (46), *61*
Smith, E. W., 101 (20), 109 (20), 116 (20, 48), *118*
Smith, G., 117 (56), *119*
Smith, K., 20 (35), 25 (45), 26 (45), 33 (74), 34 (78), 46 (113), 47 (113), *61*, *62*, *63*
Smith, S. J., 42 (99, 100), *60*, *62*
Snyder, L. E., 198 (17, 23, 25), *199*, 201 (7, 12, 17), 219 (12, 52), *274*, *275*, 333 (25), *335*
Sobelman, I., 97 (18), *118*
Sobel'man, I. I., 82 (28), 83 (30), *83*, 85 (12), *118*
Sobolev, V. V., 239 (80), *276*
Solomon, P. M., 198 (12, 29, 30), *198*, *199*, 201 (18), 206 (29), 219 (53), 221 (29, 53), 222 (29), 233 (73), *274*, *275*, *276*, 327 (4), 328 (4), 331 (4), 332 (4), *335*
Sorochenko, R. L., 83 (30), *83*, 280 (4), *283*
Souffrin, S., 342 (13), *344*
Speer, R. J., 294 (22), 295 (16), 296 (20), 298 (22), 301 (29, 30), 303 (34), 306 (34), *309*, *310*
Spitzer, L., 328 (6), 331 (16), 334 (16), *335*
Spitzer, L., Jr., 198 (8), *198*
Spruch, L., 6 (14), 18 (29, 30, 34), *60*, *61*
Staelin, D. H., 205 (26), *274*
Stebbings, R. F., 74 (9), *83*
Stein, S., 52 (127), *63*
Stevenson, D. P., 60 (148), *63*
Stewart, J. N., 295 (15), *309*
Stief, L. J., 236 (78), *276*
Stringer, T. E., 253 (88), *276*
Studier, M. H., 259 (90), 261 (90) *276*
Sugar, R., 18 (32), *61*
Sullivan, W. T., 201 (9), 213 (42), *274*, *275*
Swan, P., 16 (28), *61*
Swartz, M., 301 (26), *309*
Swenson, G. W., 241 (81), 244 (81), *276*

T

Tai, H., 22 (40), 23 (40), *61*
Tait, J. H., 29 (63), 30 (63), *61*
Takayanagi, K., 52 (122), 53 (122), *63*
Takeo, M., 85 (2), 116 (53), *117*, *119*
Talman, J. D., 109 (28), 116 (28), *118*
Tayler, R. J., 281 (8), *283*
Taylor, A. J., 20 (37), 28 (58), *61*
Taylor, H. S., 53 (134), 55 (140), 60 (147), *63*
Teller, E., 115 (47), *118*
Temkin, A., 17 (26), *61*
Thacker, D. L., 206 (33), *275*
Thackeray, A. D., 344 (24), *345*
Thaddeus, P., 198 (13, 26), *199*, 201 (19, 20), 206 (29), 219 (53), 221 (29, 53), 222 (29), 229 (71), 231 (71), *274*, *275*, *276*
Thomas, R. G., 14 (22), *60*

AUTHOR INDEX

Thornton, D. D., 198 (19, 21), *199*, 201 (5), *274*
Tigelaar, H. H., 241 (81), 244 (81), *276*
Tomasko, M. G., 328 (6), *335*
Tommasini, F., 337 (5), *340*
Tousey, R., 288 (2), 296 (19), 298 (21), *309*
Townes, C. H., 198 (19, 20, 21, 24), *199*, 201 (5, 6), 208 (40), 209 (40), 222 (40), 229 (70), 231 (70), 273 (100), *274*, *275*, *276*
Townes, D. H., 156 (4), 179 (4), *198*
Traving, G., 85 (5, 16), *117*, *118*
Tsao, 116 (54), *119*
Tucker, W. H., 319 (16), *320*
Tully, J., 31 (71), 62
Tully, J. A., 25 (47), 26 (47), *61*
Turner, B. E., 198 (15, 18), *199*, 201 (13), 206 (34), 219 (51), 226 (51), *274*, *275*
Tyren, F., 289 (7), *309*

U

Unsöld, A., 85 (1), *117*
Urnov, A. N., 82 (29), *83*

V

Vainshtein, L. A., 27 (52), 32 (73), *61*, 62, 82 (28), *83*, 97 (18), *118*
Valentine, N. A., 75 (10), 82 (10), *83*
Van de Hulst, H. C., 337 (1, 3), *340*
Van de Meydenberg, C. J. N., 337 (1), *340*
Van Regemorter, H., 31 (71), *60*, *62*, 85 (14), 104 (21), 106 (22), 111 (32), 117 (61), *118*, *119*
Van Vleck, J. H., 171 (5), *198*
Varsavsky, C., 332 (20, 21), *335*
Victor, G. A., 315 (4), *319*
Vidal, C. R., 101 (20), 109 (20), 116 (20, 48), *118*
Vinogradov, A. V., 27 (52), *61*
Vinograsov, V. P., 301 (28), *309*
Visvanathan, N., 341 (4), *344*
Vo Ky Lan, 28 (59), *61*
Volslamber, D., 116 (49), *118*
Vriens, L., 74 (23), 81 (23), *83*

W

Walker, A. B. C., 316 (6), *320*
Wallerstein, G., 218 (45, 46), *275*
Wampler, E. J., 342 (12), *344*
Wannier, G. M., 39 (92), *62*
Warner, B., 292 (12), *309*
Waters, J. E., 198 (22), *199*, 242 (82), 244 (82), *276*
Weaver, H., 201 (3), 205 (24), *274*
Wcbb, T. G. 22 (39), *61*
Weinreb, S., 201 (1, 2), *274*
Welch, W. J., 198 (19, 20, 21), *199*, 201 (5), *274*
Werner, M. W., 229 (69), 261 (69), *276*, 327 (1, 4), 328 (4), 331 (4), 332 (4), *335*, 341 (1), *344*
Wesselius, P. R., 332 (23), *335*
Westphal, J. A., 343 (17), *345*
Whitaker, W., 20 (36), *61*
Wickramasinghe, N. C., 341 (1), *344*
Wiese, W. L., 85 (13), 112 (13), *118*
Wilcock, P. D., 301 (29), *309*
Wilford, S. N., 83 (31), *83*
Williams, D. R. W., 201 (3), 205 (24), *274*
Williams, E. J., 75 (11), *83*
Williams, J. K., 55 (140), *63*
Wilson, O. C., 131 (23), *153*
Wilson, R., 294 (22), 295 (16), 298 (22), *309*, 316 (7), *320*
Wilson, R. J., 198 (9), *198*, 201 (15), *274*
Wilson, R. W., 198 (10, 12, 13, 29, 30), *198*, *199*, 201 (16, 18, 19, 20), *274*
Wilson, T. L., 147 (46), *153*, 283 (9), *283*
Wilson, T. W., 198 (26), *199*
Wilson, W. J., 205 (21, 22), 206 (33), 209 (22), 218 (21, 22), *274*, *275*
Winther, A., 76 (13), *83*
Wolf, R. A., 322 (7), *325*
Woolley, R. v. d. R., 321 (1), *325*
Wurm, K., 121 (5), *152*

Y

Yang, C., 296 (20), *309*
Yen, J. L., 206 (31), *275*

Z

Zafra, R. L., de, 235 (75), *276*
Zanstra, H., 125 (8), *152*
Zeiger, H. J., 201 (11), 234 (11), *274*
Zemach, C., 6 (10), *60*
Zirker, J. B., 321 (6), 322 (6), *325*

Zuckerman, B., 147 (44, 47), 149 (44, 50, 51), *153*, 198 (14, 15, 23, 25, 27), *199*, 201 (7, 12), 206 (30, 31, 32, 33, 34), 219 (12, 52, 58), 234 (74), 244 (58), *274, 275, 276*, 333 (25), *335*
Zuckerman, B. M., 206 (38), 233 (38), *275*

Subject Index

A

Absorption coefficient, 86, 133
Abundances of elements. 131–2, 147, 281, 286, 321 ff
Action function, 71
Adiabacity parameter, 79
Adiabatic approximation
　in line broadening, 93–4
　in molecules, 158 ff, 167, 227
Adiabatic potential, 93
　in molecules 162–3
Aerobee rocket, 298
Anharmonic oscillator, 76
Antimaser, 201, 206, 221 ff
Autocorrelation function, 94, 108
Autoionization, 49, 54, 302

B

Back reaction, 100, 106
Bethe–Born approximation, 30, 75, 79, 99
Bethe–Seaton approximation, 29
Binary encounter approximation, 72–4, 80–1, 229
Blaze angle, 303
Boltzmann law, 151
　factor, 222
Born approximation
　for potential scattering, 5 ff
　　first, 23–4, 78–80, 99, 228
　　second, 99
　for inelastic scattering, 13, 39
　　Coulomb–Born, 13–4, 26–7, 30–1
　　Coulomb–Born–Oppenheimer, 14, 27, 32
　　for e–H, 22

Born–Oppenheimer approximation for molecules, 49–50, 55
Brocklehurst–Seaton model, 280

C

Chromosphere, 291
Classical ensemble of atoms, 72
Classical frequency, 66, 68
Classical impulse approximation—*see* Binary encounter approximation
Classical theory of perturbed orbits, 76
Close coupling calculations
　theory, 16 ff
　for e–H scattering 16 ff
　for e–He scattering 26 ff
　for e–Li scattering 28 ff
　for e–molecule scattering, 53
　in line broadening, 106, 115
Collision strength, 35, 127
Controlled thermonuclear reactions, 83
Corona, 291
　UV emission, 295
　forbidden emission, 296
　abundances in, 321 ff
Correlation approximation, 20
Correspondence principle, 25, 75–6, 231
　Bohr's, 67, 71, 75–6
　for intensities, 68
　for density of states, 70, 75–6, 82
　Heisenberg's, 76
　for strongly coupled states, 77
　for cross-sections, 24, 80, 82, 130
Cross-sections
　potential scattering 1 ff
　differential, 2
　multichannel scattering 8 ff
　photodetachment, 21
　geometric, 65–6, 72
　classical ionization, 69, 72, 75

363

SUBJECT INDEX

classical excitation, 24, 70, 72, 75, **230**
  scaling law of, 71
  in astrophysics, 227 ff
Cyclotron wave, 252 ff

D

Damped oscillator, 96
de Broglie wavelength, 100
Debye radius, 100, 114, 254
Debye screened fields, 91
Density matrix, 108
Diffraction gratings
  amplitude, 302
  blazed, 303, 306
  laminar, 303, 306
  holo-, 304
Dipole approximation, 24, 79–80
  for photoionization, 40, 42, 46
  for molecules, 166 ff, 227, 245
Distorted wave approximation, 13
Doppler temperature, 138, 281
Dust clouds, 206, 218–9, 229, 235–6, 239, 322, 341 ff
  properties, 225 ff
  formation of H_2 in, 327 ff, 337

E

Effective range
  definition, 6
  approximation for multichannel scattering, 14 ff
Eikonal approximation, 7
  for e–H, 23
Electron scattering,
  by H, 15 ff
  by ions, 15 ff
  by He and He-Like ions, 26 ff
  by alkali and alkali-like ions, 31 ff
  ionization by, 38 ff
Emission
  spontaneous, 68, 87
  stimulated, 87, 189, 201 ff
  measure, 135, 145, 280
  Einstein coefficients, 150, 187, 201, 224, 266
Emissivity, line, 136
Eta Carinae, 341

Evolution operator, 101–2, 106, 108–9, 115
Exchange potential, 8, 11
  in molecular scattering, 51

F

Fine structure
  transitions between, 36 ff, 127
Fourier components, 66, 75–6
  and quantal matrix elements, 76
Fourier transform
  of line shape, 87–8, 108, 245
  spectrometers, 294
Franck–Condon principle, 55, 166, 190ff

G

Galactic centre, 341
  OH emission, 201
Gaunt factor, 111
Glauber approximation—*see* Eikonal approximation
Grains, sticking efficiency, 259, 338
  catalysis by, 261
Green function, 5

H

H II regions, 65, 67, 72, 77–8, 205–6, 218–9, 236, 277 ff, 343
  distribution of, 278
Hamilton–Jacobi theory, 82
Hartley absorption band, 287
Hartree–Fock approximation, 155
  in molecules, 160
Holtsmark distribution, 113
Hubble–Sandage system, 341
Hund's cases for molecules, 171 ff, 183–4, 193
Hyperfine doubling, 184

I

Impact parameter method, 27, 79
Infrared stars, 205–7, 209
Intercombination lines, 314 ff
Interstellar grains, 206
Isotope effect
  in dissociative attachment, 58
  in molecules, 162, 167, 192, 194

K

Kamers–Gaunt factor, 30, 41, 48
Kepler's Law, 66
Kirchoff's law, 133, 152
K-matrix,
  definition, 4
  in multichannel scattering, 11–15

L

Lambda-doubling, 165, 168, 170, 172, 174 ff, 185
  in interstellar molecules, 220, 234–6
Landé interval rule, 183, 185
Langevin spiralling, 60
Lennard–Jones potential, 117
Levinson's Theorem, 16
Lewis cutoff, 115
Line broadening
  impact, 68
  pressure, 85 ff, 116 ff
  Doppler, 41, 85, 138, 187–8, 195, 209, 213, 217, 219, 224, 235, 237, 239, 244 ff, 341
  static approximation, 88, 90–2, 95, 97, 105–9, 114–6
  Holtsmark theory, 91
  unified theories, 93, 109–110, 115
  semiclassical theories (see semiclassical theories of line broadening)
  adiabatic approximation, 93–4, 100, 103, 106, 109, 111–12
    Lindholm, 96–8, 109–11, 116–7
  quantum theory, 102 ff, 108
  nearest neighbour approximation, 88, 91, 105, 109
  of hydrogenic lines, 112, 115–6, 280
  of nonhydrogenic lines, 110–12
  in nebulae, 149
Line profile, 85, 87–88, 90, 92–3, 103–4, 187
  Doppler, 138–9, 147
  Lorentz, 96, 107, 113
  index, 44, 46
  in nebulae, 138, 142, 145, 147
Line of nodes, 156
Lippmann–Schwinger equation, 5

M

Maser action, 68
  in nebulae, 142–3
  in interstellar regions, 189, 201, 206 ff
  polarization, 250
Molecules
  rotational excitation, 50 ff, 168
  vibrational excitation, 53
  vibrational–rotational excitation, 168 ff, 205, 223, 232
  in interstellar space, 198, 202 ff, 241 ff
  formation of, 259 ff
M-matrix, 15
Monotonicity theorem, 18
Monte Carlo methods, 74–5, 79, 82
Mott scattering, 73

N

Nebulae
  planetary, 121, 342
  variable density models of, 135, 139, 145
  constant density models of, 142
  Orion, 121, 146, 207
  Orion A, 136, 144, 232, 255, 281
  Orion B, 145
  M17, 281
  Crab, 292
  microwave emitters, 202
  infrared emitters, 206–7, 342
  X-ray emitters, 292, 342
  O II sources, 207
Nebulium, 121

O

Opacity, line, 136, 333
Optical continuum radiation, 132
Optical depth, 134, 143, 220, 224, 240, 244, 246, 321
  partial, 244
Optical potential, 18, 20
Optical thickness, 134
Orbit integration, 74, 77
Orbiting Solar Observatory-1, 286
Oscillator strength, 11, 30, 41, 68, 87, 110, 150, 233, 347 ff

SUBJECT INDEX

P

Partial wave analysis, 107
  single channel, 2 ff, 228
  multichannel, 9 ff
Pauli exclusion principle, 8, 10, 16, 50
  in molecules, 160, 179
Phase shifts, 3, 232
Photodissociation, 228, 233
Photoionization, 39 ff, 329
  in nebulae, 122
Planck equation, 151
Planck intensity function, 187
Plasma, 85, 92, 102, 218, 244, 245, 288, 290, 299, 344
  He-like ions, 27, 311 ff
  frequency, 253
  length, 254
  focus, 301 ff
  theta pinch, 318
Polarizability, 51
Polarization potential, 27, 29, 106
  effective range approximation for, 6
  in ion–molecule reactions, 60
Polarized orbital approximation, 28
Predissociation, 233, 236
Principle of detailed balance, 329
Projection operators, 18
Protostar, 207, 214, 218, 270 ff, 279
Pseudo states, 22

Q

Quantum defect, 7, 15, 35
Quasars, 342

R

Radiation trapping factor, 240
Radiative transfer, equation of, 123, 133, 149
Radio continuum, 133
Raman scattering, 246, 250, 252
Recombination coefficient, 47, 49, 122, 125
  temperature dependence of, 58
  in dust clouds, 328
Recombination lines, 65, 68, 69, 130, 135 ff, 277 ff
  of He and C, 147–8, 277

Recombination, three-body, 49, 129, 331
  dielectronic, 49, 302, 316
  dissociative, 57 ff, 329
  radiative, 129, 132
  surface, 327 ff
Reflection efficiency of gratings, 288
Resonances, 13, 47, 106
  in e–H scattering, 19, 21
  in e–He scattering, 27
  of H^-, 42
  Rhydberg series of 13, 44
  in two body recombination, 49
  in molecular scattering, 53 ff
R-matrix, 14, 101
Rutherford scattering, 73

S

Saha equilibrium, 129, 151
Scalar additivity, 87
Scalar spin interaction, 183 ff
Scaling Laws, 71, 75, 86
Scattering amplitude
  definition, 1
  for multichannel scattering, 11
  for exchange, 14
Scattering length
  definition, 6
Scattering Matrix
  definition, 4
  in multichannel scattering, 11, 15
  in fine structure transitions, 37
  in line broadening, 97–8, 101, 105, 107, 109, 115
Schumann Region, 287
Selection rules in molecules, 167
Semiclassical theories,
  of collisions, 23, 82, 111
  of line broadening, 86–7, 92–3, 98–100, 109, 112, 117
Seyfert galaxies, 341 ff
Shock waves, 205, 223–4, 228, 236, 290
  chemical reactions in, 265 ff
Solar eclipse, 290, 293 ff
Stark effect, linear, 67–8, 72, 97, 112–6
  quadratic, 112
  in molecules, 160, 175
Strömgen radius, 124, 281

Strong coupling approximation, 77, 80, 82
Sudden approximation, 76–7, 79–80
Supernova, 205–6, 328

T

Tensor–spin interaction, 183 ff
T-Matrix
  definition, 4
  in multichannel scattering, 11
Transitions
  dipole, 87, 98, 103, 110, 165–6, 181 ff, 314, 347
    in interstellar molecules, 186 ff, 208 219
  quadrupole, 98, 126, 183, 185
  magnetic dipole, 126, 132, 189, 315–6, 321
  probability of radiative, 126
  electronic in molecules, 166 ff
  forbidden, 33, 125, 127–8, 226–7, 295, 321 ff
Two-quantum emission, 132, 315–7

V

Van der Waals interaction, 86, 91, 116–7, 230

Van Vleck's "trick", 171
Variational principle
  Kohn, 13
  for e–H scattering, 17–8

W

Wave packets, 71
Weak coupling approximation, 77, 79–80, 228
Weisskopf radius, 80, 89, 93, 100–1, 113
Wigner $9-j$ symbol, 37
  $3-j$ symbol, 232
W.K.B. approximation, 58

X

X-ray
  sources, 285 ff, 342
  in sun, 301
  dissociation in clouds, 332

Z

Zeeman effect, 209, 219, 250 ff